T0361406

Applications of
Epidemiological Models to Public Health Policymaking

The Role of Heterogeneity in Model Predictions

Applications of
Epidemiological Models to
Public Health Policymaking

The Role of Heterogeneity in
Model Predictions

Zhilan Feng

Purdue University, USA

World Scientific

NEW JERSEY · LONDON · SINGAPORE · BEIJING · SHANGHAI · HONG KONG · TAIPEI · CHENNAI

Published by

World Scientific Publishing Co. Pte. Ltd.

5 Toh Tuck Link, Singapore 596224

USA office: 27 Warren Street, Suite 401-402, Hackensack, NJ 07601

UK office: 57 Shelton Street, Covent Garden, London WC2H 9HE

Library of Congress Cataloging-in-Publication Data
Feng, Zhilan, 1959–
    Applications of epidemiological models to public health policymaking : the role of heterogeneity in model predictions / by Zhilan Feng, Purdue University, USA.
      pages cm
    Includes bibliographical references.
    ISBN 978-981-4522-34-2 (hardcover : alk. paper)
    1. Epidemiology--Mathematical models. 2. Communicable diseases--Epidemiology--Mathematical models. 3. Health risk assessment--Government policy. 4. Public health surveillance. I. Title.
    RA652.2.M3F46 2014
    614.4--dc23

                                                          2013050475

British Library Cataloguing-in-Publication Data
A catalogue record for this book is available from the British Library.

Printed in Singapore

To Haiyun and Henry

Preface

Mathematical models can be very useful for studying the spread and control of infectious diseases. One of the uses of these epidemiological models is to provide information that can be applied to the evaluation of public health policies related to disease control and prevention. The purpose of this book is to present several examples of such models. One of the main focuses is to compare the outcomes between simpler models (i.e., models under more simplifying assumptions by ignoring many heterogeneities) and more realistic models (i.e., models with more realistic assumptions by considering various heterogeneities).

When modeling biological systems, it is very important to maintain a good balance between the model simplicity and its ability to provide accurate information. Although in many cases simpler models can provide helpful understanding of certain biological processes, some simplifying assumptions may lead to models that generate misleading information regarding the effectiveness of disease control strategies (see, for example, sections 1.2.6 and 2.5.2). Therefore, it is critically important to have better understanding when and why more realistic assumptions (or the corresponding heterogeneities these assumptions represent) are needed in the models. Obviously, inclusion of more realisms in a model usually implies increased difficulties in the model analysis and more limited analytical results. Keeping a good balance of model simplicity and realism can be challenging.

The balancing act can be involved in several modeling aspects, two of which will be discussed in more details in this book. The first aspect we focus on is in the model assumptions about the probability distribution of the waiting time in a disease stage (e.g., the infectious stage). Most epidemiological models assume an exponentially distributed infectious stage as it will lead to a much simpler model consisting of ordinary differential equations (ODEs) in the case of continuous time. An analogous assumption in discrete time is a geometric distribution for the infectious stage, which will lead to a simpler difference equations (DEs) model. While these simple models have been very helpful for providing insights into biological questions, in some cases models with these simplifying assumptions may generate conclusions that can be biased or even misleading. It is important to understand these drawbacks of the model assumptions and consider more realistic assumptions when needed. The second aspect we will discuss in this book is concerned with various population heterogeneities including age, spatial, and time. Particularly, several examples will be presented to demonstrate the importance of considering heterogeneities such as mixing patterns among multiple sub-groups in the population, age-dependent vaccination or drug treatment, and seasonally forced disease transmission.

The models presented in this book are based primarily on academic papers that I have published together with colleagues mentioned in the Acknowledgements. This book is divided into two parts. Part I is devoted to more theoretical results of various models, while Part II focuses on the applications of the model results from Part I to the evaluation of public health policies of disease control and prevention. The control measures considered include medication use, vaccination, quarantine, and isolation. Most of the models and theoretical results presented in Part I are applicable to infectious diseases in general, the examples in Part II focus mainly on specific diseases (including influenza, SARS, TB, schistosomiasis, HIV, and HSV-2) and are formulated to address particular epidemiological questions using results from model analyses presented in Part I.

Part I is organized as follows. Chapter 1 deals with epidemic diseases, which concerns mostly with short-term behavior of the disease dynamics. Analyses of such models focus on the computation of reproduction numbers of the pathogen and final epidemic size. Both continuous-time and discrete-time epidemic models will be discussed. In Chapter 2, models for endemic diseases are considered, which are more appropriate for diseases such as tuberculosis (TB), schistosomiasis, malaria, HIV, and childhood diseases. As long-term dynamics are studied, demographic processes such as birth and death are usually included. This chapter begins with simple standard SIR and SEIR models. These models are then extended to include more heterogeneities as well as additional factors, which allow the models to be used to address biological questions that cannot be answered using the simpler models. It is also demonstrated in both chapters that assumptions made to simplify the models mathematically may sometimes lead to biased or incorrect conclusions, in which case more realistic assumptions must be considered in the models.

In Part II, results from models considered in Part I and their extensions are applied to address particular biological questions and to evaluate disease control policies. Chapter 3 includes examples of the applications to specific diseases including influenza, SARS, TB, HIV, HSV-2, and schistosomiasis. Chapter 4 demonstrates the interactive modeling tools developed using the computational software MATH-EMATICA, which are user-friendly and can be useful for assisting public health policymaking. Appendix A collects some notations and definitions, and Appendix B includes some example of MATHEMATICA codes that are used to generate the interactive modeling outcomes provided in Chapter 4.

Zhilan Feng

Acknowledgments

This book is based primarily on academic papers that I have published with valued colleagues throughout my career at Purdue University, beginning in 1996. First and foremost, I must thank those who contributed to my education in mathematics and mathematical biology: Horst Thieme, Carlos Castillo-Chavez, and Simon Levin. Without their guidance and mentorship, none of this would be possible.

Several papers in this book were collaborations with John Glasser, from whom I have learned much about formulating models and questions that are of interest to public health officials and of importance to policymaking decisions. His friendship and support have been invaluable throughout the writing of this book as well as those papers. I must also thank my Ph.D. students at Purdue for the inspiration that they have given me through teaching them and the joy I've taken in watching them succeed. Their significant contributions to more than a few of the papers referenced herein are also very important, of course. These include both graduated (Jorge Alfaro-Murillo, Cheng-Che Li, Ya Li, Lih-Ing Wu Roeger, Libin Rong, Xiaohong Wang, Yiding Yang, Pei Zhang) and my current students (Qing Han, Nancy Hernandez Ceron, Christina Lorenzo Alvey, Katia Vogt Geisse, Yiqiang Zheng).

I would like to acknowledge those who contributed to the papers referenced in this book, as their efforts led directly to its existence. They are: Fred Brauer, Carlos Castillo-Chavez, Lamwah Chow, Sara Del Valle, Meng Fan, John Glasser, Wenzhang Huang, Mimmo Iannelli, Eunok Jung, Suzanne Lenhart, Simon Levin, Maia Martcheva, F. Ellis McKenzie, Fabio Milner, Dennis Minchella, Andrew Moylan, Miriam Nuño, Zhipeng Qiu, Gregory Sandland, Eunha Shim, David Smith, Brenda Tapia-Santos, Horst Thieme, Sherry Towers, Pauline van den Driessche, Jorge Velasco-Hernandez, Dashun Xu, Ping Yan, Yingfei Yi, Haiyun Zhao, and Huaiping Zhu.

My research has been partially supported by several grants from the National Science Foundation (DMS-9720558, DMS-9974389, DMS-0314575, DMS-0719697, DMS-0920828, DMS-1022758), the James S. McDonnell Foundation, the US Centers for Disease Control and Prevention, and the Purdue Showalter Trust Award.

Finally, I dedicate this book to my children, Haiyun and Henry, for their love and support. Henry also helped edit and compile significant portions of this book.

Contents

Preface   vii

Acknowledgments   ix

Mathematical Modeling in Epidemiology   1

1. Epidemic Models   3

    1.1   Continuous-time models   3

        1.1.1   Simple SIR and SEIR epidemic models   4

        1.1.2   Reproduction number $\mathcal{R}$ and final epidemic size   5

        1.1.3   Final size relation in more complicated epidemic models .   6

    1.2   Discrete-time models   8

        1.2.1   Simple discrete SIR and SEIR models   9

        1.2.2   Incorporation of quarantine and isolation   11

        1.2.3   Models with arbitrarily distributed stage durations   15

        1.2.4   Applications of the general model to specific distributions .   18

        1.2.5   Control reproduction number $\mathcal{R}_C$ and final epidemic size .   23

        1.2.6   Influence of stage distributions on $\mathcal{R}_C$ and final size   . . .   25

        1.2.7   More examples of using the NGM approach to compute $\mathcal{R}$ 29

2. Endemic Models   41

    2.1   Classical SIR and SEIR endemic models   41

    2.2   Multi-group models and the role of mixing   43

        2.2.1   Models with random and preferential mixing   43

        2.2.2   Role of variable activity levels in STIs   48

    2.3   Multiple pathogen strains and host types   58

        2.3.1   Modeling TB and drug-resistance   58

        2.3.2   Human-schistosome-snail system   65

        2.3.3   Competitive exclusion in a multi-strain dengue model . . .   79

    2.4   Age-structured models   88

2.4.1   Optimal ages for vaccination   88
2.4.2   Infection-age-dependent transmission and progression . . .   95
2.5   Models with non-exponentially distributed stages durations   103
2.5.1   A model with general distributions for stage durations . .   108
2.5.2   The cases of gamma and exponential distributions   114
2.6   Oscillatory dynamic created by isolation   119
2.6.1   An SIQR model with isolation-adjusted incidence   119
2.6.2   A two-strain influenza model with cross immunity   138
2.7   Coupled dynamics of biological processes   143
2.7.1   Malaria epidemiology and sickle cell genetics   144
2.7.2   Coupling within- and between-host dynamics   153

Applications to Public Health Policymaking   163

3.   Applications of models to evaluations of disease control strategies   165
3.1   Influenza   165
3.1.1   Age- or group-targeted vaccination   165
3.1.2   Preferential mixing and group-targeted vaccination   173
3.1.3   Estimation of transmission parameters   178
3.1.4   Forecast for the fall wave of the 2009 H1N1 pandemic   . .   183
3.1.5   Effect of timing for implementing control programs   186
3.1.6   Herald wave   197
3.1.7   Adverse effect of antiviral use due to drug-resistance   204
3.2   SARS   212
3.3   Tuberculosis   223
3.3.1   Optimal treatment strategies for TB   223
3.3.2   Influence of HIV on TB prevalence   228
3.4   Schistosomiasis   234
3.4.1   Age-targeted drug treatment in humans   234
3.4.2   Evolution dynamics of hosts and parasites   238
3.5   Synergy between HSV-2 and HIV   246

4.   Development of interactive tools to assist public health policymaking   255
4.1   An interactive tool for policymaking in disease control   255
4.1.1   Simple examples   255
4.1.2   Notebook for the SARS model in section 3.2   258

Appendix A   Notations and definitions   269

Appendix B   Examples of MATHEMATICA codes   271

Bibliography 275

PART 1
Mathematical Modeling in Epidemiology

Chapter 1

Epidemic Models

Epidemic models are usually used to study disease dynamics during a relatively short time period (e.g., within one year). In this case, the models often ignore processes that occur on longer time scales such as demographic processes (e.g., birth and death). Consequently, the models are usually simpler and easier for mathematical analyses. In this chapter, we focus primarily on the derivation of basic and control reproduction numbers and their relation to the final epidemic size. These quantities are important for designing control strategies. For discrete-time models, we also consider how model assumptions on the distribution of infectious stage may affect the estimates of reproduction numbers and final epidemic size. Particularly, we demonstrate that the commonly used simple discrete-time SIR or SEIR models implicitly assume that the infectious stage follows a geometric distribution, which is the only distribution with the memory-less property. When compared with models that assume more realistic sojourn distributions, the models provide contradictory predictions regarding the best control strategies. We provide an explanation for the apparent cause of the discrepant assessments.

This study reveals the importance of model assumptions. While memory-less distributions may be convenient mathematically, no biological process is memory-less. Evidently the information needed to characterize sojourn distributions must be collected (and if collected, reported) so that we can make realistic models. It would be hard to overestimate the importance of this. Models with arbitrarily distributed stage durations are also considered, from which formulas for the reproduction number and final epidemic size relation are derived. These formulas are expressed in terms of the probability quantities associated with the distribution, which makes the application of these models more easily particularly when disease data cannot be fitted well to any distributions from a particular parametric family.

1.1   Continuous-time models

Most continuous-time models are formulated using differential equations, including ordinary differential equations, partial differential equations, integral differential equations, among others. Several examples of such models are presented in this and later sections.

1.1.1 *Simple SIR and SEIR epidemic models*

For the single-outbreak epidemic models presented in this section, some standard assumptions are adopted. The total population is divided into several epidemiological classes. For example, for an SIR model, the total population is divided into three epidemiological classes: susceptible (S), infectious (I), and recovered (R). Thus, in an ODE model (non-structured), the numbers of individuals in these classes at time t are $S(t)$, $I(t)$, and $R(t)$, respectively, and the total number of individuals is $N(t) = S(t) + I(t) + R(t)$. The force of infection (i.e., the number of new infections at time t) will take either the form of mass action, $\beta S I$, or the form of standard incidence, $\beta S I / N$, where β is a constant representing the infection rate per susceptible when contacting infectious individuals. This implies the assumption that the population mixing is homogeneous (i.e., no heterogeneities in age, gender, activity, susceptibility, infectiousness, etc.). If the infectious period is assumed to be exponentially distributed (see section 1.2.3 for models with more general distributions) with mean $1/\gamma$, then γ gives the per-capita rate of recovery. It is also assumed that there are no disease deaths, so that the total population size remains constant for all time. Under these assumptions, the classical SIR epidemic model has the form

$$\frac{dS}{dt} = -\beta S \frac{I}{N},$$

$$\frac{dI}{dt} = \beta S \frac{I}{N} - \gamma I, \tag{1.1}$$

$$\frac{dR}{dt} = \gamma I,$$

with initial conditions $S(0) = S_0$, $I(0) = I_0 > 0$, and $R(0) = R_0$. This model is a special case of the general Kermack-McKendrick epidemic model (see [Brauer and Castillo-Chavez (2012)] for more details). The dynamical behavior of the model (1.1) can be determined by the basic reproduction number, denoted by $\mathcal{R}_0$ and given by

$$\mathcal{R}_0 = \frac{\beta}{\gamma}, \tag{1.2}$$

which gives the average number of secondary infections generated by a typical infectious individual during his/her entire period of infection in a wholy susceptible population. Either the disease will take off if $\mathcal{R}_0 > 1$ or it will decrease to zero if $\mathcal{R}_0 < 1$. If control programs are implemented, e.g., with the recover rate being replaced by $\tilde{\gamma} > \gamma$, then the quantity $\beta/\tilde{\gamma}$ will give the control reproduction number, denoted by $\mathcal{R}_C$ (C for control). If the population has a certain level of immunity at the beginning of an epidemic due to vaccination, e.g., a fraction p of the population is immune, then the reproduction number is $\mathcal{R}_v = (1 - p)\beta/\gamma$ (v for vaccination). We will use $\mathcal{R}$ to denote the reproduction number in general.

The SIR model (1.1) assumes permanent immunity after recovery from infection. This allows for the omission of the R equation in the system (1.1), and only the two equations for S and I are needed for analysis. Other commonly used simple models include SI (no recovery), SIS (no immunity gained from infection), SIRS (temporary immunity), etc.

An extension of the SIR model (1.1) is the SEIR model, in which an exposed class E (infected but not infectious) is included:

$$
\begin{aligned}
\frac{dS}{dt} &= -\beta S \frac{I}{N}, \\
\frac{dE}{dt} &= \beta S \frac{I}{N} - \alpha E, \\
\frac{dI}{dt} &= \alpha E - \gamma I,
\end{aligned}
\tag{1.3}
$$

with appropriate initial conditions. The R equation is omitted. The parameter α is a constant representing the rate at which an exposed individual becomes infectious. This also implies the assumption that the latent period is distributed exponentially.

1.1.2 *Reproduction number $\mathcal{R}$ and final epidemic size*

A useful measure for the severity of an epidemic is the total number of people infected during the disease outbreak, which is the so-called *final epidemic size* or *final size* for simple epidemic models, it is usually possible to derive a final size relation, which links the initial condition and the basic reproduction number to the final size (see section 9.2 in [Brauer and Castillo-Chavez (2012)]).

Consider the standard SIR model (1.1) and assume that $I_0 = I(0) > 0$, $R_0 = R(0) = 0$, and $S_0 + I_0 = N$. It has been shown that

$$
I_\infty = \lim_{t \to \infty} I(t) = 0 \quad \text{and} \quad S_\infty = \lim_{t \to \infty} S(t) > 0
\tag{1.4}
$$

(more details can be found in [Brauer *et al.* (2010)]). Thus, the final epidemic size can be determined from S_0 and S_∞. Recall that the basic reproduction number for this model is $\mathcal{R}_0 = \beta/\gamma$. Then, the final size relation is given by

$$
\log \frac{S_0}{S_\infty} = \mathcal{R}_0 \left(1 - \frac{S_\infty}{N} \right).
\tag{1.5}
$$

The implementation of disease control measures (e.g., drug treatment or mask use) may reduce the transmission rate (β) and/or reduce the infectious period ($1/\gamma$), in which case the control reproduction number is $\mathcal{R}_C < \mathcal{R}_0$. Therefore, the effect of these control measures on the final epidemic size can be evaluated using the relation (1.5). However, for models involving other control measures such as quarantine and isolation, the final size relation can be more difficult to obtain.

1.1.3   *Final size relation in more complicated epidemic models*

As mentioned above, the simple models (1.1) and (1.3) implicitly assume that the disease stages are exponentially distributed. Although such an assumption can make the model analysis much easier, it is not realistic for most diseases. More importantly, in some cases, models with exponential distributed disease stages may generate incorrect conclusions (see section 2.5 for more discussions). Therefore, it is useful to also consider models which relax this assumption. In this context, more general modeling freamwork can be found in [Brauer (2008); Brauer *et al.* (2010); Feng (2007); Ma and Earn (2006); Yang and Brauer (2008)]. The example presented below is adopted from [Brauer *et al.* (2010)], in which a more general form for stage progression is considered.

Consider an epidemic with progression from S through k infected stages $I_1, I_2, \cdots, I_k$. Assume that in stage i the relative infectivity is ε_i, the distribution of stay in the I_i stage is given by P_i with

$$P_i(0) = 1, \quad \int_0^\infty P_i(t)dt < \infty, \quad P_i'(t) \leq 0.$$

Denote the total infectivity by

$$\varphi(t) = \sum_{i=1}^k \varepsilon_i I_i(t)/N.$$

Then the model reads

$$S'(t) = -\beta S(t)\varphi(t),$$

$$I_1(t) = I_0 P_1(t) + \int_0^t [-S'(s)]P_1(t-s)ds, \qquad (1.6)$$

$$I_i(t) = \int_0^t B_i(s)P_i(t-s)ds, \quad i = 2, 3, \cdots, k,$$

where

$$B_{i+1}(t) = -\int_0^t B_i(s)P_i'(t-s)ds,$$

with initial conditions

$$S(0) = 0, \quad I_1(0) = I_0 > 0, \quad I_2(0) = I_3(0) = \cdots = I_k(0) = 0.$$

The basic reproduction number for the model (1.6) is

$$\mathcal{R}_0 = \beta \sum_{i=1}^k \varepsilon_i \int_0^\infty P_i(t)dt. \qquad (1.7)$$

It is shown in [Brauer *et al.* (2010)] that the final size relation is given by

$$\ln \frac{S_0}{S_\infty} = \mathcal{R}_0 \left[1 - \frac{S_\infty}{N}\right], \qquad (1.8)$$

which has the same form as in (1.5), except that the expression of $\mathcal{R}_0$ is different.

Incorporation of treatment

In [Brauer *et al.* (2010)], two treatment models as extensions of the model (1.6) are considered. The first is treatment at the beginning of a stage, with a specified fraction of individuals entering the stage being selected for treatment and moving to a treatment compartment. The second is treatment throughout a stage, with a rate of transfer from the untreated compartment to a treatment compartment.

Treatment at the beginning of a stage

To the model given by (1.6) we add treatment at the beginning of each stage. By this, we mean that a fraction p_1 of members of S who are infected go to a treatment compartment T_1 while the remaining fraction q_1 of newly infected members go to I_1. There is a sequence $T_1, T_2, \cdots, T_n$ of treated compartments with relative infectivity δ_i and period distribution Q_i in T_i. Thus, the total infectivity is given by

$$\varphi(t) = \sum_{i=1}^{k} [\varepsilon_i I_i(t)/N + \delta_1 T_i(t)/N]. \qquad (1.9)$$

Treated members continue through the treatment stages. In addition of the members leaving an infected stage I_i, a fraction p_i enters treatment in T_{i+1} while the remaining fraction q_i continues to I_{i+1}. We let m_i denote the fraction of infected members who go through the stage I_i and n_i the fraction of infected members who go through the stage T_i. Then

$$m_1 = q_1, m_2 = q_1 q_2, \cdots, m_k = q_1 q_2 \cdots q_k$$

$$n_1 = p_1, p_2 = p_1 + q_1 p_2, \cdots, n_k = p_1 + q_1 p_2 + \cdots + q_1 q_2 \cdots q_{k-1} p_k.$$

The treatment model in this case is the same as (1.6) except that the total infectivity is replace by (1.9). The control reproduction number $\mathcal{R}_C$ is

$$\mathcal{R}_C = \beta \sum_{i=1}^{k} \left[m_i \varepsilon_i \int_0^\infty P_i(t)dt + n_i \delta_i \int_0^\infty Q_i(t)dt \right], \qquad (1.10)$$

and the final size relation takes exactly the same form as in (1.8), i.e.,

$$\ln \frac{S_0}{S_\infty} = \mathcal{R}_C \left[1 - \frac{S_\infty}{N} \right], \qquad (1.11)$$

except that the basic reproduction number $\mathcal{R}_0$ given in (1.7) is replaced by $\mathcal{R}_C$ given in (1.10).

Treatment throughout a stage

We now model treatment by moving members from I_i at proportional rate γ_i to a treatment compartment T_i with relative infectivity δ_i and duration distribution Q_i. We let $C_i(s)$ denote the input to the treatment stage T_i at time s; this input includes output from the previous treatment stage T_{i-1} if $i > 1$ as well as the input from I_i. Thus, we replace the model (1.6) by a new model containing S, I_i, T_i. The total infectivity is now given by

$$\varphi(t) = \sum_{i=1}^{k} [\varepsilon_i I_i(t) + \delta_i T_i(t)]. \tag{1.12}$$

The model in this case becomes

$$S'(t) = -\beta S(t)\varphi(t),$$

$$I_1(t) = I_0 P_1 e^{-\gamma_1 t)} + \int_0^t [-S'(s)] P_1(t-s) e^{-\gamma_1(t-s)} ds,$$

$$I_i(t) = \int_0^t B_i(s) P_i(t-s) e^{-\gamma_i(t-s)} ds, \quad i = 2, 3, \cdots, k, \tag{1.13}$$

$$T_i(t) = \int_0^t C_i(s) Q_i(t-s) ds.$$

For ease of notation, let

$$\Gamma_i = 1 - \gamma_i \int_0^\infty P_i(u) e^{-\gamma_i u} du,$$

$$\Theta_i = \gamma_i \Gamma_i \Gamma_{i-1} \cdots \Gamma_1 + \gamma_{i-1} \Gamma_{i-1} \cdots \Gamma_1 + \cdots + \gamma_1 \Gamma_1.$$

Then the control reproduction number can then be written as

$$\mathcal{R}_C = \beta N \sum_{i=1}^{k} \left[\varepsilon_i \Gamma_{i-1} \Gamma_{i-2} \cdots \Gamma_1 \int_0^\infty e^{-\gamma_i t} P_i(t) dt + \delta_i \Theta_i \int_0^\infty Q_i(t) dt \right], \tag{1.14}$$

and the final size relation is the same as (1.11).

1.2   Discrete-time models

In many instances the formulation of discrete-time models (or discrete models) are easier to follow for non-mathematicians than continuous-time models (or continuous models), which is an advantage in the public health world (compare, for example, the SEIR models with quarantine and isolation using discrete model (1.40) and using continuous model (2.88)). Discrete models may also be more easily connected to disease data than continuous models, particularly when the stage distributions of latent and infectious periods do not fit well the standard family distributions.

It should be pointed out that the discrete models considered here are not the equations from the numerical discretization of the corresponding continuous models. Instead, the discrete models are derived based on the assumption on probability distributions in the epidemiological stages. One of the key advantages of this approach is that the models will not generate negative solutions regardless of the choice of time step.

In [Brauer *et al.* (2010)], we developed a discrete-epidemic framework and highlighted, through model analyses, the similarities between comparable classical continuous-time epidemic models and the discrete-time epidemic models, particularly the expressions for the final epidemic size. These models are extended in [Hernandez-Ceron *et al.* (2013a,b)] to incorporate arbitrary distribution for the infectious period as well as quarantine and isolation. Some of these models are presented here.

We start with a standard SEIR model with a simplifying assumption on the distributions of latent and infectious stages. That is, the distribution is assumed to be geometric, which is equivalent to assuming that individuals exit a stage with a constant probability at each time step (see section 1.2.1). We show that the reproduction number $\mathcal{R}_0$ can be expressed as a product of the disease transmission rate and the mean infectious period. Although the geometric distribution is the most commonly used assumption in discrete-time disease models (as it makes the model easier to formulate and analyze), we will consider examples to illustrate that the memoryless property of the geometric distribution (an analogous property of the exponential distribution in continuous-time models) may generate biased and possibly misleading evaluations on disease control strategies, which are presented in [Feng *et al.* (2007); Hernandez-Ceron *et al.* (2013a,b)]. Thus, it is important to consider more realistic distributions, as less dispersed distributed stages seem to be more appropriate for modeling diseases with longer latent and infectious periods (see, e.g., [Lessler *et al.* (2009); Lloyd (2001b); Wearing *et al.* (2005)]).

In the models considered in section 1.2.3, the geometric distribution assumption is relaxed to allow arbitrarily distributed stage durations. The general model is then applied to the case when specific distributions such as Poisson and Binomial distributions are used. It is shown that, although the structure of the expression for $\mathcal{R}_0$ remains the same, the formula for the mean infectious period can be different. For most of the models presented in sections 1.2.1–1.2.3, the disease transmission rates are considered to be infection-age-dependent, which are shown to influence the computation of $\mathcal{R}_0$. The formulas for $\mathcal{R}_0$ are derived in two ways, one is using the next generation matrix (NGM) approach and the other is obtained from biological considerations. It is verified that the formulas are consistent. We further extend the model in section 1.2.3 to include multiple pathogen strains or heterosexual transmission for sexually transmitted infections.

1.2.1 *Simple discrete SIR and SEIR models*

We start with analogues of the continuous SIR and SEIR models, (1.1) and (1.3), respectively. Later in this section we consider more realistic assumptions on the distribution of infectious period, including models with an arbitrarily distributed infectious stage.

The simple discrete epidemic model analogous to (1.1) is:

$$S_{n+1} = S_n e^{-\beta I_n/N}$$
$$I_{n+1} = S_n[1 - e^{-\beta I_n/N}] + \delta I_n. \tag{1.15}$$

Let $G_n = e^{-\beta I_n/N}$, which represents the survival probability of a susceptible individual from being infected per unit of time given the per capita infection rate $\beta I_n/N$. This approach of using an exponential form for discrete-time models guarantees that the solutions remain nonnegative for all time, and is different from the approach of using a numerical discretization of a continuous-time version of the model (in which case the S equations reads $S_{n+1} = S_n - S_n G_n$). The parameter δ is the probability that at each time step infectious individuals remain to be infectious to the next time step. The infectious individuals recover with probability $1 - \delta$ at each time step, and move to the R class which is omitted in (1.15). It is clear that

$$0 \le G_n \le 1, \quad 0 \le \delta \le 1.$$

As $G_n \le 1$, S_n is a decreasing sequence and has a limit $S_\infty \ge 0$ as $t \to \infty$. Note that $\delta \le 1$, $S_{n+1} + I_{n+1}$ is a decreasing sequence and has a limit $S_\infty + I_\infty$ as $t \to \infty$. Also, the difference of successive terms in this sequence $-(1 - \delta)I_n$ tends to zero, and this shows that $I_\infty = 0$.

Noticing that the mean infective period is

$$1 + \delta + \delta^2 + \cdots = \frac{1}{1 - \delta},$$

the basic reproduction number is

$$\mathcal{R}_0 = \frac{\beta}{1 - \delta},$$

and the final size relation becomes

$$\ln \frac{S_0}{S_\infty} = \mathcal{R}_0 \left[1 - \frac{S_\infty}{N} \right]. \tag{1.16}$$

The following simple discrete SEIR model is considered in [Hernandez-Ceron *et al.* (2013b)]. Similarly to the model (1.15), the latent and infectious periods are assumed to follow the geometric distribution with the parameters α and δ ($\alpha < 1, \delta < 1$), respectively. This is equivalent to assuming constant transition probabilities $1 - \alpha$ and $1 - \delta$ per unit time from E to I and from I to R, respectively. Then the SEIR model with geometric distributions for the latent and infectious periods has the form

$$S_{n+1} = S_n e^{-\beta \frac{I_n}{N}},$$
$$E_{n+1} = S_n (1 - e^{-\beta \frac{I_n}{N}}) + \alpha E_n \tag{1.17}$$
$$I_{n+1} = (1 - \alpha)E_n + \delta I_n, \quad n = 1, 2, \cdots.$$

A formula for the reproduction number $\mathcal{R}$ (either $\mathcal{R}_0$ or $\mathcal{R}_C$) are derived by adopting the method used in [Allen and van den Driessche (2008)] based on the next generation matrix approach. That is, in the discrete-time case

$$\mathcal{R} = \varrho(F(I - T)^{-1}), \tag{1.18}$$

where ϱ represents the spectral radius, F is the matrix associated with new infections and T is the matrix of transitions with $\varrho(T) < 1$ (see [Allen and van den Driessche

(2008); De Jong *et al.* (1994); Lewis *et al.* (2006); Wesley *et al.* (2009)]). Here F and T are calculated on the infected variables only evaluated at the disease-free equilibrium, and the Jacobian on these variables is $F + T$, which is assumed to be irreducible.

For system (1.17), the matrices associated with new infections and transitions are

$$F = \begin{bmatrix} 0 & \beta \\ 0 & 0 \end{bmatrix} \quad \text{and} \quad T = \begin{bmatrix} \alpha & 0 \\ 1 - \alpha & \delta \end{bmatrix},$$

respectively. Then $\varrho(T) = \max\{\alpha, \delta\} < 1$,

$$(I - T)^{-1} = \begin{bmatrix} 1 - \alpha & 0 \\ -(1 - \alpha) & 1 - \delta \end{bmatrix}^{-1} = \begin{bmatrix} \frac{1}{1-\alpha} & 0 \\ \frac{1}{1-\delta} & \frac{1}{1-\delta} \end{bmatrix},$$

and

$$F(I - T)^{-1} = \begin{bmatrix} \frac{\beta}{1-\delta} & \frac{\beta}{1-\delta} \\ 0 & 0 \end{bmatrix}.$$

Finally,

$$\mathcal{R}_0 = \varrho(F(I - T)^{-1}) = \frac{\beta}{1 - \delta}. \tag{1.19}$$

1.2.2 *Incorporation of quarantine and isolation*

Consider the case when two control measures are implemented: quarantine of latent and isolation of infectious individuals. Let n denote time (time step or generation time) and let S_n represent the number of susceptible, E_n the number of exposed but not yet infectious, Q_n the number of quarantined, I_n the number of infectious, and H_n the number of isolated individuals at time n. Assume that only individuals in the I and H classes are capable of transmitting the disease; and that individuals leave the latent and infectious stages at per-capita constant rates. Let α and δ denote the fractions of individuals remaining in the latent (E or Q) and infectious (I or H) stages, respectively, at each time step with $0 < \alpha, \delta < 1$. Denote by $\alpha(1 - \gamma)$ and $\delta(1 - \sigma)$ the fractions of newly quarantined and isolated individuals, respectively, at each time step. Finally, let β denote the transmission coefficient and $\rho \in (0, 1)$ the isolation efficiency (i.e., not effective if $\rho = 0$ and perfect isolation if $\rho = 1$).

Using the diagram in Figure 1.1, we arrive at the following discrete-time single-outbreak model:

$$\begin{aligned}
S_{n+1} &= S_n G_n \\
E_{n+1} &= (1 - G_n)S_n + \alpha \gamma E_n \\
Q_{n+1} &= \alpha(1 - \gamma)E_n + \alpha Q_n \\
I_{n+1} &= (1 - \alpha)E_n + \delta \sigma I_n \\
H_{n+1} &= (1 - \alpha)Q_n + \delta(1 - \sigma)I_n + \delta H_n, \quad n = 0, 1, 2, 3, \cdots,
\end{aligned} \tag{1.20}$$

where the force of infection at generation n is defined by

$$G_n = e^{-\frac{\beta}{N}[I_n + (1-\rho)H_n]}. \tag{1.21}$$

In the E_{n+1} equation in (1.20), the first term represents new infections, while the second term $(\alpha\gamma E_n)$ denotes those individuals who were infected in the previous step (at time n) but neither became infectious (α) nor got quarantined (γ) at time $n+1$. Other equations can be explained in a similar way. The equation for the recovered class R is ignored because it does not affect any other equations in (1.20). The initial conditions are denoted by $S_0, E_0, Q_0, I_0, H_0, R_0$ with $S_0, E_0 > 0$ and $I_0 = Q_0 = H_0 = R_0 = 0$. In this model we have assumed that quarantine only captures latent individuals, but not susceptible individuals, which is reasonable if quarantined individuals are much fewer than the susceptible population.

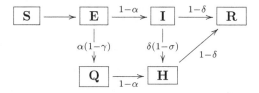

Fig. 1.1   Transition diagram for the discrete model (1.20) with constant transition rates.

The control reproduction number for model (1.20) is given by

$$\hat{\mathcal{R}}_C = \hat{\mathcal{R}}_I + \hat{\mathcal{R}}_{IH} + \hat{\mathcal{R}}_{QH}, \tag{1.22}$$

where the three quantities on the right-hand side are explicitly given by

$$\hat{\mathcal{R}}_I = \beta \frac{1-\alpha}{1-\alpha\gamma} \cdot \frac{1}{1-\delta\sigma},$$

$$\hat{\mathcal{R}}_{IH} = \beta(1-\rho)\frac{1-\alpha}{1-\alpha\gamma}\left(\frac{1}{1-\delta} - \frac{1}{1-\delta\sigma}\right), \tag{1.23}$$

$$\hat{\mathcal{R}}_{QH} = \beta(1-\rho)\frac{\alpha(1-\gamma)}{1-\alpha\gamma} \cdot \frac{1}{1-\delta}.$$

A more detailed derivation of these formulas and the biological interpretations of the factors involved in the formulas will be presented in section 1.2.3, where general distributions for the latent and infectious stages are considered (see (1.44) and (1.45)). $\hat{\mathcal{R}}_C$ gives the reproduction number restricted by the presence of control measures including quarantine and isolation. Each of the reproduction numbers in (1.23) represents the contribution (to secondary infections) by individuals stationed on distinct epidemiological state.

Because the model (1.20) does not include vital dynamics, we know that the final epidemic size can be described in terms of $N - S_\infty$ or the fraction $1 - S_\infty/N$ as described in the following result.

Theorem 1.1. *Let $\hat{\mathcal{R}}_C$ be given by Expression (1.22) then the final epidemic size generated by the dynamics of Model (1.20) satisfies the following equation*

$$\ln \frac{S_0}{S_\infty} = \left(1 - \frac{S_\infty}{N}\right) \hat{\mathcal{R}}_c. \tag{1.24}$$

Proof. From the S equation in (1.20) and Equation (1.21), we see that

$$\ln \left(\frac{S_0}{S_\infty}\right) = \frac{\beta}{N} \sum_{n=0}^{\infty} [I_n + (1 - \rho)H_n]. \tag{1.25}$$

We observe, making use of the E and S equations in (1.20), that $S_{n+1} + E_{n+1} = S_n + \alpha\gamma E_n \leq S_n + E_n$. Then, the sequence $\{S_n + E_n\}$ is decreasing and bounded by zero, thus convergent. Therefore, as $n \to \infty$

$$S_{n+1} - S_n + E_{n+1} - E_n = (\alpha\gamma - 1)E_n \to 0.$$

Similar arguments lead to the conclusion that $Q_n, I_n, H_n \to 0$ as $n \to \infty$. Moreover, the relationship

$$\sum_{n=0}^{m+1} E_n = E_0 + \sum_{n=1}^{m+1} [(1 - G_{n-1})S_{n-1} + \alpha\gamma E_{n-1}]$$

$$= E_0 + \sum_{n=1}^{m+1} (S_{n-1} - S_n) + \alpha\gamma \sum_{n=0}^{m} E_n,$$

and the fact that $S_0 + E_0 = N$, lead to

$$(1 - \alpha\gamma) \sum_{n=0}^{m} E_n = N - S_{m+1} - E_{m+1}.$$

Letting $m \to \infty$ we see that

$$\sum_{n=0}^{\infty} E_n = \frac{N - S_\infty}{1 - \alpha\gamma}. \tag{1.26}$$

From the Q equation in (1.20), it follows that

$$\sum_{n=1}^{m+1} Q_n = \alpha(1 - \gamma) \sum_{n=1}^{m+1} E_{n-1} + \alpha \sum_{n=1}^{m+1} Q_{n-1}. \tag{1.27}$$

Making use of (1.26) and (1.27), we conclude that

$$\sum_{n=1}^{\infty} Q_n = \frac{\alpha(1 - \gamma)}{1 - \alpha} \cdot \frac{N - S_\infty}{1 - \alpha\gamma} = (N - S_\infty) \left(\frac{1}{1 - \alpha} - \frac{1}{1 - \alpha\gamma}\right). \tag{1.28}$$

Similarly, using the I equation in (1.20) and Equation (1.26), we arrive at

$$\sum_{n=1}^{\infty} I_n = (N - S_\infty) \frac{1 - \alpha}{1 - \alpha\gamma} \cdot \frac{1}{1 - \delta\sigma}, \tag{1.29}$$

from where we conclude that

$$\frac{\beta}{N} \sum_{n=1}^{\infty} I_n = \left(1 - \frac{S_\infty}{N}\right) \mathcal{R}_I. \tag{1.30}$$

Finally, from the H equation in (1.20), we see that

$$\sum_{n=1}^{m+1} H_n = (1-\alpha)\sum_{n=1}^{m+1} Q_{n-1} + \delta(1-\sigma)\sum_{n=1}^{m+1} I_{n-1} + \delta\sum_{n=1}^{m+1} H_{n-1},$$

which together with equations (1.28) and (1.30), yield

$$\sum_{n=1}^{\infty} H_n = (N-S_\infty)\left[\frac{1}{1-\delta}\cdot\frac{\alpha(1-\gamma)}{1-\alpha\gamma} + \frac{1}{1-\delta}\cdot\delta(1-\sigma)\cdot\frac{1-\alpha}{1-\alpha\gamma}\cdot\frac{1}{1-\delta\sigma}\right]$$

$$= (N-S_\infty)\left[\frac{1-\alpha}{1-\alpha\gamma}\left(\frac{1}{1-\delta} - \frac{1}{1-\delta\sigma}\right) + \frac{\alpha(1-\gamma)}{1-\alpha\gamma}\cdot\frac{1}{1-\delta}\right].$$

$$(1.31)$$

Thus,

$$\frac{\beta}{N}\sum_{n=1}^{\infty}(1-\rho)H_n = \left(1-\frac{S_\infty}{N}\right)(\mathcal{R}_{IH} + \mathcal{R}_{QH}). \qquad (1.32)$$

Because our initial conditions include $Q_0 = I_0 = H_0 = 0$, the use of (1.22), (1.30), and (1.32) leads to (1.24). This last computation concludes the proof of Theorem 1.1. $\qquad\square$

The final size relation in (1.24) can be rewritten using the proportional size of the final epidemic $y = 1 - S_\infty/N$ as follows:

$$\frac{\ln S_0}{N(1-y)} = y\mathcal{R}_C. \qquad (1.33)$$

Although this last equation cannot be solved analytically, the relation between $\mathcal{R}_C$ and the final size y can, from the above expression, be numerically determined using Equation (1.33). A contour plot of the equation $\ln S_0/(N(1-y)) - y\mathcal{R}_C = 0$, Figure 1.2, supports the view that the final size y is an increasing function of $\mathcal{R}_C$.

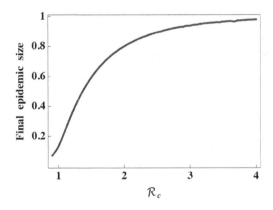

Fig. 1.2　Contour plot of the function for the final epidemic size given in (1.33).

1.2.3 *Models with arbitrarily distributed stage durations*

In [Hernandez-Ceron *et al.* (2013a,b)], we extend the discrete model (1.15) to incorporate an arbitrary distribution for the infectious stage, and illustrate that the model (1.15) is a special case when the distribution is geometric. Particularly, a final size relationship is computed and shown to involve the control reproduction number, which is a function of transmission parameters and the means of distributions used to model disease or intervention control measures. Model results and simulations highlight the inconsistencies in forecasting that emerge from the use of specific parametric distributions. Examples with the geometric, Poisson and binomial distributions are used to demonstrate the impact of the choices made in quantifying the risk posed by single outbreaks on the model predictions for the relative importance of various control measures.

A general single-outbreak discrete epidemic model involving arbitrarily distributed stage-duration distributions for the latent and the infectious stages is introduced in this section. Clarity is added to model derivation by making use of a probabilistic perspective. Hence, we let X and Y represent the time an individual spends in latent (E, Q) and infectious (I, H) classes, respectively. Similarly, denote by Z the time at which an exposed individual is quarantined (from E to Q), and W the time at which an infected individual is isolated (from I to H). X, Y, Z and W must take values on $\{1, 2, 3, \dots\}$ and their dynamics are assumed to be governed by underlying probabilistic processes. Making use of probabilistic language facilitates the interpretation and applicability of our deterministic model results. Hence, it is understood that the distribution of waiting time in each of the above classes can be described by probability distributions associated with the random variables X, Y, Z and W:

$p_i = \mathbb{P}(X > i)$ probability of remaining latent i steps after entering E,

$q_i = \mathbb{P}(Y > i)$ probability of remaining infectious i steps after entering I,

$k_i = \mathbb{P}(Z > i)$ probability of not quarantined i steps after entering E, $\qquad$ (1.34)

$l_i = \mathbb{P}(W > i)$ probability of not isolated i steps after entering I.

Notice that $p_i \geq p_{i+1}$, $\mathbb{P}(X = i) = p_{i-1} - p_i$ and $\mathbb{E}(X) = \sum_{i=0}^{\infty} p_i$, where $\mathbb{E}(X)$ is the mean or expectation of X. Similar equalities hold for q_i, k_i and l_i. Assume that $p_0 = q_0 = k_0 = l_0 = 1$, meaning that the latency, infectious, quarantine and isolation periods last at least one time step. For ease of presentation, we introduce the following notation:

$\quad A_n$ the input to E at time n (new infections),
$\quad B_n$ the input to I at time n,
$\quad C_n$ the input to Q at time n,
$\quad D_n$ the input to H from Q at time n,
$\quad F_n \ (= B_n + D_n)$ the total input to infectious class at time n.

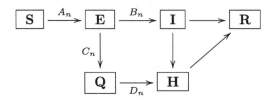

Fig. 1.3   Transition diagram for the model with arbitrarily distributed stage durations.

A transition diagram using the above notation is shown in Figure 1.3.

In the general formulation, the equation associated with changes in the susceptible population is

$$S_{n+1} = S_n G_n, \quad n = 0, 1, 2, 3, \cdots, \tag{1.35}$$

where $G_n = e^{-\frac{\beta}{N}[I_n + (1-\rho)H_n]}$. Equations for A_n are given recursively by

$$A_0 = E_0, \quad A_{n+1} = S_n(1 - G_n) = S_n - S_{n+1}, \quad n \geq 0. \tag{1.36}$$

Similarly, the equations for B_n, C_n, and D_n can be written as

$$\begin{aligned} B_{n+1} &= A_n(1 - p_1) + A_{n-1}(p_1 - p_2)k_1 + \cdots + A_1(p_{n-1} - p_n)k_{n-1} \\ &\quad + A_0(p_n - p_{n-1})k_n \end{aligned} \tag{1.37}$$

with $B_0 = 0$ and $k_0 = 1$;

$$C_{n+1} = E_n - E_{n+1} + A_{n+1} - B_{n+1}, \tag{1.38}$$

with $C_0 = 0$; and

$$D_{n+1} = Q_n - (Q_{n+1} - C_{n+1}). \tag{1.39}$$

Using the above notation the model with arbitrarily distributed stage durations can be written as:

$$\begin{aligned} S_{n+1} &= S_n G_n, \quad G_n = e^{-\frac{\beta}{N}[I_n + (1-\rho)H_n]}, \\ E_{n+1} &= A_{n+1} + A_n p_1 k_1 + \cdots + A_1 p_n k_n + A_0 p_{n+1} k_{n+1}, \\ Q_{n+1} &= A_n p_1(1 - k_1) + \cdots + A_1 p_n(1 - k_n) + A_0 p_{n+1}(1 - k_{n+1}), \\ I_{n+1} &= B_{n+1} + B_n q_1 l_1 + \cdots + B_1 q_n l_n, \\ H_{n+1} &= F_{n+1} + F_n q_1 + \cdots + F_1 q_n - I_{n+1}, \quad n = 0, 1, 2, \cdots, \end{aligned} \tag{1.40}$$

where A_n, B_n, C_n and D_n are given in (1.36), (1.37), (1.38) and (1.39). The initial conditions in model (1.40) is $S_0, E_0 > 0$ and $I_0 = Q_0 = H_0 = R_0 = 0$.

The mean sojourn time and the "quarantine adjusted" mean sojourn time in the exposed stage are given by

$$\mathcal{D}_E = \mathbb{E}(X) \quad \text{and} \quad \mathcal{D}_{E_k} = \mathbb{E}(X \wedge Z), \tag{1.41}$$

respectively, where $X \wedge Z = \min\{X, Z\}$ and

$$\mathbb{E}(X \wedge Z) = \sum_{j=0}^{\infty} \mathbb{P}(X \wedge Z > j) = 1 + \sum_{j=1}^{\infty} p_j k_j.$$

Similarly, the mean sojourn and the "isolation adjusted" mean sojourn times in the infectious stage are given by

$$\mathcal{D}_I = \mathbb{E}(Y) \quad \text{and} \quad \mathcal{D}_{I_l} = \mathbb{E}(Y \wedge W), \tag{1.42}$$

respectively, where

$$\mathbb{E}(Y \wedge W) = \sum_{j=0}^{\infty} \mathbb{P}(Y \wedge W > j) = 1 + \sum_{j=1}^{\infty} q_j l_j.$$

Denote by $\mathcal{T}_{E_k}$ the proportion of individuals in the E class who become infectious before going to the Q class, i.e., the event $E \to I$ occurs before $E \to Q$. This quantity is given by

$$\begin{aligned}
\mathcal{T}_{E_k} &= \mathbb{P}(X \le Z) = \sum_{j=1}^{\infty} \mathbb{P}(X = j, j \le Z) = \sum_{j=1}^{\infty} \mathbb{P}(X = j)\mathbb{P}(j - 1 < Z) \\
&= \sum_{j=1}^{\infty} (p_{j-1} - p_j)k_{j-1}.
\end{aligned} \tag{1.43}$$

Then

$$\mathcal{R}_I = \beta \mathcal{T}_{E_k} \mathcal{D}_{I_l} \tag{1.44}$$

represents the number of secondary infections produced in a susceptible population by an individual in the I class during his/her infectious period. All the notations used above are also listed in Table 1.1.

There are two types of individual in the H class, those who, i) entered H from I and ii) entered H from Q. From the definitions for $\mathcal{D}_I$ and $\mathcal{D}_{I_l}$ given in (1.42), the average time spent in H is $\mathcal{D}_I - \mathcal{D}_{I_l}$ for type i) individuals and is $\mathcal{D}_I$ for type ii) individuals. The proportions of type i) and type ii) individuals are $\mathcal{T}_{E_k}$ and $1 - \mathcal{T}_{E_k}$, respectively. Considering the isolation efficiency determined by ρ we know that the average numbers of secondary infections produced by type i) and type ii) individuals are

$$\mathcal{R}_{IH} = \beta(1 - \rho)\mathcal{T}_{E_k}(\mathcal{D}_I - \mathcal{D}_{I_l}) \quad \text{and} \quad \mathcal{R}_{QH} = \beta(1 - \rho)(1 - \mathcal{T}_{E_k})\mathcal{D}_I, \tag{1.45}$$

respectively, where $\mathcal{D}_I$, $\mathcal{D}_{I_l}$ and $\mathcal{T}_{E_k}$ are given in (1.42) and (1.43).

Theorem 1.2. *The control reproduction number $\mathcal{R}_C$ for the general model (1.40) can be expressed in terms of the mean $\mathcal{D}_I$, the isolation-adjusted mean $\mathcal{D}_{I_l}$, and the quarantine-adjusted probability of disease progression $\mathcal{T}_{E_k}$. That is,*

$$\mathcal{R}_c = \mathcal{R}_I + \mathcal{R}_{IH} + \mathcal{R}_{QH} \tag{1.46}$$

where $\mathcal{R}_I$, $\mathcal{R}_{IH}$ and $\mathcal{R}_{QH}$ are the stage-specific reproduction numbers defined in (1.44) and (1.45).

1.2.4   *Applications of the general model to specific distributions*

The general model (1.40) and the results from the model allows one to apply any distributions, including the use of empirical probability distributions computed from data as well as three of the most commonly used discrete distributions: the geometric, Poisson and binomial distributions (see Figures 1.4-1.6). We show first that the $\mathcal{R}_C$ formula (1.22) obtained for model (1.20) is indeed a special case of the general formula (1.46) when the distributions for the latent and infectious stages are geometric.

The case of geometric distribution

The geometric distribution, a discrete probability distribution supported on $\{1, 2, 3, \dots\}$, represents the number of independent Bernoulli trials needed to get a single success (see [Casella and Berger (2001)]). When X follows a geometric distribution with parameter α ($X \sim \text{Geom}(\alpha)$, see Figure 1.4),

$$\mathbb{P}(X = 0) = 1, \quad \mathbb{P}(X = i) = \alpha^{i-1}(1 - \alpha), \quad \mathbb{P}(X > i) = \alpha^i \quad \text{for } i = 1, 2, \dots.$$

Similar to its continuous analogue (the exponential distribution), the geometric distribution is memoryless, that is, for any $i, j \in \mathbb{N}$,

$$\mathbb{P}(X > i + j \,|\, X > i) = \mathbb{P}(X > j).$$

Consider the case where the distributions in (1.34) are geometric. Assume that $X \sim \text{Geom}(\alpha)$, $Y \sim \text{Geom}(\delta)$, $Z \sim \text{Geom}(\gamma)$ and $W \sim \text{Geom}(\sigma)$. That is,

$$p_i = \alpha^i, \quad q_i = \delta^i, \quad k_i = \gamma^i, \quad l_i = \sigma^i, \quad i = 0, 1, 2, \dots.$$

Using the geometric distribution assumption it can be verified that the equations in (1.40) for E, Q, I and H are the same as those in model (1.20).

The mean of X can be easily computed

$$\mathbb{E}(X) = \sum_{i=0}^{\infty} p_i = \sum_{i=0}^{\infty} \alpha^i = \frac{1}{1 - \alpha} > 1, \quad \alpha \in (0, 1).$$

Similarly, $\mathbb{E}(Y) = \frac{1}{1-\delta}$, $\mathbb{E}(Z) = \frac{1}{1-\gamma}$ and $\mathbb{E}(W) = \frac{1}{1-\sigma}$. Moreover, $\mathbb{P}(X \wedge Z > i) = \alpha^i \gamma^i$, so that $\mathbb{E}(X \wedge Z) = \frac{1}{1-\alpha\gamma}$. Therefore,

$$\mathcal{D}_E = \frac{1}{1 - \alpha}, \quad \mathcal{D}_I = \frac{1}{1 - \delta}, \quad \mathcal{D}_{E_k} = \frac{1}{1 - \alpha\gamma}, \quad \mathcal{D}_{I_l} = \frac{1}{1 - \delta\sigma}. \quad (1.47)$$

It follows that

$$\mathcal{T}_{E_k} = \mathbb{P}(X \le Z) = \sum_{j=1}^{\infty}(p_{j-1} - p_j)k_{j-1} = \sum_{j=1}^{\infty} \alpha^{j-1}(1 - \alpha)\gamma^{j-1} = \frac{1 - \alpha}{1 - \alpha\gamma}. \quad (1.48)$$

Combining (1.47) and (1.48), we can rewrite the components of the control reproduction number (1.46) in this case as follows:

$$\mathcal{R}_I = \beta \mathcal{T}_{E_k} \mathcal{D}_{I_l} = \beta \frac{1 - \alpha}{1 - \alpha\gamma} \cdot \frac{1}{1 - \delta\sigma},$$

$$\mathcal{R}_{IH} = \beta(1 - \rho)\mathcal{T}_{E_k}(\mathcal{D}_I - \mathcal{D}_{I_l}) = \beta(1 - \rho)\frac{1 - \alpha}{1 - \alpha\gamma} \cdot \left(\frac{1}{1 - \delta} - \frac{1}{1 - \delta\sigma} \right),$$

$$\mathcal{R}_{QH} = \beta(1 - \rho)(1 - \mathcal{T}_{E_k})\mathcal{D}_I = \beta(1 - \rho)\frac{\alpha(1 - \gamma)}{1 - \alpha\gamma} \cdot \frac{1}{1 - \delta},$$

Geometric density function

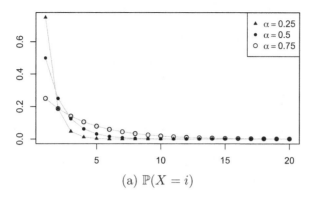

(a) $\mathbb{P}(X = i)$

Geometric survival function

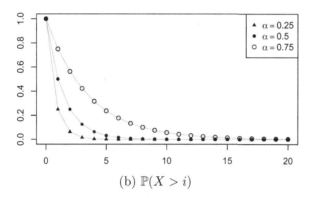

(b) $\mathbb{P}(X > i)$

Fig. 1.4 Plots of the probability mass and survival function of geometric distributions with parameters $\alpha = 0.25, 0.5$, and 0.75.

which are the same expressions as $\hat{\mathcal{R}}_I$, $\hat{\mathcal{R}}_{IH}$, and $\hat{\mathcal{R}}_{QH}$ given in (1.23).

Thus, we have verified that model (1.20) is a special case of the general model (1.40) when the stage distributions are geometric.

The control reproduction number and final epidemic size are important measures, which are often used to compare the effectiveness of control strategies like quarantine and/or isolation. In our framework, $\mathcal{R}_C$ can be used to examine the impact of the shape of the waiting time distributions as well. Here, we compare the role of three classical discrete distributions when they are used to model the latent and infectious stage durations (described by X and Y, respectively). The geometric distribution assumption (GDA) will be linked to the random variables X_g and Y_g; the Poisson distribution assumption (PDA) to X_p and Y_p; and binomial distribution assumption (BDA) to X_b and Y_b. We refer to the models corresponding to these distributions as GDM (geometric distribution model), PDM (Poisson distribution

Table 1.1   Definitions of symbols related to the discrete-time model (1.40).

Symbol	Definition
S_n	Number of susceptible individuals at time n
E_n	Number of exposed (not infectious) at time n
I_n	Number of infected (infectious) at time n
Q_n	Number of quarantined (not infectious) at time n
H_n	Number of isolated (infectious)
β	Transmission coefficient
ρ	Isolation efficiency, $\rho \in [0, 1]$
α	Fraction of individuals remaining in latent stages (E or Q) after each time step (under GDA)
δ	Fraction of individuals remaining in infectious stages (I or H) after each time step (under GDA)
$1 - \gamma$	Quarantine probability each time step (in GDM, PDM and BDM)
$1 - \sigma$	Isolation probability each time step (in GDM, PDM and BDM)
X	Random variable for the time spent in latent stages (E, Q)
Y	Random variable for the time spent in infectious stages (I, H)
Z	Random variable that indicates the moment at which quarantine occurs (from E to Q)
W	Random variable that indicates the moment at which isolation occurs (from I to H)
p_j	$\mathbb{P}(X > i)$, survival probability of X: proportion of individuals that remain latent i steps after infection
q_j	$\mathbb{P}(Y > i)$, survival probability of Y: proportion of individuals that remain infectious i steps after becoming infectious
k_j	$\mathbb{P}(Z > i)$, survival probability of Z: proportion of individuals who are not quarantined i steps after infection
l_j	$\mathbb{P}(W > i)$, survival probability of W: proportion of individuals who are not isolated i steps after becoming infectious
$\mathcal{D}_E$	Latent period ($= \mathbb{E}(X)$)
$\mathcal{D}_I$	Infectious period ($= \mathbb{E}(Y)$)
$\mathcal{D}_{I_l}$	Isolation adjusted infectious period ($= \mathbb{E}(Y \wedge Z)$)
$\mathcal{T}_E$	Quarantine adjusted probability, $\mathbb{P}(X \leq Z)$
$\mathcal{R}_C$	Control reproduction number
μ_1, μ_2	Fixed mean of X and Y, when comparing GDM, PDM and BDM
GDA, PDA, BDA	Geometric, Poisson, Binomial Distribution Assumption, respectively
GDM, PDM, BDM	Geometric, Poisson, Binomial Distribution Model, respectively

model), and BDM (binomial distribution model), respectively. For the quarantine (described by Z) and isolation (described by W) period distributions, for simplicity it will be assumed here that they follow geometric distributions with means

$$\mathbb{E}(Z) = \frac{1}{1 - \gamma} \quad \text{and} \quad \mathbb{E}(W) = \frac{1}{1 - \sigma}.$$

As a baseline case, the GDA will be assumed here with means

$$\mathbb{E}(X_g) = \frac{1}{1 - \alpha} = \mu_1 > 1 \quad \text{and} \quad \mathbb{E}(Y_g) = \frac{1}{1 - \delta} = \mu_2 > 1. \tag{1.49}$$

The GDA case will be compared to the PDA case with parameters

$$X_p - 1 \sim \text{Poiss}(\mu_1 - 1) \quad \text{and} \quad Y_p - 1 \sim \text{Poiss}(\mu_1 - 1),$$

and to the BDA case with parameters

$$X_b - 1 \sim \text{Binom}(n_1, a) \quad \text{and} \quad Y_b - 1 \sim \text{Binom}(n_2, b).$$

The support of a Poisson distributed random variable P with parameter η is the set $\{0, 1, 2, \dots\}$ with

$$\mathbb{P}(P = i) = e^{-\eta}\frac{\eta^i}{i!} \quad \text{and} \quad \mathbb{E}(P) = \eta \tag{1.50}$$

(see Figure 1.5). Therefore, X_p is a shifted Poisson random variable with support $\{1, 2, 3 \dots\}$ and mean μ_1.

Poisson density function

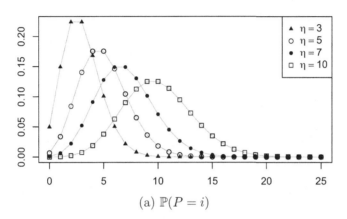

(a) $\mathbb{P}(P = i)$

Poisson survival function

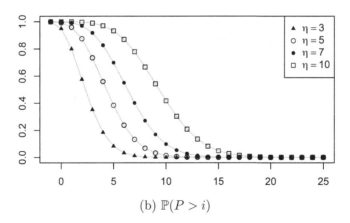

(b) $\mathbb{P}(P > i)$

Fig. 1.5 Plots of the probability mass and survival function of Poisson distributions with parameters $\eta = 3, 5, 7, 10$.

The support of a binomial distributed random variable B with integer size parameter $m > 0$ and probability parameter $p \in (0, 1)$ is the set $\{0, 1, 2, \dots, m\}$

(Figure 1.6). Its density function and expectation are given by

$$\mathbb{P}(B = i) = \binom{m}{i} p^i (1-p)^{m-i} \quad \text{and} \quad \mathbb{E}(B) = mp. \qquad (1.51)$$

Thus, the parameters n_1 and a of the shifted binomial r.v. X_b must satisfy $(n_1 - 1)a + 1 = \mu_1$. Here, n_1 is the maximum amount of time that an infected individual will remain latent (i.e., either in the E or the Q class). That is, X_b is bounded by n_1.

Binomial density function

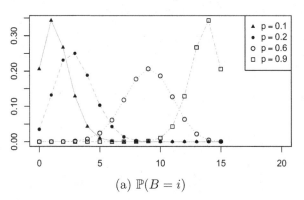

(a) $\mathbb{P}(B = i)$

Binomial survival function

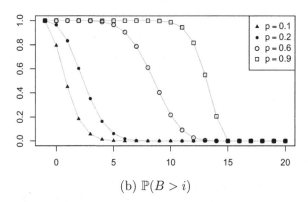

(b) $\mathbb{P}(B > i)$

Fig. 1.6   Plots of the probability mass and survival function of binomial distributions with parameters $m = 15$ and $p = 0.1, 0.2, 0.6, 0.9$.

Expected remaining sojourns

Similar to the approach used in [Feng *et al.* (2007)] for continuous-time models, we can analyze the role of stage duration distributions in the expected remaining sojourns and explore how the use of distinct distributions may lead to discrepancies in model predictions.

The expected remaining sojourn, denoted by $\mathcal{M}_U(s)$ ($U = X, Y$), represents the expected remaining time in a stage given that s units of time have elapsed in the stage. It is defined as follows. For distributions with the property that $\mathbb{P}(U > s) > 0$ for all s, where s is the time steps that have elapsed at a given stage (latent or infectious),

$$\mathcal{M}_U(s) = \sum_{n=0}^{\infty} \mathbb{P}(U > n | U > s)$$

$$= \sum_{n=0}^{\infty} \frac{\mathbb{P}(U > n + s)}{\mathbb{P}(U > s)}.$$

Thus,

$$\mathcal{M}_X(s) = \sum_{n=0}^{\infty} \frac{p_{n+s}}{p_s}, \quad \mathcal{M}_Y(s) = \sum_{n=0}^{\infty} \frac{q_{n+s}}{q_s}.$$

For bounded distributions, i.e., there exists a constant $M > 0$ such that $\mathbb{P}(U > M) = 0$,

$$\mathcal{M}_U(s) = \begin{cases} \sum_{n=0}^{\infty} \mathbb{P}(U > n | U > s) = \sum_{n=0}^{\infty} \frac{\mathbb{P}(U > n+s)}{\mathbb{P}(U > s)} & \text{if } \mathbb{P}(U > s) > 0 \\ 0 & \text{if } \mathbb{P}(U > s) = 0. \end{cases}$$

Thus, if U is bounded by M, then $\mathcal{M}_U(m) = 0$ for all $m \geq M$. Clearly $\mathcal{M}_U(0) = \mathbb{E}(U)$ for $U = X, Y$. Note that in the case of GDM, i.e., for X_g (GDA) with parameter α, we have $p_{i,g} = \alpha^j$ and

$$\mathcal{M}_{X_g}(s) = \sum_{n=0}^{\infty} \frac{\alpha^{n+s}}{\alpha^s} = \sum_{n=0}^{\infty} \alpha^n = \mathbb{E}(X_g) = \mathcal{M}_{X_g}(0),$$

which shows that the expected remaining sojourn after an individual already spent s units of time in the latent stage is independent of s. This may contribute in a significant way to the potentially biased model predictions on the effect of disease control strategies. The use of PDM and BDM may lead to more reliable assessments because of their ability to capture more accurately the description for the expected remaining sojourns. Figure 1.8 illustrates the difference among the three distribution assumptions (GDA, PDA, and BDA) by plotting the expected remaining sojourn as a function of s (the time elapsed after entering the latent stage). This figure shows that the function is constant under GDA, while the functions correspond to PDA and BDA decreases with s.

1.2.5 *Control reproduction number $\mathcal{R}_C$ and final epidemic size*

Let $\mathcal{R}_{C,g}$, $\mathcal{R}_{C,p}$ and $\mathcal{R}_{C,b}$ denote the control reproduction numbers for the GDM, PDM and BDM, respectively. For $\mathcal{R}_{C,p}$, we can first specify $\mathcal{T}_{E_k,p}$ and $\mathcal{D}_{I_l,p}$ using

Expected remaining sojourn at step s

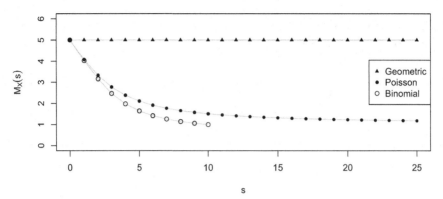

Fig. 1.7   Plots of the expected remaining sojourn time, $\mathcal{M}_X(s)$, in a disease stage (latent or infectious) when s units of time have elapsed after entering the stage. It illustrates the difference among the three distributions as specified in the legend. Because the binomial random variable X is bounded, $\mathcal{M}_X(s) = 0$ after its upper bound.

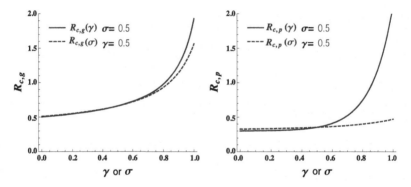

Fig. 1.8   Sensitivity of $\mathcal{R}_C$ to the variation in one of the control parameters (γ and σ) while the other is fixed. The two models considered are the GDM and PDM.

(1.50) as follows:

$$\mathcal{T}_{E_k,p} = \sum_{i=1}^{\infty} \mathbb{P}(X_p = i)\mathbb{P}(Z > i - 1) = e^{-(\mu_1-1)} \sum_{i=1}^{\infty} \frac{(\mu_1-1)^{i-1}}{(i-1)!} \cdot \gamma^{i-1}$$

$$= e^{-(\mu_1-1)(1-\gamma)}$$

$$\mathcal{D}_{I_l,p} = 1 + \sum_{i=1}^{\infty} q_{i,p}l_i = 1 + \sum_{i=1}^{\infty} \left(1 - e^{-(\mu_2-1)} \sum_{j=1}^{i} \frac{(\mu_2-1)^{j-1}}{(j-1)!}\right)\sigma^i$$

$$= 1 + \frac{\sigma}{1-\sigma} - e^{-(\mu_2-1)} \sum_{i=1}^{\infty}\sum_{j=0}^{i-1} \frac{(\mu_2-1)^j}{j!}\sigma^i = \frac{1}{1-\sigma} - e^{-(\mu_2-1)} \sum_{j=0}^{\infty}\sum_{i=j+1}^{\infty} \frac{(\mu_2-1)^j}{j!}\sigma^i$$

$$= \frac{1}{1-\sigma} - e^{-(\mu_2-1)} \sum_{j=0}^{\infty} \frac{(\mu_2-1)^j}{j!} \frac{\sigma^{j+1}}{1-\sigma} = \frac{1}{1-\sigma} - e^{-(\mu_2-1)} \frac{\sigma}{1-\sigma} \sum_{j=0}^{\infty} \frac{[(\mu_2-1)\sigma]^j}{j!}$$

$$= \frac{1-\sigma e^{-(\mu_2-1)(1-\sigma)}}{1-\sigma}.$$

Notice that, $\mathcal{D}_{I_l} = \mathcal{D}_I$ for $\sigma = 1$. Then, from (1.45)

$$
\mathcal{R}_{C,p} = \begin{cases} \beta \left[(1-p)\mu_2 + pe^{-(\mu_1-1)(1-\gamma)} \cdot \frac{1-\sigma e^{-(\mu_2-1)(1-\sigma)}}{1-\sigma} \right] & \text{if } 0 \le \sigma < 1, \\ \beta\mu_2 \left[1 - p(1 - e^{-(\mu_1-1)(1-\gamma)}) \right] & \text{if } \sigma = 1. \end{cases}
$$

Although it is computationally easy to obtain the value for $\mathcal{R}_{C,b}$ in case of a binomial distribution, its expression is not simple. For example, if $X-1 \sim \text{Binom}(n_1, a)$ then

$$
p_{i,b} = \mathbb{P}(X > i) = \sum_{k=i+1}^{n_1+1} \mathbb{P}(X = k) = \sum_{k=i+1}^{n_1+1} \binom{n_1}{k-1} a^{k-1}(1-a)^{n_1-k+1}.
$$

Similarly, $\mathcal{T}_{E_k,b}$ and $\mathcal{R}_{C,b}$ consist of finite sums that cannot be simplified.

The derivatives of $\mathcal{R}_C$ with respect to the control parameters (e.g., γ and σ) can provide useful information about the effect of controls on the reduction of $\mathcal{R}_C$. Note that $0 < \alpha, \delta, \gamma, \sigma < 1$ and $\mu_i > 1$ for $i = 1, 2$. Then the partial derivatives under GDA and PDA have the following properties:

$$
\frac{\partial \mathcal{R}_{C,g}}{\partial \gamma} = \beta\rho \cdot \frac{1-\alpha}{1-\delta\sigma} \cdot \frac{\alpha}{(1-\alpha\gamma)^2} > 0,
$$

$$
\frac{\partial \mathcal{R}_{C,g}}{\partial \sigma} = \beta\rho \cdot \frac{1-\alpha}{1-\alpha\gamma} \cdot \frac{\delta}{(1-\delta\sigma)^2} > 0,
$$

and

$$
\frac{\partial \mathcal{R}_{C,p}}{\partial \gamma} = \beta\rho(\mu_1-1) \cdot e^{-(\mu_1-1)(1-\gamma)} \cdot \frac{1-\sigma e^{-(\mu_2-1)(1-\sigma)}}{1-\sigma} > 0,
$$

$$
\frac{\partial \mathcal{R}_{C,p}}{\partial \sigma} = \beta\rho \cdot e^{-(\mu_1-1)(1-\gamma)} \cdot \frac{e^{-(\mu_2-1)(1-\sigma)}[(\mu_2-1)\sigma^2 - (\mu_2-1)\sigma - 1] + 1}{(1-\sigma)^2} > 0.
$$

To prove the last inequality, notice that $\frac{\partial \mathcal{R}_{C,p}}{\partial \sigma} > 0$ if and only if

$$
f(\sigma) = e^{-(\mu_2-1)(1-\sigma)}\left[(\mu_2-1)\sigma^2 - (\mu_2-1)\sigma - 1\right] + 1 > 0, \quad \sigma \in (0,1).
$$

From $f'(\sigma) = (\mu_2-1)e^{-(\mu_2-1)(1-\sigma)}\left[(\mu_2-1)\sigma^2 - (\mu_2-3)\sigma - 2\right]$ and $\mu_2 > 1$ (see (1.49)), we know that $f'(\sigma) = 0$ if and only if $\sigma = -\frac{2}{\mu_2-1}$ or 1. This implies that f is strictly monotone in the interval $\left[-\frac{2}{\mu_2-1}, 1 \right]$, and particularly in $(0,1)$. Then, the inequality follows from the fact that $f(0) > 0$ and $f(1) = 0$.

1.2.6 *Influence of stage distributions on $\mathcal{R}_C$ and final size*

In this section, we examine the role of model assumptions (GDA, PDA and BDA) on the impact of quarantine and isolation in reducing $\mathcal{R}_C$ and, consequently, the final epidemic size for the models (GDM, PDM and BDM). For the purpose of comparison, we take X_g, X_p, X_b with mean μ_1 and Y_g, Y_p, Y_b with mean μ_2. Figure 1.9 plots the control reproduction number $\mathcal{R}_C$ vs γ (quarantine) or σ (isolation). The left figure is for $\mathcal{R}_{c,g}$ (i.e., when X and Y are geometric distributions) and the right figure is for $\mathcal{R}_{c,p}$ (i.e., when X and Y are Poisson distributions). Figure 1.9 shows reduction in $\mathcal{R}_C$ under two control strategies: Strategy I corresponds to $\gamma = 0.5$ and $\sigma = 0.8$ (represented by a dot $\bullet$ on the curve) while Strategy II corresponds to $\gamma = 0.8$ and $\sigma = 0.5$ (represented by a solid diamond $\blacklozenge$ on the curve). We see that the two distributions (geometric and Poisson) may generate contradictory assessments. For example, it is shown in the left figure that under the geometric distribution assumption, Strategy II is more effective than Strategy I, because it leads to larger reductions in $\mathcal{R}_{c,g}$. However, it is shown in the right figure that under the Poisson distribution, Strategy I is more effective than Strategy II, because it leads to larger reductions in $\mathcal{R}_{c,p}$. The parameter values used are $\beta = 0.75, \rho = 0.95, \mu_1 = 5, \mu_2 = 10$.

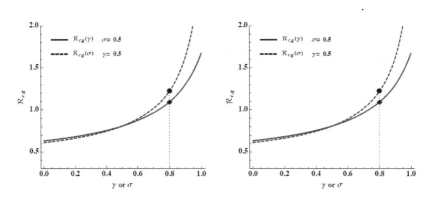

Fig. 1.9 Comparison of the GDM and PDM in terms of their evaluations on various control measures when the corresponding control reproduction numbers, $\mathcal{R}_{C,g}$ and $\mathcal{R}_{C,p}$ are used. The two control strategies considered are represented by $\gamma = 0.5$ and $\sigma = 0.8$ (labeled with a dot $\bullet$) and by $\gamma = 0.8$ and $\sigma = 0.5$ (labeled with a diamond $\blacklozenge$). The GDM (left figure) suggests that the diamond strategy is more effective (as it reduces $\mathcal{R}_C$ the most), whereas the PDM (right figure) suggests the opposite, that the circle strategy is more effective.

Similar discrepancies are also identified when the final epidemic sizes are compared (the final size can be represented by $N - S_\infty$ where N is constant population size (see (1.24)). Figure 1.10 shows curves of final epidemic sizes generated by GDM and PDM as functions of γ (for fixed σ) or σ (for fixed γ). Again, we evaluate two strategies: Strategy I represented by $\gamma = 0.5$ and $\sigma = 0.8$ and Strategy II represented $\gamma = 0.8$ and $\sigma = 0.5$. It is shown in the left figure that under GDA, Strategy II is more effective than Strategy I because it leads to larger reductions in the final size $N - S_\infty$. However, the figure on the right shows that under PDA, Strategy I is more effective than Strategy II. Parameter values used in Figure 1.10 are $\beta = 0.75, \rho = 0.95, \mu_1 = 5, \mu_2 = 10$ and $N = 1000$. Figures 1.9 and 1.10 illustrate how $\mathcal{R}_C$ and the final size vary as a function of one of the control measures when the other remains fixed.

In addition to the reproduction number and final epidemic size, we can also

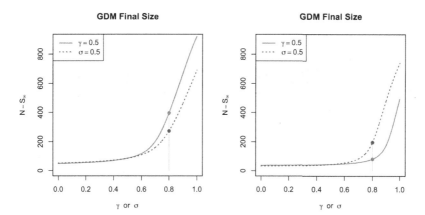

Fig. 1.10 The final size predicted by the GDM and PDM, as functions of γ (for fixed σ) or σ (for fixed γ). We observe similar discrepancies, as seen in Figure 1.9. S_∞ was computed using equation (1.24) and $N = 1000$.

examine some of the characteristics of epidemic curves generated by the three models (GDM, PDM, and BDM) and the influence on these characteristics of quarantine and isolation, as shown in Figure 1.11. The values for the control parameters are $\gamma = 0.7, \sigma = 0.8$, the initial condition is $S = 980, E_0 = 20, I_0 = Q_0 = H_0 = R_0 = 0$, and $\beta = 0.75., \rho = 0.95$. We observe that the GDM predicts a much higher peak for the initial epidemic than the other two models. However, it is not the case that the epidemic peak generated by the GDM is always higher. It is shown in Figure 1.12 that for different parameter values, the epidemic peak from the GDM may be lower than that generated by the other two models. The production of Figure 1.11 made use of the modified control parameters $\gamma = 0.9$ and $\sigma = 0.8$, the same initial conditions and $\beta = 0.4, \rho = 0.95$.

Theorem 1.3. *Let $\mathcal{R}_c$ be as given in (1.46). The final epidemic size generated by the dynamics of Model (1.40) satisfies the following final size relationship*

$$\ln \frac{S_0}{S_\infty} = \left(1 - \frac{S_\infty}{N}\right) \mathcal{R}_C. \tag{1.52}$$

Proof. From the E and Q equations in (1.40) we obtain

$$\sum_{n=0}^{\infty} E_n = \sum_{n=0}^{\infty} \left(A_n + \sum_{j=1}^{n} A_{n-j} p_j k_j\right) = A_0 + S_0 - S_\infty + \sum_{j=1}^{\infty} \sum_{n=j}^{\infty} A_{n-j} p_j k_j$$

$$= N - S_\infty + \sum_{j=1}^{\infty} \left(p_j k_j \sum_{n=0}^{\infty} A_n\right) = (N - S_\infty)\left(1 + \sum_{j=1}^{\infty} p_j k_j\right)$$

$$= (N - S_\infty)\mathbb{E}(X \wedge Z) = (N - S_\infty)\mathcal{D}_{E_k}$$

and

$$\sum_{n=1}^{\infty} Q_n = \sum_{n=1}^{\infty} \sum_{j=1}^{n} A_{n-j} p_j (1 - k_j) = \sum_{j=1}^{\infty} \sum_{n=j}^{\infty} A_{n-j} p_j (1 - k_j)$$

$$= \sum_{j=1}^{\infty} \left((p_j - p_j k_j) \sum_{n=0}^{\infty} A_n \right) = (N - S_\infty) \left(\sum_{j=1}^{\infty} p_j - \sum_{j=1}^{\infty} p_j k_j \right)$$

$$= (N - S_\infty)[\mathbb{E}(X) - \mathbb{E}(X \wedge Z)] = (N - S_\infty)[\mathbb{E}(X) - \mathcal{D}_{E_k}].$$

Because $E_n \geq 0$ and $\sum_{n=1}^{\infty} E_n < \infty$, $E_\infty = 0$. Similarly, Q_n, B_n, I_n, F_n and H_n converge to zero as $n \to \infty$. By definition of B_n (see (1.37)) we get

$$\sum_{n=1}^{\infty} B_n = \sum_{n=1}^{\infty} \sum_{j=1}^{n} A_{n-j}(p_{j-1} - p_j)k_{j-1} = \sum_{j=1}^{\infty} \left((p_{j-1} - p_j)k_{j-1} \sum_{n=j}^{\infty} A_{n-j} \right)$$

$$= (N - S_\infty) \sum_{j=1}^{\infty} (p_{j-1} - p_j)k_{j-1} = (N - S_\infty)\mathcal{T}_{E_k}.$$

Thus,

$$\sum_{n=1}^{\infty} I_n = B_1 + \sum_{n=2}^{\infty} \left(B_n + \sum_{j=1}^{n-1} B_{n-j} q_j l_j \right) = \sum_{n=1}^{\infty} B_n + \sum_{j=1}^{\infty} \left(q_j l_j \sum_{n=j+1}^{\infty} B_{n-j} \right)$$

$$= \left(\sum_{n=1}^{\infty} B_n \right) \left(1 + \sum_{j=1}^{\infty} q_j l_j \right) = (N - S_\infty)\mathcal{T}_{E_k}\mathbb{E}(Y \wedge W)$$

$$= (N - S_\infty)\mathcal{T}_{E_k}\mathcal{D}_{I_l}. \tag{1.53}$$

Notice from (1.38) and (1.39) that

$$\sum_{n=1}^{\infty} F_n = \sum_{n=1}^{\infty} (B_n + D_n) = \sum_{n=1}^{\infty} B_n + \sum_{n=1}^{\infty} (Q_{n-1} - Q_n + C_n)$$

$$= \sum_{n=1}^{\infty} B_n + Q_0 + \sum_{n=1}^{\infty} C_n = \sum_{n=1}^{\infty} B_n + \sum_{n=1}^{\infty} (E_{n-1} - E_n + A_n - B_n)$$

$$= \sum_{n=1}^{\infty} B_n + E_0 + \sum_{n=1}^{\infty} A_n - \sum_{n=1}^{\infty} B_n \tag{1.54}$$

$$= N - S_\infty.$$

Using the H equation in (1.40), together with (1.53) and (1.54), we have that

$$\sum_{n=1}^{\infty} H_n = F_1 - I_1 + \sum_{n=2}^{\infty} \left(F_n + \sum_{j=1}^{n-1} F_{n-j} q_j - I_n \right) = \sum_{n=1}^{\infty} F_n + \sum_{j=1}^{\infty} \left(q_j \sum_{n=1}^{\infty} F_n \right) - \sum_{n=1}^{\infty} I_n$$

$$= \left(\sum_{n=1}^{\infty} F_n \right) \left(1 + \sum_{j=1}^{\infty} q_j \right) - \sum_{n=1}^{\infty} I_n = \mathbb{E}(Y) \left(\sum_{n=1}^{\infty} F_n \right) - \sum_{n=1}^{\infty} I_n$$

$$= (N - S_\infty)(\mathcal{D}_I - \mathcal{T}_{E_k}\mathcal{D}_{I_l}). \tag{1.55}$$

From the S equation in (1.40), (1.53), and (1.55) we obtain

$$\ln \frac{S_0}{S_\infty} = \frac{\lambda}{N} \left[\sum_{n=1}^{\infty} I_n + (1-\rho) \sum_{n=1}^{\infty} H_n \right]$$

$$= \left(1 - \frac{S_\infty}{N}\right) \left[\lambda \mathcal{T}_{E_k} \mathcal{D}_{I_l} + \lambda(1-\rho)(\mathcal{D}_I - \mathcal{T}_{E_k} \mathcal{D}_{I_l}) \right]$$

$$= \left(1 - \frac{S_\infty}{N}\right) \mathcal{R}_C.$$

This completes the proof of Theorem 1.3. $\qquad\qquad\qquad\qquad\qquad\qquad$ □

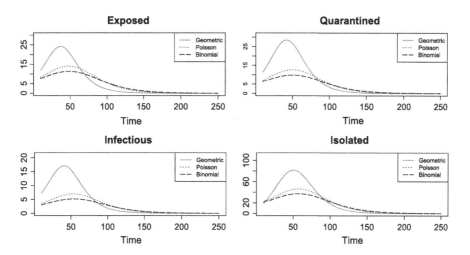

Fig. 1.11 Numerical simulations showing the epidemic curves generated by the three models: GDA, PDA, and BDA from time $n = 10$ to $n = 250$. The control parameters have values $\gamma = 0.7$ and $\sigma = 0.8$. It shows that for this set of control parameters, the GDM generates a much higher epidemic peak than the other two models.

1.2.7 *More examples of using the NGM approach to compute $\mathcal{R}$*

When models are more complicated, derivation of $\mathcal{R}$ from biological arguments can be difficult. In this case, the use of next generation matrix (NGM) might be helpful. The examples presented in this section are from [Hernandez-Ceron *et al.* (2013b)].

Arbitrary bounded distribution

Assume that the infectious period follows an arbitrary discrete (bounded) distribution represented by Y. Let again $q_i = \mathbb{P}(Y > i)$ and $\mathbb{P}(Y = i) = q_{i-1} - q_i$. It is easy to see that q_i is a decreasing function, i.e., $q_i \geq q_{i+1}$. In fact, $q_0 = 1$ and $q_m = 0$ for all $m \geq M$, where M is the maximum number of units of time that an individual takes to recover.

Because the geometric is the only memory-less discrete distribution, when other distributions are considered it is necessary to keep track of the past in order to know

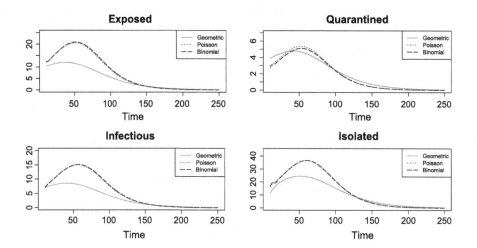

Fig. 1.12   Similar to Figure 1.11 but for a larger value of γ, at 0.95, representing a lower quarantine effort. All other parameters have the same values as in Figure 1.11. It shows that for this set of parameter values, the GDM generates a much lower epidemic peak than the other two models.

the values at the present. In fact, it is impossible to use the next generation matrix approach directly because the disease stages $(S, E, \text{ and } I)$ at time $n + 1$ cannot be written in the form

$$[\, E_{n+1}, \ I_{n+1}, \ S_{n+1}\,]^T = \mathcal{M}\left([\, E_n, \ I_n, \ S_n \,]^T\right),$$

where $\mathcal{M} : \mathbb{R}^3 \to \mathbb{R}^3$. To overcome this difficulty we can consider multiple I stages, similar to the approach known as the "linear chain trick" used in continuous models to convert a gamma distribution to a sequence of exponential distributions. Thus, we introduce the subclasses $I^{(1)}, I^{(2)}, \cdots, I^{(M)}$ (see Figure 1.13). The superscript i corresponds to the time since becoming infectious. Notice that these subclasses $I^{(i)}$ are different from those in the negative binomial model because here an individual can only stay in $I^{(i)}$ for one unit of time, and must go to either the $I^{(i+1)}$ class with probability q_i or the recovered class R with probability $1 - q_i$.

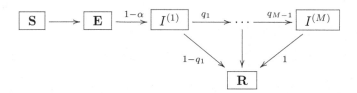

Fig. 1.13   A transition diagram for the case when the stage duration of the infectious period has an arbitrary bounded distribution with upper bound M. The superscript i is the stage age in the infectious period and individuals in $I^{(i)}$ (for all i) can enter the recovered class R with a certain probability.

From Figure 1.13 the model equations can be written as

$$S_{n+1} = S_n e^{-\sum_{i=1}^{M} \beta_i \frac{I_n^{(i)}}{N}}, \quad E_{n+1} = S_n\left[1 - e^{-\sum_{i=1}^{M} \beta_i \frac{I_n^{(i)}}{N}}\right] + \alpha E_n,$$

$$I_{n+1}^{(1)} = (1-\alpha)E_n, \quad I_{n+1}^{(2)} = q_1 I_n^{(1)}, \quad I_{n+1}^{(j)} = \frac{q_{j-1}}{q_{j-2}} I_n^{(j-1)}, \quad 3 \le j \le M,$$

(1.56)

where β_i denote the transmission rates at the infectious stage i, $1 \le i \le M$. As q_i is the probability that an infectious individual remains infectious i time units after becoming infectious, the transition probability from $I_n^{(2)}$ to $I_{n+1}^{(3)}$ is given by the probability that an infectious individual is still infectious 2 time units after becoming infectious given that the person remained infectious 1 time unit ago, i.e., q_2/q_1. This explains the $I_{n+1}^{(3)}$ equation and similarly $I_{n+1}^{(j)}$ equations for $3 \le j \le M$.

For the case when transmission rates β_i are stage-dependent, the $(M+1) \times (M+1)$ matrices for F and T are

$$F = \begin{bmatrix} 0 & \beta_1 & \beta_2 & \cdots & \beta_{M-1} & \beta_M \\ 0 & 0 & 0 & \cdots & 0 & 0 \\ 0 & 0 & 0 & \cdots & 0 & 0 \\ 0 & 0 & 0 & \cdots & 0 & 0 \\ \vdots & \vdots & \vdots & & \vdots & \vdots \\ 0 & 0 & 0 & \cdots & 0 & 0 \end{bmatrix}, \quad T = \begin{bmatrix} \alpha & 0 & 0 & \cdots & 0 & 0 \\ 1-\alpha & 0 & 0 & \cdots & 0 & 0 \\ 0 & q_1 & 0 & \cdots & 0 & 0 \\ 0 & 0 & \frac{q_2}{q_1} & \cdots & 0 & 0 \\ \vdots & \vdots & \vdots & & \vdots & \vdots \\ 0 & 0 & 0 & \cdots & \frac{q_{M-1}}{q_{M-2}} & 0 \end{bmatrix}.$$

Thus,

$$F(I-T)^{-1} = \begin{bmatrix} \sum_{i=1}^{M} \beta_i q_{i-1} & \sum_{i=1}^{M} \beta_i q_{i-1} & \cdots & \beta_M \\ 0 & 0 & \cdots & 0 \\ \vdots & \vdots & & \vdots \\ 0 & 0 & \cdots & 0 \end{bmatrix}.$$

It follows that

$$\mathcal{R}_0 = \varrho(F(I-T)^{-1}) = \sum_{i=1}^{M} \beta_i q_{i-1}.$$

(1.57)

We can also derive $\mathcal{R}_0$ from the biological definition as follows. Using the fact that the distribution Y has an upper bound M and that for a given function f

$$\sum_{m=1}^{M} \mathbb{P}(Y=m)f(m) = \mathbb{E}(f(Y)),$$

the reproduction number from the biological definition (with $f(m) = \sum_{i=1}^{m} \beta_i$) is

$$\mathcal{R}_0 = \sum_{i=0}^{M-1} \beta_{i+1} q_i = \sum_{i=0}^{M-1} \beta_{i+1} \sum_{m=i+1}^{M} \mathbb{P}(Y=m)$$

$$= \sum_{m=1}^{M} \sum_{i=0}^{m-1} \beta_{i+1} \mathbb{P}(Y=m) = \sum_{m=1}^{M} \left[\mathbb{P}(Y=m) \sum_{i=1}^{m} \beta_i\right] = \mathbb{E}\left(\sum_{i=1}^{Y} \beta_i\right).$$

(1.58)

In the case of constant transmission rate, i.e., $\beta_i = \beta$, the formula (1.58) reduces to

$$\mathcal{R}_0 = \mathbb{E}\left(\sum_{i=1}^{Y} \beta\right) = \mathbb{E}(Y\beta) = \beta\mathbb{E}(Y) = \beta\mathcal{D}_I.$$

A model with two pathogen strains

Consider the case in which two parasite strains (e.g., drug-sensitive and drug-resistant strains) compete for a single susceptible population. Assume that the infectious periods for both strains follow arbitrary discrete (bounded) distributions denoted by Y_s and Y_r, respectively. Here the subscript s stands for the sensitive strain and the subscript r stands for resistant strain. Let $q_{si} = \mathbb{P}(Y_s > i)$ and $q_{ri} = \mathbb{P}(Y_r > i)$ with $q_{s0} = q_{r0} = 1$, and the upper bounds for the two distributions (i.e., the maximum numbers of units of time that an individual takes to recover) be M_w for $w = s, r$ representing respective strains. A transition diagram is illustrated in Figure 1.14, in which we have introduced substages for ease of formulating the model equations.

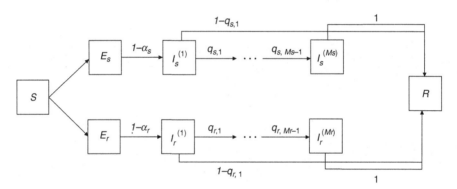

Fig. 1.14   A transition diagram for the model with two strains. The subscripts s and r denote drug-sensitive and drug-resistant strains, respectively. The infectious periods for both strains have arbitrary bounded distributions with upper bounds M_s and M_r.

Using the subscripts s and r to represent variables associated with the sensitive and resistant strains, respectively, the model equations are

$$E_{w,n+1} = S_n\left[1 - e^{-\sum_{i=1}^{M_w}\beta_{w,i}I_{w,n}^{(i)}/N}\right] + \alpha_w E_{w,n}, \quad I_{w,n+1}^{(1)} = (1-\alpha_w)E_{w,n},$$

$$I_{w,n+1}^{(2)} = q_{w,1}I_{w,n}^{(1)}, \quad I_{w,n+1}^{(j)} = \frac{q_{w,j-1}}{q_{w,j-2}}I_{w,n}^{(j-1)}, \quad 3 \leq j \leq M_w, \quad w = s, r,$$

$$S_{n+1} = S_n e^{-\left[\sum_{i=1}^{M_s}\beta_{s,i}I_{s,n}^{(i)}+\sum_{i=1}^{M_r}\beta_{r,i}I_{r,n}^{(i)}\right]/N}.$$

$$(1.59)$$

Here the constants $\beta_{w,i}$ represent the infection rates of individuals with strain w ($w = s, r$) at stage age i ($1 \leq i \leq M$). Other parameters have similar meanings as in other models. Let $(E_{s,n}, I_{s,n}^{(1)}, I_{s,n}^{(2)}, \cdots, I_{s,n}^{(M_s)}, E_{r,n}, I_{r,n}^{(1)}, I_{r,n}^{(2)}, \cdots, I_{r,n}^{(M_r)})$ be the order of variables. Then, the corresponding F and T matrices have the forms

$$F = \begin{bmatrix} F_s & 0 \\ 0 & F_r \end{bmatrix}, \quad T = \begin{bmatrix} T_s & 0 \\ 0 & T_r \end{bmatrix},$$

where F_w and T_w ($w = s, r$) are the $(M_w + 1) \times (M_w + 1)$ matrices

$$F_w = \begin{bmatrix} 0 & \beta_{w,1} & \beta_{w,2} & \cdots & \beta_{w,M_w-1} & \beta_{w,M_w} \\ 0 & 0 & 0 & \cdots & 0 & 0 \\ 0 & 0 & 0 & \cdots & 0 & 0 \\ 0 & 0 & 0 & \cdots & 0 & 0 \\ \vdots & \vdots & \vdots & & \vdots & \vdots \\ 0 & 0 & 0 & \cdots & 0 & 0 \end{bmatrix}, \quad T_w = \begin{bmatrix} \alpha_w & 0 & 0 & \cdots & 0 & 0 \\ 1-\alpha_w & 0 & 0 & \cdots & 0 & 0 \\ 0 & q_{w,1} & 0 & \cdots & 0 & 0 \\ 0 & 0 & \frac{q_{w,2}}{q_{w,1}} & \cdots & 0 & 0 \\ \vdots & \vdots & \vdots & & \vdots & \vdots \\ 0 & 0 & 0 & \cdots & \frac{q_{w,M_w-1}}{q_{w,M_w-2}} & 0 \end{bmatrix}.$$

Then

$$F(I - T)^{-1} = \begin{bmatrix} F_s(I - T_s)^{-1} & 0 \\ 0 & F_r(I - T_r)^{-1} \end{bmatrix},$$

where

$$F_w(I - T_w)^{-1} = \begin{bmatrix} \displaystyle\sum_{i=1}^{M_w} \beta_{w,i} q_{w,i-1} & \displaystyle\sum_{i=1}^{M_w} \beta_{w,i} q_{w,i-1} & \cdots & \beta_{w,M_w} \\ 0 & 0 & \cdots & 0 \\ \vdots & \vdots & & \vdots \\ 0 & 0 & \cdots & 0 \end{bmatrix}, \quad w = s, r.$$

Notice that $F_w(I - T_w)^{-1}$ reduces to a block diagonal matrix with each of the two blocks having one nonzero (positive) eigenvalue, thus ϱ is the maximum of these. It follows that

$$\mathcal{R}_0 = \varrho(F(I - T)^{-1}) = \max\left\{ \sum_{i=1}^{M_s} \beta_{s,i}\, q_{s,i-1}, \sum_{i=1}^{M_r} \beta_{r,i}\, q_{r,i-1} \right\}.$$

A model structured by gender

Consider a model that includes two sub-populations, female and male populations, with heterogeneous mixing (i.e., no sexual contacts between individuals of the same sex). Assume that the infectious periods for female and male populations follow arbitrary discrete (bounded) distributions denoted by Y_f and Y_m, respectively. Here the subscripts f and m stand for female and male, respectively. Let $q_{f,i} = \mathbb{P}(Y_f > i)$ and $q_{m,i} = \mathbb{P}(Y_m > i)$ with $q_{f,0} = q_{m,0} = 1$, and the upper bounds for the two distributions (i.e., the maximum numbers of units of time that an individual takes to recover) be M_w for $w = f, m$.

The model equations are

$$E_{w,n+1} = S_{w,n}\left[1 - e^{-\sum_{i=1}^{M_{\tilde{w}}} \beta_{\tilde{w},i} I_{\tilde{w},n}^{(i)}/N}\right] + \alpha_w E_{w,n}, \quad I_{w,n+1}^{(1)} = (1 - \alpha_w)E_{w,n},$$

$$I_{w,n+1}^{(2)} = q_{w,1} I_{w,n}^{(1)}, \quad I_{w,n+1}^{(j)} = \frac{q_{w,j-1}}{q_{w,j-2}} I_{w,n}^{(j-1)}, \quad 3 \le j \le M_w,$$

$$S_{w,n+1} = S_{w,n} e^{-\sum_{i=1}^{M_{\tilde{w}}} \beta_{\tilde{w},i} I_{\tilde{w},n}^{(i)}/N}, \quad \text{for } w = f, m.$$

Here $\tilde{w}$ represents the opposite sex of w, i.e., $\tilde{f} = m$, $\tilde{m} = f$. The constant $\beta_{\tilde{f},i}$ ($\beta_{\tilde{m},i}$) represents the infection rate to a female (male) transmitted by infectious male (female) individuals with stage age i.

Let $(E_{f,n}, I_{f,n}^{(1)}, I_{f,n}^{(2)}, \cdots, I_{f,n}^{(M_f)}, E_{m,n}, I_{m,n}^{(1)}, I_{m,n}^{(2)}, \cdots, I_{m,n}^{(M_m)})$ be the order of variables. Then the corresponding F and T matrices have the forms

$$F = \begin{bmatrix} 0 & F_m \\ F_f & 0 \end{bmatrix}, \quad T = \begin{bmatrix} T_f & 0 \\ 0 & T_m \end{bmatrix},$$

where F_w and T_w ($w = f, m$) are the $(M_w + 1) \times (M_w + 1)$ matrices

$$F_w = \begin{bmatrix} 0 & \beta_{w,1} & \beta_{w,2} & \cdots & \beta_{w,M_w-1} & \beta_{w,M_w} \\ 0 & 0 & 0 & \cdots & 0 & 0 \\ 0 & 0 & 0 & \cdots & 0 & 0 \\ 0 & 0 & 0 & \cdots & 0 & 0 \\ \vdots & \vdots & \vdots & & \vdots & \vdots \\ 0 & 0 & 0 & \cdots & 0 & 0 \end{bmatrix}, \quad T_w = \begin{bmatrix} \alpha_w & 0 & 0 & \cdots & 0 & 0 \\ 1-\alpha_w & 0 & 0 & \cdots & 0 & 0 \\ 0 & q_{w,1} & 0 & \cdots & 0 & 0 \\ 0 & 0 & \frac{q_{w,2}}{q_{w,1}} & \cdots & 0 & 0 \\ \vdots & \vdots & \vdots & & \vdots & \vdots \\ 0 & 0 & 0 & \cdots & \frac{q_{w,M_w-1}}{q_{w,M_w-2}} & 0 \end{bmatrix}.$$

Then,

$$F(I - T)^{-1} = \begin{bmatrix} 0 & F_m(I - T_m)^{-1} \\ F_f(I - T_f)^{-1} & 0 \end{bmatrix}$$

where

$$F_w(I - T_w)^{-1} = \begin{bmatrix} \sum_{i=1}^{M_w} \beta_{w,i} q_{w,i-1} & \sum_{i=1}^{M_w} \beta_{w,i} q_{w,i-1} & \cdots & \beta_{w,M_w} \\ 0 & 0 & \cdots & 0 \\ \vdots & \vdots & & \vdots \\ 0 & 0 & \cdots & 0 \end{bmatrix}, \quad w = f, m.$$

The matrix $F(I - T)^{-1}$ has rank 2 and the only two non-zero eigenvalues are

$$\pm \sqrt{\left(\sum_{i=1}^{M_f} \beta_{f,i}\, q_{f,i-1}\right)\left(\sum_{i=1}^{M_m} \beta_{m,i} q_{m,i-1}\right)}.$$

It follows that

$$\mathcal{R}_0 = \varrho(F(I - T)^{-1}) = \sqrt{\left(\sum_{i=1}^{M_f} \beta_{f,i}\, q_{f,i-1}\right)\left(\sum_{i=1}^{M_m} \beta_{m,i} q_{m,i-1}\right)}. \tag{1.60}$$

The square root in (1.60) is a consequence of the fact that the secondary infections need to be computed from one female (male) to other females (males) through the male (female) population. Let

$$\mathcal{R}_0^{(fm)} = \sum_{i=1}^{M_f} \beta_{f,i}\, q_{f,i-1}, \quad \mathcal{R}_0^{(mf)} = \sum_{i=1}^{M_m} \beta_{m,i} q_{m,i-1}.$$

Then $\mathcal{R}_0^{(fm)}$ ($\mathcal{R}_0^{(mf)}$) describes the average number of secondary infections an infectious female (male) individual can produce in a susceptible male (female) population during her (his) infectious period. Thus, $\mathcal{R}_0$ is the geometric mean of $\mathcal{R}_0^{(fm)}$ and $\mathcal{R}_0^{(mf)}$.

An SEIR model with isolation and arbitrary distribution

A transition diagram for the model with isolation is described in Figure 1.15, in which H denotes the isolated (or hospitalized) class. In this diagram, q_i and p_i denote the survival functions for individuals in the I and H classes, that is, if Y is the infectious period and W is the number of days before isolation then $q_i = \mathbb{P}(Y > i)$ and $p_i = \mathbb{P}(W > i)$. It is reasonable to assume that both distributions are bounded by M (the maximum units of time that an individual takes to recover), i.e., $q_m = p_m = 0$ for all $m \geq M$. Consider the case of constant transmission rate β and assume that isolated individuals have a reduced transmission with a reduction constant ρ, i.e., the force of infection has the form

$$\lambda(I_n, H_n) = \frac{\beta}{N}[I_n + (1 - \rho)H_n].$$

In this case, the model equations read

$$
\begin{aligned}
S_{n+1} &= S_n e^{-\lambda(I_n, H_n)}, \\
E_{n+1} &= S_n\left(1 - e^{-\lambda(I_n, H_n)}\right) + \alpha E_n, \\
I_{n+1} &= B_{n+1} + B_n q_1 p_1 + B_{n-1} q_2 p_2 + \cdots + B_1 q_n p_n, \\
H_{n+1} &= B_n q_1 (1 - p_1) + B_{n-1} q_2 (1 - p_2) + \cdots + B_1 q_n (1 - p_n),
\end{aligned}
\tag{1.61}
$$

where $B_{j+1} = (1 - \alpha)E_j$ and the initial conditions are $S_0 = N - E_0, E_0 > 0$ and $I_0 = H_0 = 0$.

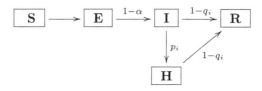

Fig. 1.15 A transition diagram for the model with isolation and an arbitrary bounded distribution for the infectious period.

As was done before, the set of equations given by (1.61) must be reformulated. Because the classes S, E, I and H at time $n + 1$ cannot be written in the form

$$[\, E_{n+1}, \ I_{n+1}, \ H_{n+1}, \ S_{n+1} \,]^T = \mathcal{M}\left([\, E_n, \ I_n, \ H_n, \ S_n \,]^T\right),$$

where $\mathcal{M} : \mathbb{R}^4 \to \mathbb{R}^4$, the next generation matrix approach cannot be directly applied. This problem is avoided by considering substages for the I and H classes, as shown in Figure 1.16. The substage $I^{(i)}$ represents individuals who have been infected for i units of time without being isolated, and $H^{(i)}$ represents isolated individuals infected for i units of time. Note that $H^{(1)} = 0$ as isolations occurs after becoming infectious. All of the I and H sub-stages are connected to R, representing the fact that infectious individuals may recover at any time. All individuals will be recovered by M units of time after becoming infectious.

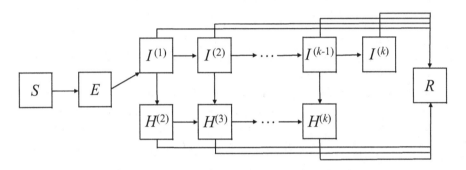

Fig. 1.16   A more detailed diagram for the model with isolation and arbitrary distributions.

Then, the E, I and H equations in (1.61) can be replaced by the following equations:

$$E_{n+1} = S_n\left(1 - e^{-\beta \sum_{i=1}^{M} [I_n^{(i)} + (1-\rho)H_n^{(i)}]/N}\right) + \alpha E_n,$$

$$I_{n+1}^{(1)} = (1-\alpha)E_n, \quad I_{n+1}^{(2)} = q_1 p_1 I_n^{(1)}, \quad I_{n+1}^{(j)} = \frac{q_{j-1} p_{j-1}}{q_{j-2} p_{j-2}} I_n^{(j-1)},$$

$$H_{n+1}^{(2)} = q_1(1-p_1)I_n^{(1)}, \quad H_{n+1}^{(j)} = \frac{q_{j-1}(p_{j-2} - p_{j-1})}{q_{j-2} p_{j-2}} I_n^{(j-1)} + \frac{q_{j-1}}{q_{j-2}} H_n^{(j-1)}, \quad 3 \leq j \leq M.$$

From the above equations, the $2M \times 2M$ matrix F associated with new infections is

$$F = \begin{bmatrix} 0 & \beta & \cdots & \beta & (1-\rho)\beta & \cdots & (1-\rho)\beta \\ 0 & 0 & \cdots & 0 & 0 & \cdots & 0 \\ 0 & 0 & \cdots & 0 & 0 & \cdots & 0 \\ \vdots & \vdots & \vdots & \vdots & \vdots & & \vdots \\ 0 & 0 & \cdots & 0 & 0 & \cdots & 0 \end{bmatrix}.$$

Let T be the $2M \times 2M$ block matrix associated with transitions. Then $I - T = \begin{bmatrix} A & 0 \\ C & D \end{bmatrix}$, where the $(M+1) \times (M+1)$ matrix A, the $(M+1) \times (M-1)$ matrix C

and the $(M-1) \times (M-1)$ matrix D are

$$
A = \begin{bmatrix}
1-\alpha & 0 & 0 & \cdots & 0 & 0 \\
-(1-\alpha) & 1 & 0 & \cdots & 0 & 0 \\
0 & -q_1 p_1 & 1 & \cdots & 0 & 0 \\
0 & 0 & -\frac{q_2 p_2}{q_1 p_2} & \cdots & 0 & 0 \\
\vdots & \vdots & \vdots & & \vdots & \vdots \\
0 & 0 & 0 & \cdots & 1 & 0 \\
0 & 0 & 0 & \cdots & -\frac{q_{M-1}p_{M-1}}{q_{M-2}p_{M-2}} & 1
\end{bmatrix}, \quad
D = \begin{bmatrix}
1 & 0 & 0 & \cdots & 0 & 0 \\
-\frac{q_2}{q_1} & 1 & 0 & \cdots & 0 & 0 \\
0 & -\frac{q_3}{q_2} & 1 & \cdots & 0 & 0 \\
\vdots & \vdots & \vdots & & \vdots & \vdots \\
0 & 0 & 0 & \cdots & 1 & 0 \\
0 & 0 & 0 & \cdots & -\frac{q_{M-1}}{q_{M-2}} & 1
\end{bmatrix},
$$

and

$$
C = \begin{bmatrix}
0 & -q_1(1-p_1) & 0 & \cdots 0 & 0 & 0 \\
0 & 0 & -\frac{q_2(p_1-p_2)}{q_1 p_1} & \cdots 0 & 0 & 0 \\
\vdots & \vdots & \vdots & \vdots & \vdots & \vdots \\
0 & 0 & 0 & \cdots 0 & -\frac{q_{M-1}(p_{M-2}-p_{M-1})}{q_{M-2}p_{M-2}} & 0
\end{bmatrix},
$$

respectively. The inverse of $I - T$ is given by

$$
\begin{bmatrix}
A^{-1} & 0 \\
-D^{-1}CA^{-1} & D^{-1}
\end{bmatrix},
$$

where

$$
A^{-1} = \begin{bmatrix}
\frac{1}{1-\alpha} & 0 & 0 & \cdots & 0 & 0 \\
1 & 1 & 0 & \cdots & 0 & 0 \\
q_1 p_1 & q_1 p_1 & 1 & \cdots & 0 & 0 \\
q_2 p_2 & q_2 p_2 & \frac{q_2 p_2}{q_1 p_1} & \cdots & 0 & 0 \\
\vdots & \vdots & \vdots & & \vdots & \vdots \\
q_{M-1}p_{M-1} & q_{M-1}p_{M-1} & q_{M-1}p_{M-1} & \cdots & \frac{q_{M-1}p_{M-1}}{q_{M-2}p_{M-2}} & 1
\end{bmatrix},
$$

$$
D^{-1} = \begin{bmatrix}
1 & 0 & 0 & \cdots & 0 & 0 \\
\frac{q_2}{q_1} & 1 & 0 & \cdots & 0 & 0 \\
\frac{q_3}{q_1} & \frac{q_3}{q_2} & 1 & \cdots & 0 & 0 \\
\vdots & \vdots & \vdots & & \vdots & \vdots \\
\frac{q_{M-2}}{q_1} & \frac{q_{M-2}}{q_2} & \frac{q_{M-2}}{q_3} & \cdots & 1 & 0 \\
\frac{q_{M-1}}{q_1} & \frac{q_{M-1}}{q_2} & \frac{q_{M-1}}{q_3} & \cdots & \frac{q_{M-1}}{q_{M-2}} & 1
\end{bmatrix},
$$

and the matrix $-D^{-1}CA^{-1}$ is given by

$$
\begin{bmatrix}
q_1(1-p_1) & * \cdots * \\
\frac{q_2}{q_1}q_1(1-p_1) + q_2(p_1-p_2) = q_2(1-p_2) & * \cdots * \\
\frac{q_3}{q_1}q_1(1-p_1) + \frac{q_3}{q_2}q_2(p_1-p_2) + q_3(p_2-p_3) = q_3(1-p_3) & * \cdots * \\
\vdots & \vdots
\end{bmatrix}.
$$

Note that only the first column of the previous matrix is relevant for computing the eigenvalues of the next generation matrix because F has rank 1 and

$$F(I-T)^{-1} = \begin{bmatrix} \beta \sum_{i=0}^{M-1} q_i p_i \; + \; \beta(1-\rho) \sum_{i=0}^{M-1} q_i(1-p_i) & * & \cdots & * \\ 0 & 0 & \cdots & 0 \\ \vdots & \vdots & & \vdots \\ 0 & 0 & \cdots & 0 \end{bmatrix}.$$

Finally, notice that, if Y is the infectious period (with survival function q_i) and W is the number of days before isolation (with survival function p_i), then

$$\sum_{i=0}^{M-1} q_i p_i = \mathbb{E}(Y \wedge W),$$

and

$$\sum_{i=0}^{M-1} q_i(1-p_i) = \sum_{i=0}^{M-1} q_i - \sum_{i=0}^{M-1} q_i p_i = \mathbb{E}(Y) - \mathbb{E}(Y \wedge W),$$

where $\mathbb{E}(Y)$ represents the mean time spent in compartments I and H, $\mathbb{E}(Y \wedge W)$ is the mean time spent in I ('isolation-adjusted' mean sojourn time) and $\mathbb{E}(Y) - \mathbb{E}(Y \wedge W)$ is the mean time spent in H. Then the control reproduction number $\mathcal{R}_C$ is given by

$$\mathcal{R}_C = \varrho(F(I-T)^{-1}) = \underbrace{\beta\mathbb{E}(Y \wedge W)}_{\mathcal{R}_I} + \underbrace{\beta(1-\rho)[\mathbb{E}(Y) - \mathbb{E}(Y \wedge W)]}_{\mathcal{R}_H}. \qquad (1.62)$$

Here $\mathcal{R}_I$ represents the number of secondary infections produced in a susceptible population by an individual in the I class during the infectious period. Similarly, $\mathcal{R}_H$ is the number of secondary infections produced by an individual in the H class.

In summary, the application of the framework presented in the section is extremely flexible because the probabilities p_i, q_i, k_i, and l_i do not have to come from a particular parametric family of discrete distributions. Model (1.40) can in fact incorporate directly empirically estimated (from the raw data) probabilities. That is, no specific assumptions on the shape of the duration-stage distribution of latent and infectious stages or on the waiting-time distributions in quarantine and/or isolation classes are required within the framework of this manuscript.

In addition, the question of what stage distribution(s) is (are) more appropriate depends on what we actually know about the epidemiological process. The specifics of each disease provide the most critical information. Researchers involved in the study of the dynamics of infectious diseases seem to prefer to work with models that make use of geometric stage-duration distributions. Needless to say, the latent or infectious stage distributions may fit better alternative distributions, for a great number of infectious diseases. Does the general use of geometric distributions matter? If the goal is to carry out a qualitative study within single-outbreak epidemic models then no but if the goal is to assess quantitatively the efficacy of control measures for specific diseases then the answer is, most likely yes.

Finally, starting with the general model (1.40) an analytic formula for the control reproduction number $\mathcal{R}_C$ (see (1.46)) is derived as well as a final epidemic size relation; completely specified by $\mathcal{R}_C$ (see (1.52)). The formula for $\mathcal{R}_C$ is expressed in terms of the means and control-adjusted means of the general stage distributions. The usefulness of the $\mathcal{R}_C$ formula given in (1.46) emerges from the fact that it was derived for general stage distributions. The expression for $\mathcal{R}_C$ allows for the investigation of its dependence on the means and control-adjusted means (e.g., $\mathcal{D}_I$, $\mathcal{D}_{I_l}$ etc.) of the stage distributions. These expressions also allow us to explore the role of control measures (quarantine and isolation) in reducing $\mathcal{R}_C$ as a function of pre-selected stage distributions. This level of specificity makes it possible and convenient to connect both with specific distributions. For simpler discrete epidemic models, the reproduction number can be derived using the next generation matrix approach (see examples in section 1.2.7). The use of model (1.40) facilitates the comparison of epidemic curves. Here, primarily for expository reasons we have focused on situations involving explicit distributions, the geometric, Poisson, and binomial. From the use of specific parametric families, we see, for example, that the GDM supports higher epidemic peaks than the PDM or BDM in some cases (see Figure 1.11) but it supports lower peaks in other cases (see Figure 1.12). Most importantly, the models with different assumptions on stage distributions may generate contradictory evaluations of disease control strategies (see Figures 1.9 and 1.10). These observations reinforce the need for caution when simple models (such as the GDM) are used in the study of infectious diseases or as the dynamic framework used to estimate parameters from data, or as the cornerstone of efforts to assess the relative effectiveness of distinct control or intervention strategies.

Chapter 2

Endemic Models

When modeling endemic diseases, i.e., diseases that are persistent in a population for a long period of time, the demographic changes (e.g., birth and death processes) need to be included in the model. These models are usually referred to as *endemic models*, and they usually have more complex dynamics than epidemic models. In this chapter, we first introduce the classical SIR and SEIR epidemic models, followed by several extensions of these simple models when various heterogeneities are incorporated.

2.1 Classical SIR and SEIR endemic models

A classical SIR endemic model reads

$$\frac{dS}{dt} = bN - \beta S\frac{I}{N} - \mu S,$$

$$\frac{dI}{dt} = \beta S\frac{I}{N} - \gamma I - \mu I, \tag{2.1}$$

$$\frac{dR}{dt} = \gamma I - \mu R,$$

where b and μ denote the per-capita birth and death rates. In the simplest case when the total population size remains constant, the birth and death rates are set to be equal, i.e., $b = \mu$. This will be the case in this chapter so that the effects of other factors on the model outcomes can be more transparent.

One of the main differences between the epidemic model (1.1) and the endemic model (2.1) is that in the epidemic model disease always dies out, whereas for the endemic model there is a possible non-trivial steady state (an endemic equilibrium) at which the fraction of infected individuals, denoted by I^*/N, is positive. The existence of this endemic equilibrium depends on the basic reproduction number of the model (2.1), which is given by

$$\mathcal{R}_0 = \frac{\beta}{\gamma + \mu}. \tag{2.2}$$

The disease will die out if $\mathcal{R}_0 < 1$, and when $\mathcal{R}_0 > 1$ the disease will stabilize at the endemic level

$$\frac{I^*}{N} = \frac{\mu}{\gamma + \mu}\left(1 - \frac{1}{\mathcal{R}_0}\right). \tag{2.3}$$

Figure 2.1 depicts a typical time plot (see (a)) and phase portrait (see (b)) of the system (2.1).

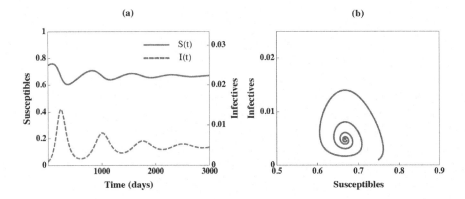

Fig. 2.1   Time plot (a) and phase portrait (b) of the simple SIR model (2.1).

For diseases that have a relatively long latent period or if quarantine of latent individuals need to be considered, it is necessary to include a latent class in the model. The classical SEIR endemic model has the following form:

$$\frac{dS}{dt} = bN - \beta S \frac{I}{N} - \mu S,$$

$$\frac{dE}{dt} = \beta S \frac{I}{N} - \alpha E - \mu E,$$

$$\frac{dI}{dt} = \alpha E - \gamma I - \mu I, \tag{2.4}$$

$$\frac{dR}{dt} = \gamma I - \mu R,$$

where α is the rate at which a latent individual becomes infectious. The basic reproduction number from this model is

$$\mathcal{R}_0 = \frac{\beta}{\gamma + \mu}\frac{\alpha}{\alpha + \mu}. \tag{2.5}$$

Comparing with the $\mathcal{R}_0$ in (2.2), the expression (2.5) has an extra factor, $\alpha/(\alpha+\mu)$, which represents the probability that a latent individual will enter the infectious class (i.e., survive and become infectious). Besides the difference in $\mathcal{R}_0$, the dynamical outcomes of the two models are similar. That is, the disease will die out if $\mathcal{R}_0 < 1$, and when $\mathcal{R}_0 > 1$, the disease will stabilize at the endemic level

$$\frac{I^*}{N} = \frac{\mu\alpha}{(\gamma + \mu)(\alpha + \mu)}\left(1 - \frac{1}{\mathcal{R}_0}\right). \tag{2.6}$$

The factor $\alpha/(\alpha + \mu)$ in the expression of (2.5) $\mathcal{R}_0$ and the endemic level (2.6) represents the probability that a latent individual survives and becomes infectious, and $1/(\alpha+\mu)$ denotes the death-adjusted latent period. It has also been shown that $\mathcal{R}_0$ can help determine the global dynamics of a system under certain conditions (see [Castillo-Chavez *et al.* (2002)]).

It worth pointing out that the birth and death terms in the models (2.1) and (2.4) can also be interpreted as school-enter and exit rates, respectively when a school population is considered, or rates of entering and exiting sexually active population when a saxually transmitted disease is studied.

Both the SIR model (2.1) and SEIR model (2.4) assume that the mixing between individuals is homogeneous and ignore heterogeneities such as age-dependent activity levels, infectivity, and susceptibility. However, in many cases, it is necessary to consider one or more types of heterogeneities and therefore, extensions of the simple models (2.1) and (2.4) need to be studied. Several examples are presented in the following section.

2.2 Multi-group models and the role of mixing

The classical models (2.1) and (2.4), as well as many other endemic models assume that the mixing among individuals in the entire population is homogeneous. In many cases, the total population consists of distinct groups of subpopulations based on age, spatial location, gender, activity levels, etc. There heterogeneities can be critical to be considered in the models, which are to be used to study certain biological questions. Some of the heterogeneities can be incorporated in the models by further dividing the population into subgroups according to the particular heterogeneities interested. In the section, several types of heterogeneities are discussed including chronological age, infection age, activity level, and gender.

2.2.1 *Models with random and preferential mixing*

The results presented in this section are mainly from [Glasser *et al.* (2012); Chow *et al.* (2011)]. The focus is on developing a modeling framework to incorporate preferential mixing between subgroups of the population, and examine how preferential mixing may affect the threshold conditions that can be used for evaluating disease control strategies.

One of the commonly used mixing functions is the convex combination of contacts within and among groups of similarly aged individuals, respectively termed preferential and proportionate mixing. Recently published face-to-face conversation and time-use studies suggest that parents and children and co-workers also mix preferentially. As indirect effects arise from the off-diagonal elements of mixing matrices, these observations are exceedingly important. Accordingly, we refined the formula published by Jacquez *et al.* [Jacquez *et al.* (1988)] to account for these newly-observed patterns and estimated age-specific fractions of contacts with each preferred group. As the ages of contemporaries need not be identical nor those of parents and children to differ by exactly the generation time, we also estimated the variances of the Gaussian distributions with which we replaced the Kronecker delta commonly used in theoretical studies. Our formulae reproduce observed patterns and can be used, given contacts, to estimate probabilities of infection on contact, infection rates, and reproduction numbers. As examples, we illustrate these calculations for influenza based on infection levels from a prospective household study during the 1957 pandemic and for varicella based on cumulative incidence estimated from a cross-sectional serological survey conducted from 1988-94, together with contact rates from the several face-to-face conversation and time-use studies. Susceptibility to infection on contact generally declines with age, but may be elevated among adolescents and adults with young children.

While pathogens spread via interpersonal contacts, transmission may be modeled within and between groups of similar individuals. Appropriate levels of aggregation depend on questions of interest and observations available. Given suitable expressions for heterogeneous mixing, this mean field approach yields dynamic networks whose nodes are ever changing sub-populations defined by age, location, or other strata. Recently, there has been an explosion of models in which network structure defines social contacts among individuals (see, e.g., [Newman (2003)]). Epidemic-control measures have been evaluated using both approaches.

Consider first the case when there are n groups, which differ in their activity levels (e.g., the average per capita contact rates), denoted by a_i, and population size, denoted by N_i, for $i = 1, 2, \cdots, n$. Busenberg and Castillo-Chavez [Busenberg and Castillo-Chavez (1991)] define c_{ij} as proportions of contacts that members of group i have with group j, given that i has contacts. Their criteria that mixing models should meet are:

(i) $c_{ij} \geq 0$,

(ii) $\sum_{j=1}^{k} c_{ij} = 1$,

(iii) $a_i N_i c_{ij} = a_j N_j c_{ji}, \quad i, j = 1, 2, \cdots, n$.

One example of the mixing functions satisfying the constraints (i)-(iii) is the proportionate (or random) mixing for which

$$c_{ij} = \frac{a_i N_i}{\sum_{j=1}^{n} a_j N_j}. \tag{2.7}$$

Jacquez *et al.* [Jacquez *et al.* (1988)] developed a more general mixing function to take into account the preferential contact between individuals of similar ages. Such preference has been observed in data from various populations (reflected by the lighter area along the main diagonal). In their formulation, it is assumed that proportion ε_i of i-group contacts is reserved for others in group i (called preferences), and that the complement $1 - \varepsilon_i$ is distributed among all groups, including i, via the proportionate mixing formula shown in (2.7). The mixing function reads

$$c_{ij} = \varepsilon_i \delta_{ij} + (1 - \varepsilon_i) \frac{(1 - \varepsilon_j) a_j N_j}{\sum_{l=1}^{n} (1 - \varepsilon_l) a_l N_l} \tag{2.8}$$

where δ_{ij} is the Kronecker delta (i.e., $\delta_{ij} = 1$ if $i = j$ and $\delta_{ij} = 0$ if $i \neq j$). Jacquez *et al.* [Jacquez *et al.* (1988)] obtained the first of these preferential mixing expressions by allowing the fraction of within-group contacts, ε, to vary between groups in Nold's [Nold (1980)] preferred mixing model. Hethcote's [Hethcote (1996)] equation (4.14) is the same as hers with epsilon and its complement reversed. Similar ideas are evident in Barbour's [Barbour (1978)] modeling of schistosomiasis or the extensive HIV modeling at the beginning of the pandemic (see, e.g., [Castillo-Chavez *et al.* (1989b)] and references therein). Castillo-Chavez *et al.* [Castillo-Chavez *et al.* (1991)], and Blythe and Castillo-Chavez [Blythe and Castillo-Chavez (1989)] use the log-normal distribution and an arbitrary continuous function, respectively, for their main diagonals.

Recently collected data (e.g., [Mossong *et al.* (2008)]) reveals also preferential mixing between parents and children in addition to that among contemporaries (see the left plot in Figure 2.2). In [Glasser *et al.* (2012)], we further extend the preferential mixing function (2.8) to include contacts between parents and children (the sub- and super-diagonals in Figure 2.2) and among co-workers as well as contemporaries (the main diagonal in Figure 2.2):

$$c_{ij} = \phi_{ij} + \left(1 - \sum_{l=1}^{4} \varepsilon_{li}\right) f_j, \quad f_j = \frac{\left(1 - \sum_{l=1}^{4} \varepsilon_{li}\right) a_j N_j}{\sum_{k=1}^{n} \left(1 - \sum_{l=1}^{4} \varepsilon_{lk}\right) a_k N_k}. \tag{2.9}$$

When $\vec{\varepsilon}_2 = \vec{\varepsilon}_3 = \vec{\varepsilon}_4 = 0$ (where $\vec{\varepsilon}_l = (\varepsilon_{l1}, \cdots, \varepsilon_{ln})$), the expression (2.9) reduces to the formula (2.8) of Jacquez *et al.* [Jacquez *et al.* (1988)]. Figure 2.2 illustrates contacts among contemporaries (ε_1) as well as between parents and children (ε_2 and ε_3) ($\vec{\varepsilon}_4 = 0$ in this figure).

Because the sub- and super-diagonals extend over ages $i \geq G$ and $i \leq L - G$, respectively, where G is the generation time (i.e., average age at which women bear children), L is longevity (i.e., average expectation of life at birth), and $L > G$, we define ϕ_{ij} as:

$$\phi_{ij} = \begin{cases} \delta_{ij}\varepsilon_{1i} + \delta_{i(j+G)}\varepsilon_{2i} + I_W(i,j)\frac{\varepsilon_{4i}}{W_{\max} - W_{\min}}, & i \geq G, \\ \delta_{ij}\varepsilon_{1i} + \delta_{i(j-G)}\varepsilon_{3i} + I_W(i,j)\frac{\varepsilon_{4i}}{W_{\max} - W_{\min}}, & i \leq L - G. \end{cases} \tag{2.10}$$

Only people whose ages equal or exceed G can have children, and only those whose ages equal to or less than $L - G$ can have parents, but people aged at least G but not more than $L - G$ can have both children and parents. In these inequalities, we mix indices and real numbers, but if age classes are 0–4, 5–9, $\cdots$ and $G = 25$ years, for example, by $i > G$ we mean $i >$ class 5. Other new variables are $W_{\min}$ and $W_{\max}$, average ages at entry to and exit from the workforce (Figure 2.2, right), ε_{1i}-ε_{4i}, fractions of contacts reserved for contemporaries, children $(j - G)$, parents $(j + G)$,

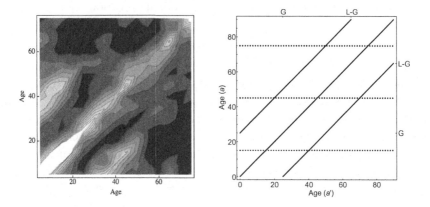

Fig. 2.2   The figure on the left shows observed patterns of contacts between age-groups [Mossong *et al.* (2008)]. The lighter areas correspond to higher level of contacts. The figure on the right shows schematic contact matrices illustrating the main and off-diagonals representing contacts among contemporaries, between children and parents, and vice versa. In this figure and the text, $G = 25$ and $L = 90$ years.

and co-workers (if $W_{\min} \geq i,\ j \leq W_{\max}$), respectively, and the corresponding delta or indicator function. That is,

$$\delta_{i(j \pm G)} = \begin{cases} 1 & \text{if } i = j \pm G, \\ 0 & \text{otherwise.} \end{cases} \qquad \text{and} \qquad I_W(i,j) = \begin{cases} 1 & \text{if } i,j \in W, \\ 0 & \text{otherwise} \end{cases}$$

where $W = [W_{\min}, W_{\max}]$. Notice that the non-zero elements of $\vec{\varepsilon}_2$ and $\vec{\varepsilon}_3$ are related. If $G = 25$ years, for example, then $a_i N_i \varepsilon_{2i} = a_j N_j \varepsilon_{3j}$, for $i = 6, 7, \cdots, j = i - 5$. Notice also that $0 \leq \sum_{l=1}^{4} \varepsilon_{li} < 1$ and the mixing among co-workers does not depend on age provided that $i \geq W_{\min}$ and $j \leq W_{\max}$.

While delta formulations are undeniably heuristic, contemporaries need not be exactly the same age [Hethcote (1996)], nor need the ages of parents and children to differ by exactly the generation time. Accordingly, we reformulate θ_{ij} to incorporate this more realistic feature. Let α and α' denote the ages of susceptible and infected individuals, respectively. Further, let $a(\alpha)$ denote the average number of contacts per person aged α per unit of time and $N(\alpha)$ denote the number of people aged α. The continuous analogue of c_{ij} can be formulated as:

$$c(\alpha, \alpha') = \phi(\alpha, \alpha') + \left[1 - \sum_{l=1}^{4} \varepsilon_l(\alpha)\right] f(\alpha'),$$

$$f(\alpha') = \frac{[1 - \sum_{l=1}^{4} \varepsilon_1(\alpha')] a(\alpha') N(\alpha')}{\int_0^\infty [1 - \sum_{l=1}^{4} \varepsilon_1(u)] a(u) N(u) du}, \tag{2.11}$$

where

$$\phi(\alpha, \alpha') = \begin{cases} g_1(\alpha, \alpha') \varepsilon_1(\alpha) + g_2(\alpha, \alpha') \varepsilon_2(\alpha) + I_W(\alpha, \alpha') \dfrac{\varepsilon_4(\alpha)}{W_{max} - W_{min}}, & \alpha \geq G, \\[2ex] g_1(\alpha, \alpha') \varepsilon_1(\alpha) + g_3(\alpha, \alpha') \varepsilon_3(\alpha) + I_W(\alpha, \alpha') \dfrac{\varepsilon_4(\alpha)}{W_{max} - W_{min}}, & \alpha \leq L - G, \end{cases}$$

with

$$g_1(\alpha, \alpha') = \frac{1}{\sqrt{2\pi}\sigma_1(\alpha)} e^{-\frac{(\alpha'-\alpha)^2}{2[\sigma_1(\alpha)]^2}},$$

$$g_2(\alpha, \alpha') = \frac{1}{\sqrt{2\pi}\sigma_2(\alpha)} e^{-\frac{[\alpha'-(\alpha-G)]^2}{2[\sigma_2(\alpha)]^2}},$$

$$g_3(\alpha, \alpha') = \frac{1}{\sqrt{2\pi}\sigma_3(\alpha)} e^{-\frac{[\alpha'-(\alpha+G)]^2}{2[\sigma_3(\alpha)]^2}},$$

where the $g_k(\alpha, \alpha')$ ($k = 1, 2, 3$) are Gaussian kernels with standard deviations $\sigma_k(\alpha)$ and

$$I_W(\alpha, \alpha') = \begin{cases} 1 & \text{if } \alpha, \alpha' \in [W_{min}, W_{max}] \\ 0 & \text{otherwise.} \end{cases}$$

Besides the above-mentioned relationship between $\varepsilon_2(\alpha)$ and $\varepsilon_3(\alpha)$, for each $\alpha, 0 \leq \sum_{l=1}^{4} \varepsilon_1(\alpha) < 1$. Figure 2.3 corresponds to the horizontal lines on Figure 2.2 (left).

We fit a hybrid of these discrete and continuous formulations of our model (i.e., our discrete formulation with Gaussian kernels instead of deltas) to observations from the four above-mentioned empirical studies, all of which are discrete, using the FindMinimum function in MATHEMATICA. This amounts to choosing ε_{li} and σ_{ki}, as well as G, L and the W_s, that minimize an objective function, here the mean squared error. With one starting value for each variable, FindMinimum uses BFGS quasi-Newton methods. When there are constraints, FindMinimum uses interior point methods. We found it necessary to constrain $\vec{\varepsilon}_1$ and $\vec{\sigma}_k$ lest the main diagonal dominate, and fixed G, L, and the W_s after convergence. We ensured that solutions were robust by using different initial conditions for simple models and solutions of simpler models (e.g., with off-diagonals identical, without contacts among co-workers, ...) for more complex ones.

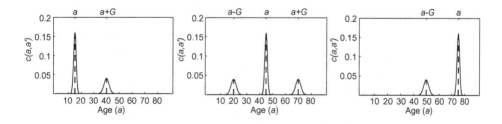

Fig. 2.3 Left to right, the panels correspond to horizontal lines on Figure 2.2 (left). People of all ages contact contemporaries preferentially, but children aged $< G$ also contact their parents, adults aged $> L - G$ their children, and those in between both. The heights of the Gaussian curves are determined by $\varepsilon(\alpha)$ and their widths by $\sigma(\alpha)$. A similar line through Figure 2.2 (right) anywhere between $y = W_{\min}$ and $y = W_{\max}$ would equal ε_4 between $x = W_{\min}$ and $x = W_{\max}$, but be 0 otherwise.

Applications of these mixing functions will be presented in Chapter 3 (see section 3.1.1).

2.2.2   *Role of variable activity levels in STIs*

For sexually transmitted infections (STIs), critical heterogeneities to consider include gender and levels of sexual activity. Their consideration also leads to multigroup models. The example presented in this section considers the synergy between HIV and HSV-2. One of the questions that is of interest for public health officials is how treatment of HSV-2 may influence the prevalence and control of HIV.

In [Feng *et al.* (2013)], we analyzed a model that includes both HIV and HSV-2 infections. The model considers one male population and multiple female populations based on their activity levels with variable male preferences to different female groups. Results from the model demonstrate that the heterogeneity in activity levels and male preference in mixing may play an important role in model outcomes. More details of the model analysis are presented below.

Several mathematical models have been developed to investigate the transmission dynamics of HSV-2 ([Blower *et al.* (1998); Foss *et al.* (2009); Newton and Kuder (2000); Schinazi (1999)] and references therein) and HIV ([Cohen *et al.* (2008); Blower and Dowlatabadi (1994); Blower *et al.* (2001, 2000); Mukandavire and Garira (2007a,b)] and references therein). To our knowledge, however, there have only been a few modeling studies of the epidemiological synergy between HSV-2 and HIV. Using the individual-based model STDSIM, White *et al.* [White *et al.* (2008)] studied the population-level effect of HSV-2 therapy on the incidence of HIV in sub-Saharan Africa. Foss *et al.* [Foss *et al.* (2011)] developed a dynamic HSV/HIV model to estimate the contribution of HSV-2 to HIV transmission from clients to female sex workers in southern India and the maximum potential impact of 'perfect' HSV-2 suppressive therapy on HIV incidence. Blower and Ma [Blower and Ma (2004)] used a transmission model that specifies the dynamics of HIV and HSV-2 to predict the effect of a high-prevalence HSV-2 epidemic on HIV incidence. Abu-Raddad *et al.* [Abu-Raddad *et al.* (2008)] constructed a deterministic compartmental model to describe HIV and HSV-2 transmission dynamics and their interaction. However, the model studied in [Blower and Ma (2004)] does not include heterogeneity in sexual activity and assumes that individuals mix randomly, whereupon each infective individual is equally likely to spread the disease to all others. Also, gender is not incorporated into the models studied by either [Blower and Ma (2004)] or [Abu-Raddad *et al.* (2008)]. The models in [Foss *et al.* (2011)] and [White *et al.* (2008)] incorporate various heterogeneities, including gender and/or age, but not sexual activity, and only numerical simulations are conducted.

The motivation for our model to include different levels of sexual activity is based on the consideration that individuals differ in number of sexual partners, and thereby risk of being infected and infecting others. Thus, heterogeneity in sexual activity may have important implications for the transmission and control of STDs. Mixing between different activity groups has been shown theoretically to influence the rate and pattern of infection in defined communities. If mixing is assortative by activity, for example, infection will be largely restricted to a core of highly active individuals with only occasional transmission to others. The pathogen will spread more slowly, but affect a higher fraction of the total community if mixing is random [Anderson (1990)]. Most importantly, the well-mixed assumption of unstratified models is particularly inappropriate for sexual activity as the number of contacts that each individual has is considerably smaller than the population size [Keeling and Eames (2005)]. Finally, gender may be an important factor in modeling the epidemiological synergy between HSV-2 and HIV as shown in the meta-analysis of several studies that male parameters differ from the corresponding female parameters. For example, the male-to-female HSV-2 transmission probability is greater than the female-to-male transmission probability [Corey *et al.* (2004); Wald *et al.* (2001)], and thus the risk of female-to-male transmission per sex act is less than the risk of male-to-female transmission [Mukandavire and Garira (2007a,b)]. Thus, to fully understand the epidemiological synergy between HSV-2 and HIV and to investigate measures for controlling these sexually-transmitted diseases, it is important to analyze models that consider heterogeneities in sexual activity, mixing within and between different activity groups and genders.

The model in [Feng *et al.* (2013)] incorporates multiple sexual activities within both female and male populations, and heterogeneous mixing between females and males from different activity groups. We derive several reproduction numbers for HIV and HSV-2 (either a disease alone or invasion) that determine the dynamical outcomes of the interaction between HIV and HSV-2. We provide a systematic qualitative analysis of the model including the existence of equilibria, and their local and global behavior. The model is also used to explore the epidemiological synergy between HSV-2 and HIV by investigating the contribution of HSV-2 to HIV prevalence, as well as by evaluating the potential population-level impact of HSV-2 therapy on HIV control.

Consider a population consisting of sexually active female and male individuals. We consider the case in which the female population is divided into sub-groups based on levels of sexual activity (e.g., number of partners) with a low-risk group (e.g., members of the general population) and a high-risk group (e.g., sex workers), while all individuals in the male population have the same activity level. These sub-populations are labeled by the subscripts f_1, f_2, m, which denote low- and high-risk females and males, respectively. Let N_i denote the population sizes of groups $i, i = m, f_1, f_2$. The population in each group is assumed to be homogeneous in the sense that individuals have the same infectious period, duration of immunity, contact rate, and so on. We divide the progression of HIV into two stages, acute infection and AIDS. Similarly, HSV-2 is represented by acute and latent infection stages. Because individuals infected with HSV-2 alone or HSV-2 alone can become co-infected with both HIV and HSV-2, each group i ($i = m, f_1, f_2$) is further divided into seven epidemiological classes or subgroups: susceptible, infected with acute HSV-2 only (A_i), infected with latent HSV-2 only (L_i), infected with HIV only (H_i), infected with HIV and acute HSV-2 (P_i), infected with HIV and latent HSV-

$2(Q_i)$ and AIDS (D_i). A transition diagram between these epidemiological classes within group i is depicted in Figure 2.4.

For each sub-population i ($i = f_1, f_2, m$) there is a per capita recruitment rate μ_i into the susceptible group. For all classes there is a constant per capita rate μ_i of exiting the sexually active population. Thus, the total population N_i in group i remains constant for all time. Susceptible people in group i acquire infection with HSV-2 or HIV at the rate $\lambda_i^A(t)$ or $\lambda_i^H(t)$, respectively. Upon being infected with HSV-2, people in group i enter the class A_i (infected with acute HSV-2 only). These individuals become latent L_i at the constant rate ω_i^A (an average duration in A_i is $1/\omega_i^A$). Following an appropriate stimulus in individuals with latent HSV-2, reactivation may occur [Blower *et al.* (1998)]. We assume that people with latent HSV-2 only reactivate at the rate γ_i^L. Individuals with HIV are assumed to develop AIDS at the rate d_i^H. Let δ_i^A and δ_i^L denote the enhanced susceptibility to HIV infection for individuals in group i with acute or latent HSV-2 infection. Classes P_i and Q_i are similar to A_i and L_i, respectively, except that A_i and L_i denote individuals with HSV-2 only whereas P_i and Q_i denote individuals with co-infections. The difference in stage durations are indicated by the superscripts (e.g., $1/\gamma_i^L$ for the L class and $1/\gamma_i^Q$ for the Q class). Finally, the anti-viral treatment rates for the A_i and P_i individuals are denoted by θ_i^A and θ_i^Q, respectively. Because anti-viral medications will also suppress reactivation of latent HSV-2, we assume that the reactivation rate of people with latent HSV-2 γ_i^L (or γ_i^Q) is a decreasing function of θ_i^A (or θ_i^P), denoted by $\gamma_i^L(\theta_i^A)$ (or $\gamma_i^Q(\theta_i^P)$). Table 2.1 includes the definition of frequently used variables and parameters. The sources for most of the parameter values are from [Foss *et al.* (2009); Abu-Raddad *et al.* (2008)] (see [Feng *et al.* (2013)] for more details).

Based on Figure 2.4, the model is described by the following system of differential equations for $i = m, f_1, f_2$:

$$\frac{dS_i}{dt} = \mu_i N_i - (\lambda_i^A(t) + \lambda_i^H(t))S_i - \mu_i S_i,$$

$$\frac{dA_i}{dt} = \lambda_i^A(t)S_i + \gamma_i^L(\theta_i^A)L_i - \delta_i^A \lambda_i^H(t)A_i - (\omega_i^A + \theta_i^A + \mu_i)A_i,$$

$$\frac{dL_i}{dt} = (\omega_i^A + \theta_i^A)A_i - \delta_i^L \lambda_i^H(t)L_i - (\gamma_i^L(\theta_i^A) + \mu_i)L_i,$$

$$\frac{dH_i}{dt} = \lambda_i^H(t)S_i - \delta_i^H \lambda_i^A(t)H_i - (\mu_i + d_i^H)H_i, \qquad (2.12)$$

$$\frac{dP_i}{dt} = \delta_i^A \lambda_i^H(t)A_i + \delta_i^H \lambda_i^A(t)H_i + \gamma_i^Q(\theta_i^P)Q_i - (\omega_i^P + \theta_i^P + \mu_i + d_i^P)P_i,$$

$$\frac{dQ_i}{dt} = \delta_i^L \lambda_i^H(t)L_i + (\omega_i^P + \theta_i^P)P_i - (\gamma_i^Q(\theta_i^P) + \mu_i + d_i^Q)Q_i,$$

where the functions $\lambda_i^j(t)$ represent the forces of infection given below. Let b_i ($i = m, f_1, f_2$) be the rate at which individuals in group i acquire new sexual partners (also referred to as contact rates), and let c_j denote the probability that a male chooses a female partner in group j $(j = f_1, f_2)$. Then $c_1 + c_2 = 1$. For ease of notation, let

$$c_1 = c, \quad c_2 = 1 - c.$$

Overall, the number of female partners in groups j $(j = f_1, f_2$ that males acquire should be equal to the number of male partners that females in groups j acquire. These observations lead to the following balance conditions:

$$b_m c N_m = b_{f_1} N_{f_1}, \qquad b_m(1 - c)N_m = b_{f_2} N_{f_2}. \tag{2.13}$$

To ensure that constraints in (2.13) are satisfied, we assume in numerical simulations that b_m and c are fixed constants with b_{f_1} and b_{f_2} being varied according to N_m, N_{f_1} and N_{f_2}.

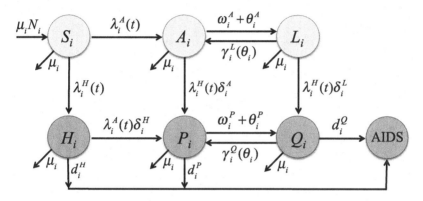

Fig. 2.4 Transition diagram of the coupled dynamics between HIV and HSV-2. The top row includes classes infected with HSV-2 only, and the bottom row includes classes infected with either HIV only or coinfected with HIV and HSV-2.

The force of infection functions can be expressed as

$$
\lambda_m^H(t) = \sum_{i=1}^{2} b_m c_i \beta_{f_i m}^H \frac{H_{f_i} + \delta_{f_i}^P P_{f_i} + \delta_{f_i}^Q Q_{f_i}}{N_{f_i}},
$$

$$
\lambda_{f_j}^H(t) = b_{f_j} \beta_{m f_j}^H \frac{H_m + \delta_m^P P_m + \delta_m^Q Q_m}{N_m}, \quad j = 1, 2,
$$

$$
\lambda_m^A(t) = \sum_{i=1}^{2} b_m c_i \beta_{f_i m}^A \frac{A_{f_i} + \sigma_{f_i}^P P_{f_i}}{N_{f_i}},
$$

$$
\lambda_{f_j}^A(t) = b_{f_j} \beta_{m f_j}^A \frac{A_m + \sigma_m^P P_m}{N_m}, \quad j = 1, 2,
$$

$$\tag{2.14}$$

where

$$N_i = S_i + A_i + L_i + H_i + P_i + Q_i, \quad i = m, f_1, f_2$$

denotes the total population size of group i. In (2.14), β_{im}^H (β_{mi}^H), $i = f_1, f_2$ are the HIV transmission probabilities per partner between females infected with HIV in group i and susceptible males (between males infected with HIV and susceptible females in group i); β_{im}^A (β_{mi}^A), $i = f_1, f_2$ are the HSV-2 transmission probabilities per partner between females infected with acute HSV-2 in group i and susceptible males (between males infected with acute HSV-2 and susceptible females in group i); δ_i^P and δ_i^Q ($i = m, f_1, f_2$) are the enhanced HIV infectiousness of co-infected individuals, and σ_i^P ($i = m, f_1, f_2$) are the enhanced HSV-2 infectiousness of co-infected individuals. All variables and parameters are summarized in Table 2.1.

Table 2.1   Frequently used symbols and parameter values used in simulations.

Symbol	Description	Value
m, f_1, f_2	Male, Low-risk female, High-risk female	
N_i	Total population size of $(i = m, f_1, f_2)$	
S_i	Number of susceptibles	
A_i, L_i	Number with acute, latent HSV-2	
H_i	Number with HIV only of group i	
P_i, Q_i	Number with HIV and acute, latent HSV-2	
D_i	Number with AIDS	
μ_i	per capita recruitment rate of group i	(1/312,1/204)
$\frac{1}{\mu_i}$	Average sexual life span	312, 204
$\gamma_i^L(0)$	Baseline reactivation rate of latent HSV-2	0.436, 0.339
$\gamma_i^Q(0)$	Baseline reactivation rate of latent HSV-2	0.469, 0.365
$\omega_i^A,$	Rate of acute HSV-2 becoming latent $(j = m, f_1, f_2)$	2.678, 2.083
ω_i^P	Rate of acute HSV-2 becoming latent	1.875, 1.458
$\theta_i^A,\ \theta_i^P$	Treatment rate of acute HSV-2	Varied
$\delta_m^A,\ \delta_{f_1}^A = \delta_{f_2}^A$	Enhanced susceptibility to HIV due to acute HSV-2	2.7, 3.1
δ_i^L	Enhanced susceptibility to HIV due to latent HSV-2	1.0
δ_i^H	Enhanced susceptibility to HSV-2 due to HIV	1.0
$\delta_i^P = \delta_i^Q$	Enhancement of HIV infectiousness in co-infected	1.0
σ_i^P	Enhancement of HSV-2 infectiousness in co-infected	1.0
$d_i^H = d_i^P = d_i^Q$	Rate of progression from HIV to AIDS	0.0104
b_m	Rate of male sexual contacts with females	Varied
b_{f_1}	Rate of group 1 female sexual contact of with males	$= b_m c \frac{N_m}{N_{f_1}}$
b_{f_2}	Rate of group 2 female sexual contact of with males	$= b_m c_2 \frac{N_m}{N_{f_2}}$
$c_1 = c$	Fraction of male contacts with female group 1	Varied
$c_2 = 1 - c$	Fraction of male contacts with female group 2	Varied
$\beta_{f_1 m}^H,\ \beta_{f_2 m}^H$	Probability of HIV infection per male contact	0.1, 0.02
$\beta_{m f_1}^H,\ \beta_{m f_2}^H$	Probability of HIV infection per female contact	0.2, 0.04
$\beta_{f_1 m}^A,\ \beta_{f_2 m}^A$	Probability of HSV-2 infection per male contact	1.0, 0.2
$\beta_{m f_1}^A,\ \beta_{m f_2}^A$	Probability of HSV-2 infection per female contact	1.0, 0.4

In this table, $i = m, f_1, f_2$. The time unit is months.

Reproduction numbers

For each of the two diseases, we can compute the reproduction number in the absence of the other disease. Let $\mathcal{R}_0^A$ and $\mathcal{R}_0^H$ denote these reproduction numbers for HSV-2 and HIV, respectively. Due to the loop between the symptomatic and asymptomatic stages of HSV-2, the derivation of analytical expression for $\mathcal{R}_0^A$ for model (2.12) is not straightforward. A detailed derivation of the following formula for $\mathcal{R}_0^A$ can be found in [Feng *et al.* (2013)]:

$$\mathcal{R}_0^A = \sqrt{\left(\mathcal{R}_{mf_1m}^A\right)^2 + \left(\mathcal{R}_{mf_2m}^A\right)^2}, \qquad (2.15)$$

where

$$\mathcal{R}_{mf_jm}^A = \sqrt{\frac{b_{f_j}\beta_{mf_j}^A}{\omega_m^A + \theta_m^A + \mu_m} \cdot P_m^A \cdot \frac{b_m c_j \beta_{f_jm}^A}{\omega_{f_j}^A + \theta_{f_j}^A + \mu_{f_j}} \cdot P_{f_j}^A}, \quad j = 1, 2$$

with P_i^A ($i = m, f_1, f_2$) representing the probability that an individual of group i is in the acute stage (A), which is given by

$$P_i^A = \frac{(\omega_i^A + \theta_i^A + \mu_i)(\gamma_i^L(\theta_i^A) + \mu_i)}{[\gamma_i^L(\theta_i^A) + \omega_i^A + \theta_i^A + \mu_i]\mu_i}, \quad i = m, f_1, f_2. \qquad (2.16)$$

The formulas for P_i^A in (2.16) can be explained as follows. Let

$$p = \frac{\omega_i^A + \theta_i^A}{\omega_i^A + \theta_i^A + \mu_i}, \quad q = \frac{\gamma_i^L}{\gamma_i^L + \mu_i},$$

where p represents the probability that an individual moves from the acute stage (A) to the latent stage (L), and q represents the probability that an individual moves from L to A. Thus, the probability that an individual is in the acute stage within the $A \rightleftharpoons L$ loop is

$$\sum_{k=1}^{\infty}(pq)^k = \frac{(\omega_i^A + \theta_i^A + \mu_i)(\gamma_i^L + \mu_i)}{(\gamma_i^L + \omega_i^A + \theta_i^A + \mu_i)\mu_i} = P_i^A.$$

Notice that in the formula for $\mathcal{R}_0^A$ the balance conditions in (2.13) have been used. Other factors in $\mathcal{R}_{mf_im}^A$ ($i = 1, 2$) also have clear biological interpretations:

- $b_{f_j}\beta_{mf_j}^A$ is the number of new infections that a male will cause in females of group j ($j = 1, 2$) per unit of time;
- $b_m c_j \beta_{f_jm}^A$ is the number of new infections that a female in group j ($j = 1, 2$) will cause in males per unit of time;
- $\frac{1}{\omega_i^A + \theta_i^A + \mu_i}$ ($i = m, f_1, f_2$) represents the mean time that an individual in group i remains infected (i.e., in either A or L).

Thus, $\sqrt{\mathcal{R}_{mf_jm}^A}$ represents the average secondary HSV-2 male infections by one male individual through females in group j ($j = 1, 2$) while in the infectious stage (A) in a completely susceptible population. The square root is associated with the fact that we need to consider both the male-to-female and female-to-male processes to obtain the number of secondary infections. The overall reproduction number $\mathcal{R}_0^A$ is an average of $\mathcal{R}_{mf_im}^A$ $(i = 1, 2)$.

Let $\mathcal{R}_0^H$ denote the basic reproduction number for HIV in the absence of HSV-2. Then

$$\mathcal{R}_0^H = \sqrt{\left(\mathcal{R}_{mf_1m}^H\right)^2 + \left(\mathcal{R}_{mf_2m}^H\right)^2},$$

where

$$\mathcal{R}_{mf_jm}^H = \sqrt{\frac{b_{f_j}\beta_{mf_j}^H}{d_m^H + \mu_m} \cdot \frac{b_m c_j \beta_{f_jm}^H}{d_{f_j}^H + \mu_{f_j}}}, \quad j = 1, 2.$$

The biological meanings of $\mathcal{R}_{mf_1m}^H$ and $\mathcal{R}_{mf_2m}^H$ can be explained in the similar way as those of $\mathcal{R}_{mf_1m}^A$ and $\mathcal{R}_{mf_2m}^A$. It is clear that $\mathcal{R}_0^H$ represents the average secondary HIV male infections by one male individual (through both female groups) during the whole HIV infectious period in a completely susceptible population.

Invasion reproduction numbers

Let $\mathcal{R}_A^H$ denote the invasion reproduction number for HIV in a population where the HSV-2 infection is already established at the endemic equilibrium, which is denoted by E_{∂}^A. The nonzero components of E_{∂}^A are $S_i^0, A_i^0,$ and L_i^0, representing the density of susceptible, acute HSV-2, and HSV-2 latent, respectively, in group i. Let $N_i^0 = S_i^0 + A_i^0 + L_i^0$. For ease of notation, let

$$\lambda_m^{A0} = b_m \sum_{i=1}^2 c_i \beta_{f_jm}^A \frac{A_{f_j}^0}{N_{f_j}^0}, \qquad \lambda_{f_j}^{A0} = b_{f_j}\beta_{mf_j}^A \frac{A_m^0}{N_m^0}, \quad j = 1, 2$$

and

$$\vec{\delta}_i = \left(1, \delta_i^P, \delta_i^Q\right), \qquad \vec{x}_i^0 = \left(S_i^0, \delta_i^A A_i^0, \delta_i^L L_i^0\right)^T, \quad i = m, f_1, f_2.$$

Note that the system (2.12) has 9 infected variables with HIV ($H_i, P_i, Q_i, i = m, f_1, f_2$). Consider the HIV-free equilibrium E_{∂}^A of system (2.12). The matrices $\mathcal{F}^H$ and $\mathcal{V}^H$ (corresponding to the new infection and remaining transfer terms, respectively) are given by

$$\mathcal{F}^H = \begin{pmatrix} 0 & F_{f_1m}^H & F_{f_2m}^H \\ F_{mf_1}^H & 0 & 0 \\ F_{mf_2}^H & 0 & 0 \end{pmatrix}, \qquad \mathcal{V}^H = \begin{pmatrix} V_m^H & 0 & 0 \\ 0 & V_{f_1}^H & 0 \\ 0 & 0 & V_{f_2}^H \end{pmatrix}, \qquad (2.17)$$

where

$$F_{f_jm}^H = b_m c_j \beta_{f_jm}^H \frac{\vec{x}_m^0}{N_m^0}\vec{\delta}_{f_j}, \qquad F_{mf_j}^H = b_{f_j}\beta_{mf_j}^H \frac{\vec{x}_{f_j}^0}{N_{f_j}^0}\vec{\delta}_m, \quad j = 1, 2$$

and

$$
V_i^H =
\begin{pmatrix}
(\mu_i + d_i^H + \delta_i^H \lambda_i^{A0}) & 0 & 0 \\
-\delta_i^H \lambda_i^{A0} & \omega_i^P + \theta_i^P + \mu_i + d_i^P & -\gamma_i^Q(\theta_i^P) \\
0 & -(\omega_i^P + \theta_i^P) & \gamma_i^Q(\theta_i^P) + \mu_i + d_i^Q
\end{pmatrix},
\tag{2.18}
$$

for $i = m, f_1, f_2$. Then, the next generation matrix for HIV, denoted by K_H, can be expressed by

$$
K_H = \mathcal{F}^H (\mathcal{V}^H)^{-1}
$$

$$
=
\begin{pmatrix}
0 & F_{f_1 m}^H (V_{f_1}^H)^{-1} & F_{f_2 m}^H (V_{f_2}^H)^{-1} \\
F_{m f_1}^H (V_m^H)^{-1} & 0 & 0 \\
F_{m f_2}^H (V_m^H)^{-1} & 0 & 0
\end{pmatrix}
\tag{2.19}
$$

$$
:= (k_{ij})_{9 \times 9}
$$

where the entries k_{ij} of the matrix K_H can be found in Appendix A of [Feng *et al.* (2013)].

Noting that $\text{Rank}(K_H) = 2$ and that the sum of the diagonal elements in matrix K_H is zero, it follows from Vieta's formulas that if the numbers of susceptible people and those with acute and latent HSV-2 in group i are S_i^0, A_i^0, L_i^0, respectively, the reproduction number for HIV infection is given by

$$
\mathcal{R}_A^H = R_A^H(S_i^0, A_i^0, L_i^0, 0, 0, 0) := \rho(K_H) = \sqrt{-E_2(K_H)}
$$

$$
= \sqrt{\sum_{i=1}^{3} \sum_{j=4}^{9} k_{ij} k_{ji}},
\tag{2.20}
$$

where $\rho(K_H)$ represents the spectral radius of the matrix K_H and $E_2(K_H)$ is the sum of all the 2×2 principal minors of matrix K_H (see Appendix A in [Feng *et al.* (2013)] for the definition of k_{ij}). It is shown in [Feng *et al.* (2013)] that invasion is possible if and only if $\mathcal{R}_A^H > 1$.

The results on the existence and local stability of the boundary equilibria are summarized in Table 2.2, and corresponding sketches of the local dynamics of system (2.12) are shown in Figure 2.5.

Figures 2.6 and 2.7 illustrate results of numerical simulations, which show how disease dynamics may depend on the reproduction numbers. They also help to confirm and extend the analytic results illustrated in Figure 2.5. For example, the left plot in Figure 2.6 is for the case when enhancement of HIV by HSV-2 is relatively strong and co-infection by HSV-2 has little effect on the rate of progression from the HIV stage to AIDS or HIV mortality. In this case, $\mathcal{R}_0^A > 1, \mathcal{R}_0^H < 1$

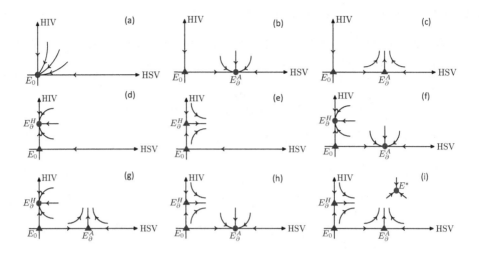

Fig. 2.5   Sketches of the local dynamics of the system (2.12). The three boundary equilibria, E_0, $E_{\partial_H}^A$ and $E_{\partial_A}^H$, are labeled by either a ▲ (if it is unstable) or a • (if it is stable).

Table 2.2   Existence and stability of equilibria of system (2.12).

		E_0	E_δ^A	E_δ^H	Corresponding sketches (see Figure 2.5)
$\mathcal{R}_0^A < 1$	$\mathcal{R}_0^H < 1$	LS	DNE	DNE (conditional)	(a)
$\mathcal{R}_0^A > 1$	$\mathcal{R}_A^H < 1$	US	LS	DNE (conditional)	(b)
$\mathcal{R}_0^H < 1$	$\mathcal{R}_A^H > 1$	US	US	DNE (conditional)	(c)
$\mathcal{R}_0^A < 1$	$\mathcal{R}_H^A < 1$	US	DNE	LS (conditional)	(d)
$\mathcal{R}_0^H > 1$	$\mathcal{R}_H^A > 1$	US	DNE	US	(e)
	$\mathcal{R}_H^A < 1$ $\mathcal{R}_A^H < 1$	US	LS	LS (conditional)	(f)
$\mathcal{R}_0^A > 1$	$\mathcal{R}_H^A < 1$ $\mathcal{R}_A^H > 1$	US	US	LS (conditional)	(g)
$\mathcal{R}_0^H > 1$	$\mathcal{R}_H^A > 1$ $\mathcal{R}_A^H < 1$	US	LS	US	(h)
	$\mathcal{R}_H^A > 1$ $\mathcal{R}_A^H > 1$	US	US	US	(i)

DNE: does not exist; US: unstable; LS: locally stable.

and $\mathcal{R}_A^H > 1$. Our analytic results suggest that, under these conditions i) HIV cannot establish itself in a population when HSV-2 is absent; ii) HIV can invade a population in which HSV-2 is endemic; and iii) the HSV-2 equilibrium E_∂^A is stable for the HSV-2 subsystem, but unstable for the full system (2.12). These are confirmed by the simulations illustrated in Figure 2.6 (left). We observe that while HIV can invade and persist in the presence of HSV-2 (the dashed curve), it dies out in the absence of HSV-2 (the solid curve), suggesting that HSV-2 infection can favor the invasion of HIV. The right plot in Figure 2.6 is for the case when enhancement factors are relatively weak and co-infection by HSV-2 has a significant effect on the rate of progression from the HIV stage to AIDS or HIV mortality. In this case, $\mathcal{R}_0^A > 1, \mathcal{R}_0^H > 1$ and $\mathcal{R}_A^H < 1$. From our analytic results, the HSV-2 equilibrium E_{∂}^A is stable for the full system (2.12). Thus, while HIV can establish itself in the absence of HSV-2 (as $\mathcal{R}_0^H > 1$), it cannot invade a population in which HSV-2 is endemic. This is confirmed by simulations presented in Figure 2.6 (right). We observe that HIV can invade and persist in the absence of HSV-2 (the solid curve), and that it dies out in the presence of HSV-2 (the dashed curve), suggesting that HSV-2 infection may not always favor the invasion of HIV.

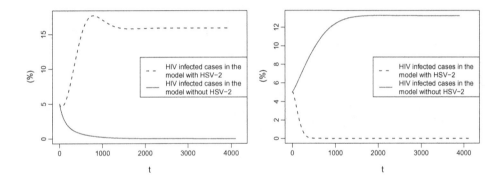

Fig. 2.6 Numerical solutions of the system (2.12) for parameters in different regions corresponding to the cases when the enhancement of HIV by HSV-2 is relatively stronger (left) or weaker (right). In both plots, the dashed and solid curves represent levels of HIV infection with and without HSV-2 present, respectively. For the left plot, the reproduction numbers are $\mathcal{R}_0^A > 1, \mathcal{R}_A^H < 1$ and $\mathcal{R}_A^H > 1$; whereas for the right plot, the reproduction numbers are $\mathcal{R}_0^A > 1, \mathcal{R}_0^H > 1$ and $\mathcal{R}_A^H < 1$.

Simulations presented in Figure 2.7 demonstrate the role of the HSV-2 invasion reproduction number $\mathcal{R}_H^A$. For these simulations, parameter values were chosen such that $\mathcal{R}_0^A > 1$, $\mathcal{R}_0^H > 1$ and $\mathcal{R}_A^H > 1$. The plots in (a) and (b) are for the cases when $\mathcal{R}_H^A < 1$ and $\mathcal{R}_H^A > 1$, respectively. Our analytic results suggest that, in case (a) the HIV equilibrium E_{∂}^H is l.a.s. (see Table 2.2). This is indeed confirmed by the simulations, which show that HIV converges to a positive level while HSV-2 goes to extinction. The simulations shown in (b) extend our analytical results. While we do not have results on the existence and stability of a coexistence equilibrium (i.e., one where both HIV and HSV-2 are present), our analytical results show that under the condition for (b) all boundary equilibria (E_0, E_{∂}^A and E_{∂}^H) are unstable, suggesting that a coexistence equilibrium may exist. This is indeed the case as exhibited in Figure 2.7(b), which shows that both HIV and HSV-2 are stabilized at

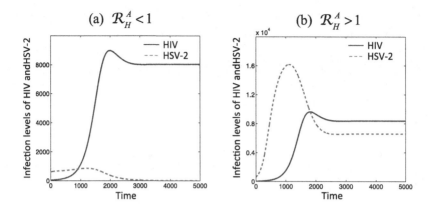

Fig. 2.7 Time plots of the system (2.12) illustrating that the invasion reproduction number $\mathcal{R}_H^A$ determines whether or not HSV-2 can invade a population in which HIV is already established. In both plots $\mathcal{R}_0^H > 1$, $\mathcal{R}_0^A > 1$ and $\mathcal{R}_A^H > 1$, in which case both HSV-2 and HIV can persist in the absence of the other pathogen. The plot in (a) is for the case when $\mathcal{R}_H^A < 1$ and it shows that HSV-2 cannot invade. The plot in (b) is for the case when $\mathcal{R}_H^A > 1$ and we observe that HSV-2 is able to invade and both pathogens stabilize at endemic levels (coexistence).

positive levels.

These stability analyses and threshold conditions are helpful in the investigation of the effect of HSV-2 therapy on the prevalence and control of HIV. More details are presented in Chapter 3 (see section 3.5).

2.3 Multiple pathogen strains and host types

This section includes several models dealing with various biological systems involving multiple pathogen strains and host types. The results are adopted from [Castillo-Chavez and Feng (1997)] on multiple drug-resistant strains of TB, [Xu et al. (2012); Yang et al. (2012); Zhang et al. (2007b)] on the human-schistosome-snail system, and [Feng and Velasco-Hernandez (1997)] on dengue.

2.3.1 *Modeling TB and drug-resistance*

Models presented in this section are based on the two-strain TB model studied in [Castillo-Chavez and Feng (1997)]. Despite its sociological and historical importance, the study of TB transmission using statistical and mathematical models has not received enough attention until very recent years. Tuberculosis is caused by *Mycobacterium tuberculosis*. The disease is most commonly transmitted from a person suffering from infectious (active) tuberculosis to other persons by infected droplets created when the person with active TB coughs or sneezes. Among generally healthy persons, infection with TB is highly likely to be asymptomatic. Data from a variety of sources suggest that the lifetime risk of developing clinically evident TB after being infected is approximately 10%, with 90% likelihood of the infection remaining latent [Hopewell (1994)]. Individuals who have a latent TB infection are not clinically ill nor capable of transmitting TB ([Miller (1993)]). At greater ages, the immunity of persons who have been previously infected may wane, and they may be then at risk of developing active TB as a consequence of either exogenous reinfection (i.e., acquiring a new infection from another infectious individual) or endogenous reactivation of latent bacilli (i.e., re-activation of a pre-existing dormant infection) ([Styblo (1991)], [Smith and Moss (1994)]).

The epidemiology of TB disease is not simple. For the purpose of this study we only provide a superficial view, which we believe is sufficient for a rough understanding of the dynamics of TB transmission at the population level. General sources of information on TB dynamics suggest that TB is hard to transmit. Transmission (it is said) occurs only when there is prolonged close contact between a susceptible person and a person who has an active case of TB. Nonetheless, under the right conditions a single person with active TB can infect many other people ([Salyers and Whitt (1994)]). For example, it seems that about 13 persons were infected with TB per year by one source of infection in a Netherlands community in the period 1921-1938 [Styblo (1991)]. However, it is not clear that TB is in fact hard to transmit. Recent documented cases of TB transmission during lengthy plane trips [Kolata (1995); MMWR (1999)]) seem to indicate that transmission may be highly facilitated in a modern society. It is not at all unlikely that the risk of infection may be quite high in public places where there are actively infected TB individuals present. Mathematical models have been developed to estimate the probability of transmission of TB in close public environments. These models support the view that the acquisition of TB infection may not be as difficult as previously thought [Nardell (1995)]. A naive look at the fact that one third of the world population is actually infected suggests that either the tubercle bacillus is easy to acquire, or that in many parts of the world exposure and re-exposure to TB is extremely persistent, or both. Current epidemiological studies strongly support the claim that latent individuals are unable to transmit the tubercle bacillus and that only individuals with "active" TB are capable of spreading this bacteria. Therefore, latent individuals provide a tremendously large reservoir for the tubercle bacillus but as latent carriers of this bacillus they are uncapable of transmission. What are the epidemiological consequences of this situation in a world where populations become closer and closer? Here lies one of the central issues associated with the study of TB dynamics.

Latent TB individuals may remain in this latent stage for variable periods of time (in fact, many die without ever developing active TB). Apparently, the longer that we carry this bacteria the less likely we are to develop active TB unless our immune system becomes seriously compromised by other diseases. Consequently, age of infection as well as chronological age are important factors in disease progression. How important are these factors as predictors or measures of spread at the population level? Because it has been estimated that 10% of those infected with TB actually develop active TB during their lifetime, the 10% rule has become a useful measure for rough and immediate public health measures. This rule is useful but at the same time it is also superficial. It is well known that TB progression is not uniform but in fact is closely linked to various other factors such as nutritional status and/or access to decent medical care and living conditions [Bloom (1994)]. The good news is that latent and active TB can be treated with antibiotics. The bad news is that its treatment has side effects (sometimes quite serious) and takes a long time. Carriers of the tubercle bacillus who have not developed TB disease can be treated with a single drug *INH*; unfortunately, it must be taken religiously for 6-9 months. Treatment for those with active TB requires the simultaneous use of three drugs for a period of about 12 months. Lack of compliance with these drug treatments (a very serious problem) not only may lead to a relapse but to the development of antibiotic resistant TB–one of the most serious public health problems facing society today.

TB remains the leading cause of death by an infectious disease in the world. TB is also the most prevalent infection in the world [Bloom (1994); Miller (1993)]. As stated before, a third of the world's population is a carrier of tuberculosis and is at risk for developing active TB. It is estimated that there are between 8 and 10 million new cases per year, of which about 3 million people die [Kochi (2001)]. In the United States, the estimated total number of TB infections lies between 10-15 million persons [Miller (1993)]. However, dramatic increases in the incidence of TB (new cases per year) have occurred within the United States over the past few years. From 1985 to 1991, the number of reported cases of TB has increased 18% with 26,283 cases reported in 1991 [Kent (1993)]. In 1991, a large California prison with 5,421 inmates and 1,500 staff members had 18 cases of active TB [Salyers and Whitt (1994)]. Against the backdrop of an increasing incidence of TB in the United States there is a second problem, namely that of multi-drug resistant TB (MDR-TB). Resistant TB develops when the treatment of a TB patient is inadequate or incomplete, thereby allowing some of the stronger/resistant bacilli to survive and prosper. Outbreaks of MDR-TB in the United States have begun to alarm doctors and public health officials. Over 80% of the patients in these outbreaks have died, often within weeks of being diagnosed as having tuberculosis. These problems are compounded by economics, as the cost of treating a patient with MDR-TB can exceed $250,000: nearly 100 times the cost of treating most other TB cases (Press release WHO /89 Nov. 1994). The emergence of the HIV epidemic has dramatically increased the risk of developing clinical TB in infected persons, substantially increasing TB rates globally [Miller (1993)].

The host population is divided into the following epidemiological class or subgroups: Susceptibles (S), Latent with sensitive stain (L_1, infected but not infectious), Infectious with sensitive strain (I_1), and (effectively) Treated (T) individuals with sensitive strain. N denotes the total population. For the resistant strain, L_2 (latent) and I_2 (infectious) have similar meanings as L_1 and I_1. Because it is very hard to cure a patient with resistant TB we ignore the treatment of the resistant strain. Furthermore, we assume that I_2 individuals can infect S, L_1, and T individuals. From the disease transmission diagram (see Figure 2.8) we can write the following system of ordinary differential equations:

$$\dot{S} = \Lambda - \beta_1 S \tfrac{I_1}{N} - \beta_2 S \tfrac{I_2}{N} - \mu S,$$

$$\dot{L}_1 = \beta_1 S \tfrac{I_1}{N} - (\mu + k_1)L_1 - r_1 L_1 + pr_2 I_1 + \beta_1^* T \tfrac{I_1}{N} - \beta_2 L_1 \tfrac{I_2}{N},$$

$$\dot{I}_1 = k_1 L_1 - (\mu + d_1)I_1 - r_2 I_1,$$

$$\dot{L}_2 = qr_2 I_1 - (\mu + k_2)L_2 + \beta_2(S + L_1 + T)\tfrac{I_2}{N}, \qquad (2.21)$$

$$\dot{I}_2 = k_2 L_2 - (\mu + d_2)I_2,$$

$$\dot{T} = r_1 L_1 + (p + q)r_2 I_1 - \beta_1^* T \tfrac{I_1}{N} - \beta_2 T \tfrac{I_2}{N} - \mu T,$$

where $N = S + L_1 + I_1 + T + L_2 + I_2$. Λ is the recruitment rate, β_1 and β_2 are the rates at which susceptible individuals become infected by one infectious individual with sensitive and resistant strains, respectively; $\beta_1^* \leq \beta_1$ is for the treated individuals who may have a reduced susceptibility to the sensitive strain; μ is the per-capita natural death rate; k_i is the rate at which an individual leaves the latent class by becoming infectious with strain i; d_i is the per-capita disease induced death rate for individuals infected with strain i, and r_1 and r_2 are per-capita treatment rates for sensitive latent and infections individuals, respectively. $p + q$ the proportion of those treated infectious individuals who did not complete their treatment. The proportion p modifies the rate that departs from the latent class; $qr_2 I$ gives the rate at which individuals develop resistant-TB because they did not complete the treatment of active TB. Therefore $p \geq 0, q \geq 0$ and $p + q \leq 1$.

The system (2.21) has four possible equilibria denoted by E_0 (infection-free), E_1 (only the sensitive strain is present), E_2 (only the resistant strain is present), and E^* (coexistence of both strains). The existence of these equilibria depend on the reproductive numbers for the sensitive and resistant strains, which are given by

$$\mathcal{R}_S = \left(\frac{\beta_1 + pr_2}{\mu + d_1 + r_2}\right)\left(\frac{k_1}{\mu + k_1 + r_1}\right)$$

and

$$\mathcal{R}_R = \left(\frac{\beta_2}{\mu + d_2}\right)\left(\frac{k_2}{\mu + k_2}\right),$$

respectively. It is shown that these reproduction numbers help determine the competitive outcomes of the two parasite strains. The dynamics can be very different for the case of $q = 0$ (no treatment failure) and $q > 0$, particularly in terms of the coexistence of resistant strain with the sensitive strain. In addition to the reproduction numbers, there are two functions, $\mathcal{R}_R = f(\mathcal{R}_S)$ and $\mathcal{R}_R = g(\mathcal{R}_S)$, which

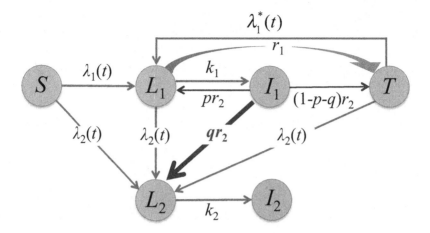

Fig. 2.8   A diagram for the two-strain TB model showing transitions between the epidemiological classes. The thicker arrow and the rate qr_2 represent the development of resistant TB due to treatment of infections with the sensitive strain. The functions for the force of infection are $\lambda_i(t) = \beta_i I_i(t)/N$ for $i = 1, 2$, and $\lambda_1^*(t) = \beta_1^* I_1(t)/N$.

divide the parameter region in the $(\mathcal{R}_S, \mathcal{R}_R)$ plane into subregions for the stability of equilibria:

$$f(\mathcal{R}_S) = \frac{1}{1 + \frac{1 - \mathcal{R}_S}{(\mathcal{R}_S - AB)(1 + 1/B)}}$$

$$g(\mathcal{R}_S) = \tfrac{1}{C}\left(AB + C - 1 \pm \sqrt{(AB + C - 1)^2 + 4(\mathcal{R}_S - AB)C}\right),$$

(2.22)

for $\mathcal{R}_S \geq 1$ where

$$A = \frac{pr_2}{\mu + k_1 + r_1}, \quad B = \frac{k_1}{\mu + d_1 + r_2}, \quad C = \frac{\mu}{\mu + k_1 + r_1}.$$

The properties of f and g include

$$f(1) = g(1) = 1, \quad f(\mathcal{R}_S) < g(\mathcal{R}_S) \quad \text{for } \mathcal{R}_S > 1.$$

The stability results for the system (2.21) are summarized in the following theorems.

Theorem 2.1. *Assume that $q = 0, d_1 = d_2 = 0$. Then*

    (a) *The disease-free equilibrium E_0 is g.a.s. if $\mathcal{R}_S < 1$ and $\mathcal{R}_R < 1$.*

    (b) *If $\mathcal{R}_S > 1$, then the boundary equilibrium E_1 is l.a.s. if $\mathcal{R}_R < f(\mathcal{R}_S)$ and unstable if $\mathcal{R}_R > f(\mathcal{R}_S)$.*

    (c) *If $\mathcal{R}_R > 1$, then the boundary equilibrium E_2 is l.a.s. if $\mathcal{R}_S < 1$ or if $\mathcal{R}_S > 1$ and $\mathcal{R}_R > g(\mathcal{R}_S)$. E_2 is unstable if $\mathcal{R}_S > 1$ and $\mathcal{R}_R < g(\mathcal{R}_S)$.*

    (d) *The coexistence equilibrium E^* exists and is l.a.s. if $\mathcal{R}_S > 1$ and $f(\mathcal{R}_S) < \mathcal{R}_R < g(\mathcal{R}_S)$.*

Figure 2.9(a) depicts the stability of the equilibria for the system (2.21) for the case of $q = 0$. Although the analytic results in Theorem 2.1 assumes that there is no disease related mortality $d_1 = d_2 = 0$, the numerical simulations suggest that the results hold also for $d_1 > 0$ and $d_2 > 0$ (see [Castillo-Chavez and Feng (1997)]. Figure 2.10 shows some simulation results of the model, illustrating the disease outcomes for $(\mathcal{R}_S, \mathcal{R}_R)$ in different regions as shown in Figure 2.9(a). In this figure, the parameter values used for the simulations are $\mu = 0.143, \beta_1 = 13, k_1 = 1, q = 0, p = 0.5, r_1 = 1, r_2 = 2, k_2 = 1, \Lambda = 35, d_1 = d_2 = 0$. For this set of values, $\mathcal{R}_S = 3.45$. Figures 2.10(a)–(c) correspond to different $\mathcal{R}_R$ (or equivalently β_2).

When $q > 0$, the boundary equilibrium E_1 is never stable, and the coexistence region III is much larger than that in the case of $q = 0$, as stated in the following theorem and illustrated in Figure 2.9(b).

Theorem 2.2. *Assume that $q > 0, d_1 = d_2 = 0$. Then*

(a) *The disease-free equilibrium E_1 is g.a.s. if $\mathcal{R}_S < 1$ and $\mathcal{R}_R < 1$.*

(b) *If $\mathcal{R}_R > 1$, then the boundary equilibrium E_2 is l.a.s. if $\mathcal{R}_S < 1$ or if $\mathcal{R}_S > 1$ and $\mathcal{R}_R > g(\mathcal{R}_S)$. E_2 is unstable if $\mathcal{R}_S > 1$ and $\mathcal{R}_R < g(\mathcal{R}_S)$.*

(c) *The equilibrium E_3 exists and is l.a.s. iff $\mathcal{R}_S > 1$ and $\mathcal{R}_R < g(\mathcal{R}_S)$.*

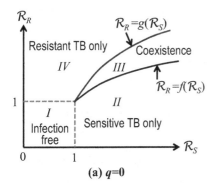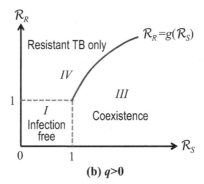

(a) q=0   **(b) q>0**

Fig. 2.9   (a) A bifurcation diagram for the system in the case $q = 0$. There are four regions I, II, III, and IV in the parameter space $(\mathcal{R}_S, \mathcal{R}_R)$. In region I, E_0 is a global attractor and other equilibria are unstable when they exist. In regions II and IV, E^* does not exist while E_1 and E_2 are l.a.s., respectively. In region III, E^* exists and is l.a.s. (b) A bifurcation diagram for the system in the case $q > 0$. There are three regions I, III, and IV in the parameter space $(\mathcal{R}_S, \mathcal{R}_R)$ (E_1 does not exist), in which E_0, E_2, and E^* are stable, respectively.

These results demonstrate that lack of drug treatment compliance by TB patients may have an important implication for the maintenance of antibiotic resistant strains. To make the role of antibiotic resistance transparent, we first studied a special version of our two-strain model with two competing strains of TB: the typical strain plus a resistant strain that was not the result of antibiotic resistance ($q = 0$).

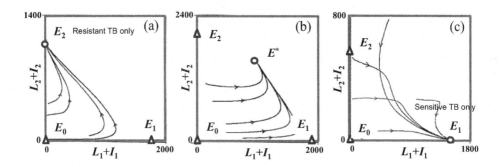

Fig. 2.10   Phase portraits of solutions to (2.21). This choice of parameter values gives a fixed value $\mathcal{R}_S = 3.45$. In (a) $\mathcal{R}_R = 2$ and hence $(\mathcal{R}_S, \mathcal{R}_R) \in$ IV. In (b) $\mathcal{R}_R = 24.4$ and hence $(\mathcal{R}_S, \mathcal{R}_R) \in$ III. In (c) $\mathcal{R}_R = 1.2$ and hence $(\mathcal{R}_S, \mathcal{R}_R) \in$ II. Circle indicates a stable equilibrium, and triangle indicates an unstable equilibrium.

In this last situation, we found that co-existence is possible but rare while later we noticed that co-existence is almost certain when the second strain is the result of antibiotic resistance. In our two-strain model there is a superinfection-like term $\beta_2 L_1 I_2 / N$. Is this necessary to obtain the co-existence result because it is well known that superinfection can cause co-existence (see [Levin and Pimentel (1981); Nowak and May (1994)])? The answer is no. In fact it can be shown that in the absence of the superinfection-like term co-existence is still almost the rule when the second strain is the result of antibiotic resistance (see Figure 2.9).

Our results show that coexistence of naturally resistant strains is limited. Deterministic models such as the one we have used here in fact suggest that competitive exclusion is the preferred outcome and not coevolution. Our results do not differ significantly from those developed by [Levin and Pimentel (1981)] in which it was found that coexistence of two strains of maxoma were only possible within a window of opportunity in parameter space. The recent work on super-infection by [Nowak and May (1994); May and Nowak (1994)] suggest that super-infection would enhance coexistence and pressumably coevolution. However, their models assume that the population size of the competing hosts are constant. If this assumption is removed then again the size of the region of coexistence shrinks and may even become a disconnected set ([Castillo-Chavez and Velasco-Hernandez (1998); Mena-Lorca *et al.* (1999)]).

In this study, we have just looked at the basic transmission dynamics of TB on a homogeneously mixing population (the null model). Our results are clear: such populations cannot support pathogen diversity, that is, competitive exclusion seems to be the preferred outcome. Furthermore, we found that antibiotic resistant (just as pesticide resistance) enhances – one may even say promotes) – coexistence. This is not surprising but reminds us of the challenges facing public health officials. Resistant TB will remain a serious threat to our communities as long as many members of our society do not have regular access to medical care.

One of the typical features for TB is its long and variable period of latency (more details can be found in section 2.4.2). This feature is neglected in model (2.21), which assumed exponentially distributed latent period so that the system consists of ordinary differential equations. A model taking into account a more realistic distribution of latency is discussed in section 2.4.2.

2.3.2 *Human-schistosome-snail system*

Although in many cases, simple models are sufficient for addressing speific biological questions, sometimes, for the model to be more readily linked to specific biological processes, more complexities and components of the biological system need to be included in the models. The purpose of this section is to demonstrate the situation in which models with increased complexities may generate additional insights that are impossible for simpler models. It is important to keep in mind that these models need to remain tractable mathematically.

The results presented in this section are adopted from [Yang *et al.* (2012)] for the case of multiple types of intermediate host and a single parasite strain, and [Xu *et al.* (2012)] for the case of multiple parasite strains and a single type of intermediate host.

Studying particular host-parasite interactions can often provide valuable insights into general patterns of disease establishment and persistence in natural systems. Addressing these systems using mathematical models not only provides us with information about the importance of particular disease parameters, it also allows us to follow the outcomes of such interactions over evolutionary time. Currently, most coevolutionary models of host-parasite interactions involve microparasites with direct life cycles (e.g.,[Anderson and May (1982a, 1991); Boots and Bowers (1999); Boots and Haraguchi (1999); Bowers *et al.* (1994); Frank (1992); May and Anderson (1983); May and Nowak (1995); Miller *et al.* (2005); Nowak and May (1994)]). There have been few theoretical studies investigating host-parasite coevolution using indirectly transmitted parasites (e.g., [Dobson (1988)]). Indirect life cycles involve multiple species of hosts (definitive and intermediate), all of which are required for the parasite to complete its development from larvae to an adult. Variable selective forces imposed by different host species (or strains) can have important consequences for parasite fitness at a number of points throughout the life cycle (see [Davies *et al.* (2001); Gower and Webster (2004, 2005)]).

Empirical work has demonstrated that parasites may face different host strategies for mitigating or preventing infection at different points in the life cycle [Minchella (1985)]. In some cases, hosts may actively resist parasite attack by altering morphological, physiological, or immunological factors [Sandland and Minchella (2003a)]. However, these strategies can be costly and often result in trade-offs with other host fitness parameters such as growth, reproduction, and survival [Beck *et al.*

(1984); Boots and Bowers (1999); Bowers *et al.* (1994); Sandland and Minchella (2003b)]. Alternatively, hosts may express strategies that allow parasite infection, but reduce fitness costs associated with invasion through reproductive enhancement (termed "fecundity compensation") [Minchella and Loverde (1981); Sandland and Minchella (2004)] or suppression of resistance responses, which themselves can cause host damage and pathogenicity (termed "tolerance"). Recently, a study by Miller *et al.* [Miller *et al.* (2005)] investigated the evolutionary consequences of hosts employing two different strategies in the face of infection: control, in which hosts reduced infection pathogenicity by actively reducing parasite replication rates, and tolerance, where hosts accommodated infection but did not reduce pathogen replication rates. Differences in the strategies exhibited by different species or strains of host, can significantly influence parasite evolution [Zhang *et al.* (2007b)]. Although this scenario is of great interest from an evolutionary standpoint, it is challenging mathematically.

This study investigates the evolutionary outcomes of interactions between a complex life-cycle parasite (*S. mansoni*) and its hosts (humans and snails). Schistosomes are dioecious helminth parasites with indirect life cycles. Although several species of parasite compose the genus *Schistosoma*, we focus on *Schistosoma mansoni*, a parasite of great human concern as it causes morbidity and mortality in over 200 million people worldwide [Chitsulo *et al.* (2000)]. *S. mansoni* uses snails along with humans to complete its life cycle. Adult (female) worms within human hosts produce eggs, which, hatch into miracidia when they come in contact with water. These larvae (miracidia) can then infect snails. After 5 weeks, parasites (cercariae) are released from snails and are infective to human hosts (see Figure 2.11, which is adopted from [Yang *et al.* (2012)] based on the information from CDC). The complexities of the model come from several sources. i) parasite load; ii) two hosts; iii) age-dependent transmission and targeted treatment of humans; iv) multiple parasite strains to consider evolution of parasite characteristics.

Multiple strains of parasites

Consider the case in which the human host is structured by choronological age denoted by a. Let $n(t,a)$ denote the density function of human hosts of age a at time t, and let $p(t,a)$ denote the density function of parasites carried by human hosts of age a at time t. Based on the models in [Anderson and May (1978); Dobson (1988); Hadeler (1982); Hadeler and Dietz (1983); Zhang *et al.* (2007a)], the equations for $n(t,a)$ and $p(t,a)$ take the forms (a derivation of these equations can be found in the appendix of [Xu *et al.* (2012)] or [Yang *et al.* (2012)]):

$$\frac{\partial}{\partial t}n(t,a) + \frac{\partial}{\partial a}n(t,a) = -\mu_h(a)n(t,a),$$

$$\frac{\partial}{\partial t}p(t,a) + \frac{\partial}{\partial a}p(t,a) = \beta(a)n(t,a)C(t) - \Big(\mu_h(a) + \mu_p + f(\sigma(a))\Big)p(t,a),$$

$$n(t,0) = \Lambda_h, \ n(0,a) = n_0(a), \ p(t,0) = 0, \ p(0,a) = p_0(a).$$

Schistosomiasis

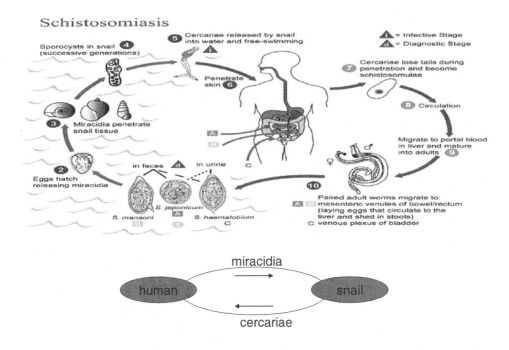

Fig. 2.11 A diagram showing the schistosome life cycle. The parasite uses both the human (denitive) host and the snail (intermediate) host to complete its development.

The parameters $\mu_h(a)$ and μ_p denote the per capita natural death rates of human hosts of age a and the adult parasites within human hosts, respectively; $f(\sigma(a))$ is the age-dependent killing rate of parasites due to effective drug treatment strategy $\sigma(a)$ for humans; $\beta(a)$ is the per capita infection rate of human hosts by cercariae. $C(t)$ is the density of free-swimming cercariae released by infected snail hosts, whose form will be specified later. The boundary conditions are $n(t,0) = \Lambda_h$ (birth rate of human hosts) and $p(t,0) = 0$ (humans are born uninfected), and initial conditions are $n(0,a) = n_0(a)$ and $p(0,a) = p_0(a)$ for given bounded functions with compact supports. Note that in the $n(t,a)$ equation the parasite-induced death rate has been ignored. This is because the disease mortality in humans is low and the main focus of this study is on the evolution of parasite-resistance in the snail host. The p equation for the case of multiple parasite strains is presented in [Xu *et al.* (2012)].

For the intermediate snail population, the model includes the possibility of coinfection of snails by different strains of parasites. Assume that the double infection can occur only after an infection by a single strain of parasites and that the order in which coinfection occurs is irrelevant ([Gower and Webster (2005)]). For simplicity, coinfections with the same strain of parasites are ignored, i.e., the coinfection rates $\rho_{ij} = 0$ if $i = j$ [Mosquera and Adler (1998)]. Under these assumptions, the snail population can be divided into four epidemiological classes: uninfected snail hosts

(S), hosts infected only by parasites of strain i (I_i, $i = 1, 2$), and hosts co-infected with both strains (I_{12}). Following an approach similar to that [Mosquera and Adler (1998)], we can describe the population dynamics for the intermediate snail host using the following set of differential and integral equations, which, together with the equations for the equations for humans n and two strains of adult parasites p_i, lead to the following full model for the human-schistome-snail system:

$$
\left(\frac{\partial}{\partial t} + \frac{\partial}{\partial a} \right) n(t, a) = -\mu_h(a) n(t, a),
$$

$$
\left(\frac{\partial}{\partial t} + \frac{\partial}{\partial a} \right) p_i(t, a) = \beta_i(a) n(t, a) C_i - \left(\mu_h(a) + \mu_p + f_i(\sigma(a)) \right) p_i(t, a),
$$

$$
\frac{d}{dt} S = \Lambda_s - (\rho_1 M_1 + \rho_2 M_2) S - \mu_s S,
$$

$$
\frac{d}{dt} I_i = \rho_i M_i S - \rho_{ji} M_j I_i - (\mu_s + \delta_i) I_i, \tag{2.23}
$$

$$
\frac{d}{dt} I_{12} = \rho_{12} M_1 I_2 + \rho_{21} M_2 I_1 - (\mu_s + \delta_{12}) I_{12},
$$

$$
C_i = c_i I_i + c_i' I_{12}, \quad M_i = \gamma_i \int_0^\infty p_i(t, a) da, \quad 1 \le i \ne j \le 2
$$

with the initial and boundary conditions

$$
n(t, 0) = \Lambda_h, \quad n(0, a) = n_0(a), \quad p_i(t, 0) = 0, \quad p_i(0, a) = p_{i0}(a), \quad i = 1, 2.
$$

To study the dynamics of system (2.23) on the evolutionary (long) time scale, we can simplify the mathematical analysis by ignoring the transient dynamics and considering only the limiting system that captures the asymptotical behaviors of system (2.23). It is shown in Appendix B of [Xu *et al.* (2012)] that the limiting system of (2.23) is given by

$$
\frac{d}{dt} S = \Lambda_s - (\rho_1 M_1 + \rho_2 M_2) S - \mu_s S,
$$

$$
\frac{d}{dt} I_i = \rho_i M_i S - \rho_{ji} M_j I_i - \mu_{si} I_i,
$$

$$
\frac{d}{dt} I_{12} = \rho_{12} M_1 I_2 + \rho_{21} M_2 I_1 - \mu_{s12} I_{12}, \tag{2.24}
$$

$$
M_i = \int_0^\infty \left(c_i I_i(t - a) + c_i' I_{12}(t - a) \right) R_{hi}(a) da, \quad i, j = 1, 2, \quad i \ne j,
$$

together with given initial value $S(0) = S^0$ and initial functions $I_i^0(s)$ and $I_{12}^0(s)$. In the system (2.24), we have used the following short-hand notations

$$
\mu_{si} = \mu_s + \delta_i, \quad \mu_{s12} = \mu_s + \delta_{12}, \quad \pi_h(a) = e^{-\int_0^a \mu_h(w) dw},
$$

$$
\mu_{hi}(a) = \mu_h(a) + \mu_p + f_i(\sigma(a)), \quad \pi_{hi}(a, \tau) = e^{-\int_a^{a+\tau} \mu_{hi}(w) dw},
$$

$$
R_{hi}(\tau) = \Lambda_h \gamma_i \int_0^\infty \beta_i(a) \pi_h(a) \pi_{hi}(a, \tau) da, \quad \mathcal{R}_{hi} = \int_0^\infty R_{hi}(\tau) d\tau.
$$

These notations have clear biological meanings. For example, $\pi_h(a)$ is the survival probability of human hosts of age a while $\pi_{hi}(a, \tau)$ is the survival probability of an adult parasite in a human host of age $a+\tau$ who was infected τ time units ago (i.e., τ is the infection age of the host). Therefore, $R_{hi}(\tau)$ gives the total number of strain i miracidia produced by adult parasites within all human hosts with infection age τ due to one cercaria, and $\mathcal{R}_{h\,i}$ gives the total number of strain i miracidia produced by adult parasites in all human hosts due to one cercaria.

Notice that susceptible snail hosts with the population level fixed at Λ_s/μ_s totally release $\frac{\Lambda_s \rho_i c_i}{\mu_s \mu_{si}}$ number of cercariae due to one miracidium of strain i. The basic reproduction number of strain i parasites is given by

$$\mathcal{R}_i = \frac{\Lambda_s \rho_i c_i}{\mu_s \mu_{si}} \mathcal{R}_{hi}, \quad i = 1, 2. \tag{2.25}$$

It is shown in Appendix C of [Xu *et al.* (2012)] that when only a single strain of parasites (e.g., strain i) is present in the population, the population dynamics is determined completely by the basic reproduction number $\mathcal{R}_i$. That is, the disease-free equilibrium is globally asymptotically stable if $\mathcal{R}_i < 1$, and it is unstable if $\mathcal{R}_i > 1$, in which case a unique endemic equilibrium Q_i ($i = 1, 2$) exists and is stable. The endemic equilibria Q_1 and Q_2 are given by

$$Q_1 = (S_1^*, I_1^*, 0, 0), \quad \text{which exists if and only if } \mathcal{R}_1 > 1,$$

and

$$Q_2 = (S_2^*, 0, I_2^*.0), \quad \text{which exists if and only if } \mathcal{R}_2 > 1,$$

where

$$S_i^* = \frac{\Lambda_s}{\mu_s} \frac{1}{\mathcal{R}_i}, \quad I_i^* = \frac{\Lambda_s}{\mu_{si}} \left(1 - \frac{1}{\mathcal{R}_i} \right), \quad i = 1, 2.$$

The reproduction number of an invader cercaria (strain 2) is given by

$$\mathcal{R}_{21} = \frac{c_2 \rho_2 \mathcal{R}_{h2} S_1^*}{\mu_{s2} + \rho_{12} c_1 \mathcal{R}_{h1} I_1^*} + \frac{\rho_{12} c_1 \mathcal{R}_{h1} I_1^* c_2' \rho_2 \mathcal{R}_{h2} S_1^*}{(\mu_{s2} + \rho_{12} c_1 \mathcal{R}_{h1} I_1^*) \mu_{s12}} + \frac{c_2' \rho_{21} \mathcal{R}_{h2} I_1^*}{\mu_{s12}}, \tag{2.26}$$

which determines whether strain 1 parasites can invade the population in which strain 1 is at the equilibrium Q_1.

Note that a typical invader cercaria can produce miracidia in the number $\mathcal{R}_{h2}$ through the definitive human host. These miracidia can infect snails and consequently generate cercariae via two potential paths: (i) by infecting the uninfected snails S_1^* and (ii) by co-infecting the infected snails I_1^*. In path (i), the number of snails infected by the miracidia $\mathcal{R}_{h2}$ is given by $\rho_2 \mathcal{R}_{h2} S_1^*$. These infected snails may or may not become co-infected by resident parasites. Suppose that an infected snail (by an invader miracidium) remains singly infected for an average time T_2 and co-infected for an average time T_{12}. Then the probability that the infected snail dies while being singly infected is $\mu_{s2} T_2$, and the probability of dying while being co-infected is $\mu_{s12} T_{12}$. Therefore,

$$\mu_{s2} T_2 + \mu_{s12} T_{12} = 1. \tag{2.27}$$

Notice that T_2 can be determined directly by the total rate $\mu_{s2} + \rho_{12}c_1 \mathcal{R}_{h1} I_1^*$ (at which infected snails I_2 leave the class I_2 either dying or being co-infected), i.e.,

$$T_2 = \frac{1}{\mu_{s2} + \rho_{12}c_1 \mathcal{R}_{h1} I_1^*}. \tag{2.28}$$

Then T_{12} can be determined from (2.27) as

$$T_{12} = \frac{\rho_{12}c_1 \mathcal{R}_{h1} I_1^*}{\mu_{s2} + \rho_{12}c_1 \mathcal{R}_{h1} I_1^*} \frac{1}{\mu_{s12}}. \tag{2.29}$$

Therefore, when the number of susceptible snails available to invading miracidia is S_1^*, a typical invader cercaria can reproduce the following number of cercariae

$$(c_2 T_2 + c_2' T_{12})\rho_2 \mathcal{R}_{h2} S_1^*. \tag{2.30}$$

In path (ii), the number of snails infected by resident parasites is I_1^*. These infected snails can be co-infected by the invading miracidia $\mathcal{R}_{h2}$ produced by adult parasites due to one invader cercaria, and remain being co-infected for an average time $1/\mu_{s12}$. Thus, through path (ii) a typical invader cercaria can reproduce the following number of cercariae:

$$\frac{c_2' \rho_{21} \mathcal{R}_{h2} I_1^*}{\mu_{s12}}. \tag{2.31}$$

Therefore, from (2.30) and (2.31) with T_2 and T_{12} being replaced by (2.28) and (2.29).

It is clear from the deduction of $\mathcal{R}_{21}$ that if $\mathcal{R}_{21} > 1$, a small number of invaders can start a growing population and hence the invasion will be successful in the environment Q_1 set by the residents. Therefore, the quantity $\mathcal{R}_{21}$ can be used as a measure of the fitness of invading parasites. We call $\mathcal{R}_{21}$ the invasion reproduction number (or invasion fitness) for strain 2 parasites. The following results are proved in [Xu *et al.* (2012)].

Theorem 2.3. *Let $\mathcal{R}_{21}$ be as defined in (2.26). The equilibrium Q_1 is stable if $\mathcal{R}_{21} < 1$ and it is unstable if $\mathcal{R}_{21} > 1$.*

From the symmetry an invasion reproduction number $\mathcal{R}_{12}$ can be derived, and a similar stability result holds for the equilibrium Q_2. These results allow us to use the invasion reproduction numbers $\mathcal{R}_{ij}$ $(i, j = 1, 2, i \neq j)$ to investigate the evolutionarily stable strategies for parasite's traits, as demonstrated in section 3.4.

Multiple types of hosts

The model considered in Yang *et al.* [Yang *et al.* (2012)] includes two types of intermediate (snail) host:

$$\frac{\partial}{\partial t}n(t,a) + \frac{\partial}{\partial a}n(t,a) = -\mu_h(a)n(t,a),$$

$$\frac{\partial}{\partial t}p(t,a) + \frac{\partial}{\partial a}p(t,a) = \beta(a)n(t,a)C(t) - \left(\mu_h(a) + \mu_p + f(\sigma(a))\right)p(t,a),$$

$$\frac{d}{dt}S_k(t) = \frac{b_{1k}S_k(t)}{b_{2k} + \sum_{i=1,2}[S_i(t) + I_i(t)]} - \rho_k M(t)S_k(t) - \mu_s S_k(t),$$

$$\frac{d}{dt}I_k(t) = \rho_k M(t)S_k(t) - (\mu_s + \delta_k)I_k(t), \quad k = 1, 2,$$

(2.32)

$$C(t) = \sum_{k=1,2} c_k I_k(t), \quad M(t) = \gamma \int_0^\infty p(t,a)da,$$

$$n(t,0) = \Lambda_h, \quad n(0,a) = n_0(a), \quad p(t,0) = 0, \quad p(0,a) = p_0(a).$$

All parameters used in the model (2.32) are listed in Table 2.3.

Table 2.3   Definition of parameters used in the continuous-time model (2.32).

Symbol	Definition
$\mu_h(a)$	per capita natural death rate of human hosts of age a
$\sigma(a)$	effective drug-treatment rate of human hosts of age a
θ	drug-resistance level of adult parasite within human hosts
Λ_h	birth rate of human hosts
$\beta(a)$	per capita infection rate of human hosts by free-living cercariae
γ	per capita effective egg-production rate of adult parasites
b_{1k}	saturation constant for the growth rate of snails
b_{2k}	scaling constant for the growth rate of snails
ρ_k	per capita infection rate of type k susceptible snail hosts
c_k	cercaria releasing rate of an infected snail hosts of type k
μ_s	per capita natural death rate of snail hosts
μ_p	per capita natural death rate of adult parasites
δ_k	per capita disease-induced death rate of infected snail hosts of type k

Note: $k = 1, 2$.

Because the $n(t,a)$ equation in system (2.32) is independent of other variables, we can solve it by integrating the equation along the characteristic lines:

$$n(t,a) = \begin{cases} \Lambda_h \pi_h(a), & t \geq a, \\ n_0(a-t)\dfrac{\pi_h(a)}{\pi_h(a-t)}, & t < a, \end{cases}$$

(2.33)

where $n_0(a)$ is a given function and

$$\pi_h(a) = e^{-\int_0^a \mu_h(w)dw}$$

(2.34)

represents the survival probability of a definitive host of age a. Substituting the solution (2.33) for $n(t,a)$ in the $p(t,a)$ equation in (2.32), and integrating the equation we get

$$p(t,a) = \begin{cases} p_1(t,a), & t \geq a, \\ p_2(t,a), & t < a \end{cases}$$

(2.35)

with

$$p_1(t,a) = \Lambda_h \int_0^a \beta(w)C(t+w-a)\pi_h(w)\frac{\pi_\sigma(a)}{\pi_\sigma(w)}dw,$$

$$p_2(t,a) = p_0(a-t)\frac{\pi_\sigma(a)}{\pi_\sigma(a-t)} + \int_{a-t}^a \beta(w)C(t+w-a)n_0(a-t)\frac{\pi_\sigma(a)}{\pi_\sigma(w)}\frac{\pi_h(w)}{\pi_h(a-t)}dw,$$

where

$$\pi_\sigma(a) = e^{-\int_0^a [\mu_h(w)+\mu_p+f(\sigma(w))]dw} \tag{2.36}$$

represents the survival probability of an adult parasite in a definitive host of age a.

Our approach is to replace the partial differential equations in (2.32) by integral equations and then analyze the limiting system as $t \to \infty$ of the resulting equations. For this purpose, we rewrite the function $M(t)$ as

$$M(t) = \gamma \int_0^\infty p(t,a)da = \gamma \int_0^t p_1(t,a)da + \gamma \int_t^\infty p_2(t,a)da.$$

Notice that the solutions of the system (2.32) are bounded. Thus, the last integral in the above expression goes to zero as $t \to \infty$. Let $\tilde{M}(t)$ be the limiting function of $M(t)$, i.e.,

$$\tilde{M}(t) = \gamma \int_0^\infty p_1(t,a)da.$$

Then, by changing the orders of integrations we obtain

$$\begin{aligned}
\tilde{M}(t) &= \gamma \int_0^\infty p_1(t,a)da \\
&= \gamma \int_0^\infty \Lambda_h \int_0^a \beta(w)C(t+w-a)\pi_h(w)\pi_\sigma(a)\pi_\sigma^{-1}(w)dwda \\
&= \gamma\Lambda_h \int_0^\infty C(t-\tau)\int_0^\infty \beta(w)\pi_h(w)\pi_\sigma(w+\tau)\pi_\sigma^{-1}(w)dwd\tau \\
&= \sum_{k=1,2} c_k \int_0^\infty I_k(t-\tau)R_h(\tau)d\tau,
\end{aligned}$$

where

$$R_h(\tau) = \gamma\Lambda_h \int_0^\infty \beta(w)\pi_h(w)\pi_\sigma(w+\tau)\pi_\sigma^{-1}(w)dw. \tag{2.37}$$

Substituting $\tilde{M}(t)$ for $M(t)$ in the S_k and I_k equations in system (2.32), we get the following limiting system for S_k and I_k that are independent of the variables $n(t,a)$ and $p(t,a)$:

$$\frac{d}{dt}S_k(t) = \frac{b_{1k}S_k(t)}{b_{2k}+\sum_{i=1,2}(S_i(t)+I_i(t))} - \rho_k\tilde{M}(t)S_k(t) - \mu_s S_k(t),$$

$$\frac{d}{dt}I_k(t) = \rho_k\tilde{M}(t)S_k(t) - (\mu_s+\delta_k)I_k(t), \quad k=1,2, \tag{2.38}$$

$$\tilde{M}(t) = \sum_{k=1,2} c_k \int_0^\infty I_k(t-a)R_h(a)da.$$

From (2.33), (2.35), and $C(t) = \sum_{k=1,2} c_k I_k(t)$ we know that as $t \to \infty$,

$$n(t,a) \to \Lambda_h \pi_h(a),$$

$$p(t,a) \to \Lambda_h \int_0^a \beta(w) \left[\sum_{k=1,2} c_k I_k(t+w-a) \right] \pi_h(w) \frac{\pi_\sigma(a)}{\pi_\sigma(w)} dw,$$

where $\pi_h(a)$ and $\pi_\sigma(a)$ are given in (2.34) and (2.36). Thus, the asymptotic behaviors of $n(t,a)$ and $p(t,a)$ can be determined once the behaviors of $I_k(t)$ ($k = 1,2$) are known. Thus, it is sufficient to analyze the limiting system (2.38). Furthermore, the invasion criteria can be derived by studying the boundary equilibrium points of (2.38), at which only one type of snail host exists (both uninfected and infected snails can be present).

In the case when only one type (type 1 or 2) of snail host is present, the system (2.38) reduces to

$$\frac{d}{dt} S_k(t) = \frac{b_{1k} S_k(t)}{b_{2k} + S_k(t) + I_k(t)} - \rho_k \tilde{M}_k(t) S_k(t) - \mu_s S_k(t),$$

$$\frac{d}{dt} I_k(t) = \rho_k \tilde{M}_k(t) S_k(t) - (\mu_s + \delta_k) I_k(t), \qquad (2.39)$$

$$\tilde{M}_k(t) = c_k \int_0^\infty I_k(t-a) R_h(a) da, \quad k = 1 \text{ or } 2.$$

Notice that the only difference between the systems (2.38) and (2.39) is between $\tilde{M}(t)$ (which is a sum of two terms) and $\tilde{M}_k(t)$ (which has a single term). Due to the mathematical symmetry between the two types of snail hosts, the analysis of system (2.39) for $k = 1$ and for $k = 2$ will be identical. Consider the order of variables of the system (2.39) to be (S_k, I_k). Because the carrying capacity of the type k snail host in the absence of infection is $\frac{b_{1k}}{\mu_s} - b_{2k}$, we need to assume that

$$\frac{b_{1k}}{\mu_s} - b_{2k} > 0, \quad k = 1,2 \qquad (2.40)$$

so that the snail population size is positive in the absence of infection. The system (2.39) always has the parasite-free equilibrium $E_{k0} = (S_{k0}, 0) = (b_{1k}/\mu_s - b_{2k}, 0)$.

The existence of an interior equilibrium (i.e., all components are positive) depends on the parasite reproduction number $\mathcal{R}_k$ within the type k snail host and the quantity

$$\mathcal{R}_h = \int_0^\infty R_h(\tau) d\tau$$

and $R_h(\tau)$ is given in (2.37). An expression of $\mathcal{R}_k$ is given by

$$\mathcal{R}_k = \frac{\rho_k c_k \mathcal{R}_h}{\mu_s + \delta_k} \left(\frac{b_{1k}}{\mu_s} - b_{2k} \right). \qquad (2.41)$$

Notice from the expression of $R_h(\tau)$ (see (2.36)) that

$$\beta(w) \Lambda_h \pi_h(w)$$

represents the average number of adult parasites within human hosts of age w produced by one cercariae, and

$$\pi_\sigma(w+\tau)\pi_\sigma^{-1}(w)$$

is the survival probability of an adult parasite in a human host of age $w+\tau$ who was infected τ time units ago while at age w (i.e., τ is the infection age of a human host of age $w+\tau$). Thus, the quantity

$$\int_0^\infty \beta(w)\Lambda_h\pi_h(w)\pi_\sigma(w+\tau)\pi_\sigma^{-1}(w)dw$$

gives the total number of adult parasites within human hosts with infection age τ produced by one cercariae, and $\mathcal{R}_h(\tau)$ gives the total number of miracidia produced by adult parasites within human hosts due to one cercariae. Finally, from the definitions of ρ_k (per capita infection rate of snails of type k), c_k (cercaria releasing rate by one infected snail of type k), $\mu_s+\delta_k$ ($1/(\mu_s+\delta_k)$ is the duration of infection of a snail of type k), and $b_{1k}/\mu_s - b_{2k}$ (size of snails of type k in a susceptible population), we know that $\mathcal{R}_k$ represents the total number of secondary cercaria produced by one cercariae. That is, $\mathcal{R}_k$ indeed is the parasite's reproduction number associated with the type k snail hosts.

Let $E_k^* = (S_k^*, I_k^*)$ denote an interior equilibrium of the system (2.39). Then

$$S_k^* = \frac{\mu_s+\delta_k}{\rho_k c_k \mathcal{R}_h} = \frac{1}{\mathcal{R}_k}\left(\frac{b_{1k}}{\mu_s} - b_{2k}\right), \tag{2.42}$$

and I_k^* satisfies the equation

$$\frac{b_{1k}}{b_{2k} + S_k^* + I_k^*} - \rho_k c_k \mathcal{R}_h I_k^* - \mu_s = 0. \tag{2.43}$$

Equation (2.43) can be rewritten as

$$F(I_k^*) = 0, \tag{2.44}$$

where $F(x)$ is a quadratic function defined by

$$F(x) = \rho_k c_k \mathcal{R}_h x^2 + \left[\rho_k c_k \mathcal{R}_h(b_{2k} + S_k^*) + \mu_s\right]x + \mu_s(b_{2k} + S_k^*) - b_{1k},$$

and S_k^* is given in (2.42). Notice that

$$F(0) = \mu_s(b_{2k} + S_k^*) - b_{1k} = \frac{\mu_s(\mu_s+\delta_k)}{\rho_k c_k \mathcal{R}_h}(1 - \mathcal{R}_k).$$

Clearly, $F(0) < 0 \; (> 0)$ when $\mathcal{R}_k > 1 \; (< 1)$. Notice also that $F(x)$ has a minimum at

$$x_{\min} = -\frac{\rho_k c_k \mathcal{R}_h(b_{2k} + S_k^*) + \mu_s}{2\rho_k c_k \mathcal{R}_h} < 0.$$

This implies that the equation $F(x) = 0$ has a unique positive solution for $\mathcal{R}_k > 1$ and no positive solution for $\mathcal{R}_k < 1$. Thus, we have the following result:

Theorem 2.4. *There is a unique interior equilibrium E_k^* when $\mathcal{R}_k > 1$, and E_k^* does not exist when $\mathcal{R}_k < 1$.*

The next result shows that the reproduction number $\mathcal{R}_k$ and the threshold value $\mathcal{R}_k = 1$ also determine the stabilities of the equilibrium points E_{k0} and E_k^* of the system (2.39).

Theorem 2.5. *Assume that the condition (2.40) holds.*

(a) *The parasite-free equilibrium $E_{k0} = \left(\frac{b_{1k}}{\mu_s} - b_{2k}, 0 \right)$ is locally asymptotically stable if $\mathcal{R}_k < 1$, and it is unstable if $\mathcal{R}_k > 1$.*

(b) *When $\mathcal{R}_k > 1$, there exists a threshold value $b_{1k} = \tilde{b}_{1k}$ (with all other parameters being fixed), such that the unique endemic equilibrium $E_k^* = (S_k^*, I_k^*)$ determined in Theorem 2.4 is locally asymptotically stable if $b_{1k} < \tilde{b}_{1k}$.*

Moreover, numerical simulations show that a Hopf bifurcation may occur at some critical point $b_{1k} = b_{1k}^* > \tilde{b}_{1k}$, leading to the appearance of stable periodic solutions.

The analytic proofs for Part (a) and Part (b) of Theorem 2.5 can be found in the Appendix of [Yang *et al.* (2012)]. The existence of a Hopf bifurcation and the threshold point b_{1k}^* is confirmed via numerical simulations. This is illustrated in Figure 2.12, which shows numerical solutions of the fraction $\frac{I_k(t)}{S_k(t)+I_k(t)}$ for two values of b_{1k} below or above b_{1k}^* ($k = 1$ or 2). The solution converges to the equilibrium E_k^* when $b_{1k} < b_{1k}^*$ (see (a)), and a stable periodic solution exists for $b_{1k} > b_{1k}^*$ (see (b)). Other parameter values used in the simulations are: $\Lambda_h = 8$, $\mu_h = 0.014$, $\mu_s = 0.5$, $\delta_k = 1.8$, $\mu_p = 0.2$, $\beta_0 = 2 \times 10^{-5}$, $\rho_k = 10^{-4}$, $\gamma_0 = 3$, $b_{2k} = 600$ and $c_k = 20000$ ($k = 1, 2$). The time unit is year. Most of the values are from [Feng *et al.* (2004a)].

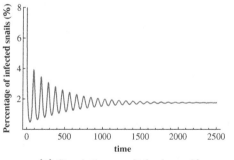
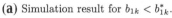
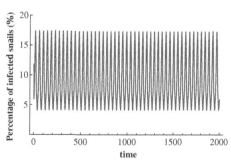

(a) Simulation result for $b_{1k} < b_{1k}^*$. (b) Simulation result for $b_{1k} > b_{1k}^*$.

Fig. 2.12 Time plots of the fraction of infected snail hosts $\frac{I_k(t)}{S_k(t)+I_k(t)}$ for the reduced system (2.39) with a single type of snail host ($k = 1$ or 2). The plot in (a) is for the case when the parameter values are chosen such that $\mathcal{R}_k > 1$ and $b_{1k} = 330 < b_{1k}^*$. It shows that the equilibrium E_k^* is stable. The plot in (b) is for the case when $\mathcal{R}_k > 1$ and $b_{1k} = 450 > b_{1k}^*$. It shows that a stable periodic orbit exists. Other parameter values are given in the text.

Next, we consider the case when both types of snail hosts are included. For the full system (2.38) in which both types of snail hosts are included, an interior equilibrium of the reduced system (2.39) corresponds to a boundary equilibrium of the full system (2.38). Again, we omit the extreme case when both snail populations are absent. Using the order of variables $U = (S_1, I_1, S_2, I_2)$ we know that the system (2.38) has the following non-trivial boundary equilibrium points:

$$U_{10} = \left(\frac{b_{11}}{\mu_s} - b_{21}, 0, 0, 0\right), \quad U_{02} = \left(0, 0, \frac{b_{12}}{\mu_s} - b_{22}, 0\right),$$

$$U_1^* = (S_1^*, I_1^*, 0, 0), \qquad U_2^* = (0, 0, S_2^*, I_2^*),$$

where S_k^* and I_k^* are determined in (2.42)-(2.44). The subscript "10" in U_{10} represents the equilibrium at which only type 1 snail host is present and there is no infection. Similarly, the subscript "02" in U_{02} represents the equilibrium with type 2 snail host only and without infection. The system (2.38) may also have other boundary equilibria. However, for the biological questions we are interested in, we will only consider the ones listed above.

An interior equilibrium of system (2.38), denoted by $U^\circ = (S_1^\circ, I_1^\circ, S_2^\circ, I_2^\circ)$ with all components positve, satisfies the following system of four equations

$$\frac{b_{1k}S_k^\circ}{b_{2k} + \sum_{i=1,2}(S_i^\circ + I_i^\circ)} - \rho_k M^\circ S_k^\circ - \mu_s S_k^\circ = 0,$$

$$\rho_k M^\circ S_k^\circ - (\mu_s + \delta_k)I_k^\circ = 0, \quad k = 1, 2,$$

where $M^\circ = \sum_{i=1,2} c_i I_i^\circ \mathcal{R}_h$. It is very difficult to solve analytically the above system for U°. This is the main reason that we will study questions related to the coexistence equilibria via numerical simulations.

To avoid a degenerate case to occur (i.e., the characteristic equation at the parasite-free equilibrium U_{10} or U_{02} has a zero eigenvalue), we assume that the two carrying capacities are not equal. Without loss of generality, assume that the type 1 snail host has a higher carrying capacity, i.e., $\frac{b_{11}}{\mu_s} - b_{21} > \frac{b_{12}}{\mu_s} - b_{22} > 0$. Under this condition it can be shown that U_{10} is locally asymptotically stable when $\mathcal{R}_1 < 1$ and unstable when $\mathcal{R}_1 > 1$; whereas U_{02} is always unstable.

The more interesting results are about the stabilities of U_1^* and U_2^*, as these results are directly related to the invasion properties of a population. For example, the stability of the equilibrium E_1^* for the reduced system (2.39) indicates that the snail population of type 1 has stabilized at the equilibrium U_1^* for the full system (2.38) in the absence of snails of type 2. When a small number of new species of snails (type 2) is introduced into this environment, they will be able to invade into the population of host type 1 only if the equilibrium U_1^* for the full system (2.38) is unstable. In such a case, we refer to the snail host of type 1 as the "resident host" and the snail host of type 2 as the "mutant host", and refer to this invasion as invasion of U_1^* by the mutant host.

We first derive the invasion condition using the invasibility analysis (see [Bowers (1999); Bowers and Turner (1997)]). A mathematical proof of the result will be provided at the end of this section. In order for the mutant snails to successfully invade U_1^*, they must have a positive growth rate. In the process of invasion by a susceptible mutant snail, the snail may either remain uninfected or become infected if it did not die. Let T_U denote the average duration in which the snail is alive and uninfected. Then

$$T_U = \frac{1}{\mu_s + \rho_2 c_1 I_1^* \mathcal{R}_h}.$$

Let T_I denote the average duration of being infected (and did not die). Note that the lifespan of an uninfected (infected) snail of type 2 is $1/\mu_s$ ($1/(\mu_s + \delta_2)$), and that

$$\frac{T_U}{1/\mu_s} + \frac{T_I}{1/(\mu_s + \delta_2)} = 1. \tag{2.45}$$

From equation (2.45) we have

$$T_I = \frac{\rho_2 c_1 I_1^* \mathcal{R}_h}{(\mu_s + \delta_2)(\mu_s + \rho_2 c_1 I_1^* \mathcal{R}_h)}.$$

Let ϕ_U and ϕ_I denote the net contribution of the snail to the growth rate of the population during the periods T_U and T_I respectively. Recall the assumption that infected snails do not reproduce. Thus,

$$\phi_U = \frac{b_{12}}{b_{22} + S_1^* + I_1^*} - \mu_s, \qquad \phi_I = -(\mu_s + \delta_2).$$

Then, the total contribution of the snail to the population growth of the mutant type is

$$g_2 = \phi_U T_U + \phi_I T_I$$

$$= \left(\frac{b_{12}}{b_{22} + S_1^* + I_1^*} - \mu_s \right) \frac{1}{\mu_s + \rho_2 c_1 I_1^* \mathcal{R}_h} - \frac{(\mu_s + \delta_2)\rho_2 c_1 I_1^* \mathcal{R}_h}{(\mu_s + \delta_2)(\mu_s + \rho_2 c_1 I_1^* \mathcal{R}_h)} \tag{2.46}$$

$$= \frac{\frac{b_{12}}{b_{22} + S_1^* + I_1^*}}{\mu_s + \rho_2 c_1 I_1^* \mathcal{R}_h} - 1.$$

Clearly, g_2 must be positive for the mutant type to successfully invade the resident population. From the condition $g_2 > 0$ we obtain the invasion reproduction number $\mathcal{R}_{21}$ and the invasion condition:

$$\mathcal{R}_{21} = \frac{\frac{b_{12}}{b_{22} + S_1^* + I_1^*}}{\mu_s + \rho_2 c_1 I_1^* \mathcal{R}_h} > 1. \tag{2.47}$$

The interpretation of $\mathcal{R}_{21}$ is clear. The fraction $\frac{b_{12}}{b_{22} + S_1^* + I_1^*}$ gives the number of new mutant snails produced by one susceptible snail of type 2 while the type 1 snail host is at the equilibrium U_1^*, and $\frac{1}{\mu_s + \rho_2 c_1 I_1^* \mathcal{R}_h} = T_U$ is the duration in which the mutant snail is productive, i.e., alive and uninfected). Thus, $\mathcal{R}_{21}$ represents the net reproduction number of the mutant host in the environment where the resident host is at the equilibrium U_1^*.

By symmetry, we can also obtain the condition under which the type 1 host (as the mutant) can invade U_2^* is

$$\mathcal{R}_{12} = \frac{\frac{b_{11}}{b_{21} + \hat{S}_2^* + I_2^*}}{\mu_s + \rho_1 c_2 I_2^* \mathcal{R}_h} > 1, \tag{2.48}$$

where $\mathcal{R}_{12}$ is the invasion reproduction number for the type 1 snail host.

Remark: The quantity $\mathcal{R}_k$ (for $k = 1, 2$) is a measure of the parasite's reproduction associated with the type k snail host, whereas the quantity $\mathcal{R}_{ji}$ ($1 \leq i \neq j \leq 2$) is a measure of the snail host's reproduction. More precisely, $\mathcal{R}_{ji}$ describes the reproductive ability of type j snails when introduced to an environment where the type i host is at the endemic equilibrium with $S_i^* > 0$ and $I_i^* > 0$. We need to point out that equation (2.47) also provides an equivalent version of fitness for the mutant host. As discussed in Hoyle and Bowers [Hoyle and Bowers (2008)], the fitness is defined as being the per capita growth rate of a rare mutant invader (see [Metz *et al.* (1992)]). A sign-equivalent version of it is given by $\mathcal{R}_0 - 1 > 0$, where $\mathcal{R}_0$ is the dominant eigenvalue of the next generation matrix as defined in [Diekmann and Heesterbeek (2000)]. The quantity $\mathcal{R}_{21}$ given in (2.47) has a similar meaning.

Using these invasion reproduction numbers we can describe the stability results for U_k^* ($k = 1, 2$) as follows.

Theorem 2.6. *Let $b_{11} < \tilde{b}_{11}$ (this will exclude the possibility of periodic orbits in the reduced system (2.39), see Theorem 2.5).*

(a) *The equilibrium U_1^* exists if the basic reproduction number $\mathcal{R}_1 > 1$.*
(b) *U_1^* is locally asymptotically stable if the invasion reproduction number $\mathcal{R}_{21} < 1$ and unstable if $\mathcal{R}_{21} > 1$.*

A symmetric result also holds:

Theorem 2.7. *Let $b_{12} < \tilde{b}_{12}$.*

(a) *The equilibrium U_2^* exists if the basic reproduction number $\mathcal{R}_2 > 1$.*
(b) *U_2^* is locally asymptotically stable if the invasion reproduction number $\mathcal{R}_{12} < 1$ and unstable if $\mathcal{R}_{12} > 1$.*

A proof of Theorem 2.6 can be found in [Yang *et al.* (2012)].

Theorems 2.6 and 2.7 imply that the coexistence of both types of snail hosts with a positive infection level will be expected when both invasion reproduction numbers exceed 1, i.e., $\mathcal{R}_{21} > 1$ and $\mathcal{R}_{12} > 1$. This is illustrated in Figure 2.13. However, these are only sufficient but not necessary conditions. This is because in Theorems 2.6 and 2.7, there are other conditions such as $\mathcal{R}_1 > 1$ or $\mathcal{R}_2 > 1$ (they ensure the existence of U_1^* or U_2^*). In the case when U_1^* and U_2^* both exist and are stable, coexistence cannot occur if one of the invasion reproduction numbers is below 1. This is illustrated in Figure 2.14.

We remark also that although the outcome of competitive-exclusion between the two types of snail hosts is expected without the shared parasite, it may not be the case in the presence of parasites. This is why the condition $\mathcal{R}_k > 1$ is required, which guarantees the establishment of the parasite population in the type k snail host. Notice that the conditions for coexistence depend on the invasion reproduction numbers $\mathcal{R}_{12}$ and $\mathcal{R}_{21}$, which depend on the levels of infection in the two types of snail hosts.

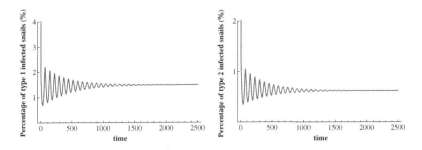

Fig. 2.13   Time plots of the fractions of infected snail hosts $\frac{I_k(t)}{\sum_{i=1,2}(S_i(t)+I_i(t))}$ of type k $(k = 1, 2)$ for the full system (2.32). Both $\mathcal{R}_{12}$ and $\mathcal{R}_{21}$ are greater than 1. The system converges to the endemic equilibrium $U^\diamond$. All parameter values are the same as in Figure 2.12 except that $\rho_2 = 3 \times 10^{-5}$.

2.3.3   *Competitive exclusion in a multi-strain dengue model*

This section includes the model studies for dengue virus presented in [Feng and Velasco-Hernandez (1997)]. It focuses on the competition between multiple strains of dengue virus and the conditions for coexistence.

Dengue fever is a viral disease endemic in many areas of the world that is invading and recolonizing regions where either it was absent or it had been eradicated. The dengue virus has 4 different serotypes. We construct and analyze a mathematical model for its transmission dynamics. The model is a system of differential equations that incorporates variable population size in both host and mosquito populations, two co-circulating strains and frequency-dependent biting rates. The model constitutes a framework to discuss conditions for coexistence or competitive exclusion of closely related pathogen strains.

Most diseases are produced by a spectrum of closely related pathogens rather than by single strains and dengue is clearly an example of this assertion. In dengue an analogous phenomenon to superinfection (Nowak and May [Nowak and May (1994)], May and Nowak [May and Nowak (1994)]) occurs. One strain invades the host population, produces a brief period of temporary immunity to other strains but when the immunity is lost the host becomes susceptible to reinfection with another strain. In dengue, before reinfection can occur, there is a period where the host is

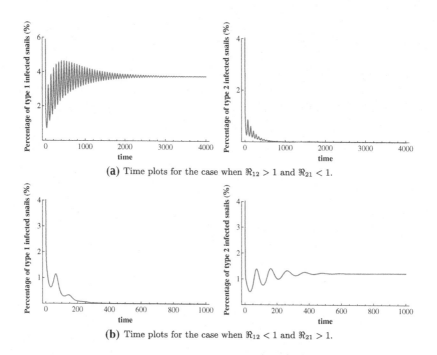

(a) Time plots for the case when $\Re_{12} > 1$ and $\Re_{21} < 1$.

(b) Time plots for the case when $\Re_{12} < 1$ and $\Re_{21} > 1$.

Fig. 2.14 Similar to Figure 2.13 but for the case when both U_1^* and U_2^* are existent and stable ($\mathcal{R}_1$ and $\mathcal{R}_2$ are both greater than 1). Plots in (a) show that the system converges to the boundary equilibria U_1^* when $\mathcal{R}_{12} = 1.13 > 1$ ($b_{12} = 317$) and $\mathcal{R}_{21} = 0.985 < 1$, whereas plots in (b) show that the system converges to the boundary equilibria U_2^* when $\mathcal{R}_{12} = 0.95 < 1$ ($b_{12} = 335$) and $\mathcal{R}_{21} = 1.04 > 1$.

resistant, in varying degrees, to all strains [Monath *et al.* (1989)]. In dengue fever we are thus confronted with a vector-transmitted disease, co-circulating strains, certain degree of cross-immunity or even increased susceptibility to infection, and a variable host population size. Under these conditions one important theoretical problem that we address here is that of the coexistence of all strains or the eventual extinction of some of them. A similar problem has been theoretically explored by several authors [Levin and Pimentel (1981); Bremermann and Thieme (1989); Castillo-Chavez *et al.* (1995, 1996)].

There are numerous published results discussing the problem of coexistence in pathogen-host interactions. Levin and Pimentel [Levin and Pimentel (1981)] constructed a mathematical *SI* model where the population in the absence of disease grows exponentially. Two strains with different virulences compete with each other. The most virulent strain can 'takeover' hosts already infected with the less virulent strain. With these assumptions a globally stable equilibrium is possible where both strains may coexist [Capasso (1993)]. The stability of the positive equilibrium is only guaranteed for certain range of values of superinfection. Outside this range one of the boundary equilibria is asymptotically stable.

Bremermann and Thieme [Bremermann and Thieme (1989)] postulate a competitive exclusion principle in an SIR model with a variable population size. Several strains compete for a single host population. The pathogens differ in their virulence. In this model virulence is a strictly convex function of the transmission rate, implying that the evolution of virulence leads to a transmission rate that maximizes the basic reproductive number of the pathogen [Bremermann and Thieme (1989)].

Castillo-Chavez *et al.* [Castillo-Chavez *et al.* (1995)] find, for a SIS two-sex model with variable population size, that competitive exclusion is the norm: the strain with the highest reproduction number persists in both host types. The model in [Mena-Lorca *et al.* (1999)] is used to study the effect of variable population, virulence and density-dependent population regulation. In this model too, coexistence is feasible only in certain window of parameter space.

Consider a human population settled in a region where a mosquito population of the genus *Aedes* is present and carrier of the dengue virus. Model equations then stand as follows ($' = d/dt$):

$$S'(t) = h - (B_1 + B_2)S - uS,$$

$$I_1'(t) = B_1 S - \sigma_2 B_2 I_1 - uI_1,$$

$$I_2'(t) = B_2 S - \sigma_1 B_1 I_2 - uI_2,$$

$$Y_1'(t) = \sigma_1 B_1 I_2 - (e_1 + u + r)Y_1, \tag{2.49}$$

$$Y_2'(t) = \sigma_2 B_2 I_1 - (e_2 + u + r)Y_2,$$

$$R'(t) = r(Y_1 + Y_2) - uR,$$

and

$$M'(t) = q - (A_1 + A_2)M - \delta M,$$

$$V_i'(t) = A_i M - \delta V_i, \quad i = 1, 2. \tag{2.50}$$

$N = S + I_1 + I_2 + Y_1 + Y_2 + R$ and $T = M + V_1 + V_2$ are the host and vector total population sizes respectively. Primary infections in human hosts are produced by either of the two strains at rates

$$B_i = \beta_i V_i / (c + \omega_h N)$$

for $i = 1, 2$ (in vector to host transmission), while primary infections in vectors are produced at rates

$$A_i = \alpha_i (I_i + Y_i) / (c + \omega_v N),$$

where c, ω_h and ω_v are scaling constants. These function forms describe frequency-dependent disease transmission. Both are special cases of the Holling type II functional response [Castillo-Chavez *et al.* (1994)] and are also generalizations the contact rates of the Ross-Macdonald model for Malaria [Aron and May (1982)] and for

Chagas disease [Velasco-Hernandez (1994)]. The parameters h and q are recruitment rates, u and δ are death rates.

We assume that once a mosquito is infected it never recovers and it cannot be reinfected with a different strain of virus. Secondary infections, therefore, may take place only in the host. Two cases can occur: either previously I_1 individuals are infected by strain 2, through contact with infected mosquitoes V_2, becoming Y_2 hosts, or previously I_2 individuals are infected with strain 1, through contact with V_1 mosquitoes, to become Y_1 infected hosts, at rates $\sigma_1 B_1 I_2$ and $\sigma_2 B_2 I_1$, respectively. Here, σ_i is a positive real number that may mimic either cross-immunity ($\sigma_i < 1$) or increased susceptibility ($\sigma_i > 1$) by immune enhancement. This type of dynamics is analogous to superinfection (cf. [Nowak and May (1994)]). In dengue, the immunity developed after infection is a factor that does not appear in superinfection models. In dengue, either of the primary infected populations can be reinfected with the other strain. General results on the effects of cross-immunity in SIS and SI models respectively [Nowak and May (1994); Mena-Lorca *et al.* (1999)], indicate that for certain values of σ coexistence of competing strains is possible. As we will show, the existence of cross-immunity together with the induction of specific permanent immunity, and frequency-dependent contact rates, prevent coexistence. The generic outcome of our model is competitive exclusion although, in some cases, in a very long time scale.

To summarize, if $\sigma_i < 1$, primary infections confer partial immunity to strain i; if $\sigma_i = 1$ secondary infections with strain i take place as if they were primary infections, and if $\sigma_i > 1$ primary infections increase susceptibility to strain i. Once an individual has suffered from both infections it gets immunity to both strains at a rate r independent of the sequence of infections.

Assume that in the absence of disease the total vector population has reached the corresponding asymptotical constant $T = q/\delta$. Then the M equation can be removed. Let

$$\Omega = \{(S, I_1, I_2, Y_1, Y_2, R, V_1, V_2) : \ S + I_1 + I_2 + Y_1 + Y_2 + R \le \frac{h}{u}, \ V_1 + V_2 \le \frac{q}{\delta}\}$$

be the set bounded by the total host and vector population in the absence of disease. We can immediately identify three equilibrium solutions to (2.49)–(2.50), the disease-free equilibrium $E_0^* = (S^*, 0, 0, 0, 0, 0, 0, 0)$ and the two (boundary) equilibria

$$E_1^* = (S_1^*, I_1^*, 0, 0, 0, 0, V_1^*, 0),$$

where only strain 1 survives, and

$$E_2^* = (S_2^*, 0, I_2^*, 0, 0, 0, 0, V_2^*)$$

where only strain 2 survives.

The basic reproduction number is therefore

$$\mathcal{R}_0 = \max\{\sqrt{\mathcal{R}_1}, \ \sqrt{\mathcal{R}_2}\}$$

with

$$\mathcal{R}_i = \frac{\alpha_i \beta_i h q / \delta u}{u \delta (c + h \omega_h / u)(c + h \omega_v / u)}.$$

This formula is a generalization of the Ross-Macdonald basic reproduction number to the case of multiple strains, frequency-dependent contact rates and variable population size in both host and vector. As in most epidemiological models, the disease is able to invade the host population if $\mathcal{R}_0 > 1$, whereas the virus eventually disappears from the host population if $\mathcal{R}_0 \leq 1$ (local result).

For the stability of the boundary equilibria E_1^* and E_2^*, it turns out that it is possible for E_1^* and E_2^* to be both locally asymptotically stable. Consider the superinfection coefficients (σ_1 and σ_2) as the bifurcation parameters. Let f and g be the respective functions of σ_1 and σ_2 given by

$$f(\sigma_1) = \frac{\delta(u+r)}{a_2 b_2 I_1^* (\hat{T} - V_1^*)} \left(1 - \frac{u}{u + \sigma_1 b_1 V_1^*} \frac{\mathcal{R}_2}{\mathcal{R}_1} \right),$$

$$g(\sigma_2) = \frac{\delta(u+r)}{a_1 b_1 I_2^* (\hat{T} - V_2^*)} \left(1 - \frac{u}{u + \sigma_2 b_2 V_2^*} \frac{\mathcal{R}_1}{\mathcal{R}_2} \right),$$

$$(2.51)$$

where I_i^* and V_i^* are the corresponding components of the boundary equilibria E_i^* ($i = 1, 2$). Then for the case when $\mathcal{R}_1 > \mathcal{R}_2 > 1$, the following result holds.

Theorem 2.8. *Let f and g be the functions defined in (2.51).*

(1) E_1^* is locally asymptotically stable if $\sigma_2 < f(\sigma_1)$ for every $\sigma_1 > 0$, and unstable if $\sigma_2 > f(\sigma_1)$.

(2) E_2^* is locally asymptotically stable if $\sigma_2 > g^{-1}(\sigma_1)$ for every $\sigma_1 > 0$, and unstable if $\sigma_2 < g^{-1}(\sigma_1)$.

(3) E_1^* and E_2^* are locally asymptotically stable if $g^{-1}(\sigma_1) < \sigma_2 < f(\sigma_1)$.

Note that it is possible that both E_1^* and E_2^* are locally asymptotically stable. This conclusion indicates, at the very least, that there are situations where the asymptotic dynamics of our model depends on the initial conditions. In this case, we can draw a bifurcation diagram in the (σ_1, σ_2) plane as shown in Figure 2.15.

Although the analytic result for the stability of coexistence equilibrium is difficult to obtain due to the disease-induced death, the following results were found through numerical simulations of model (2.49)–(2.50) when $e_i > 0$:

(1) If $\mathcal{R}_i > 1 > \mathcal{R}_j$ then the boundary equilibrium E_j^* and the interior endemic equilibrium do not exist and the boundary equilibrium E_i^* is globally asymptotically stable (see Figure 2.16).

(2) Whenever $\mathcal{R}_i > 1$ for $i = 1, 2$ there exists an equilibrium point in the interior of Ω. This point has a local unstable and a local stable manifold of positive dimension (see Figure 2.17).

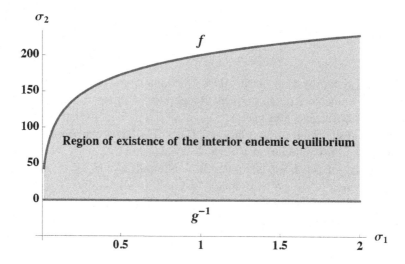

Fig. 2.15   Region of parameter space (σ_1, σ_2) where both boundary equilibria are locally asymptotically stable. Fixed parameter values are $r = 0.71/\text{day}$, $\mu = 0.000039/\text{day}$, $\delta = 0.71/\text{day}$, $h = 0.9775$, $\alpha_1 = 0.002$, $\alpha_2 = 0.015$, $\beta_1 = 0.001$, $\beta_2 = 0.001$, $c = 10$, $\hat{T} = 50000$, $\hat{N} = 25000$. The corresponding basic reproduction numbers are $R_1 = 2.4$ and $R_2 = 2.08$.

(3) When $\mathcal{R}_i > \mathcal{R}_j > 1$, the superinfection coefficients σ_1 and σ_2 may change the asymptotic behavior of the system, rendering strain j as the winner over strain i (which would be the winner if $\sigma_1 = \sigma_2 = 1$, see Figure 2.18).

(4) When $\mathcal{R}_i > \mathcal{R}_j > 1$ and both boundary equilibria are locally asymptotically stable, there exists a separatrix that cuts Ω into two disjoint basins of attraction (one for each boundary equilibrium, see Figure 2.17).

Model simulations show that in the host population there is no long-term persistence of both strains. However, the unusual nature of the endemic equilibrium (a 'saddle' point) produces a relatively prolonged (years of duration) quasi-steady state when both $\mathcal{R}_i$ are greater than one. Given the inherent time-scale of the disease (months), this quasi-steady state would look as a stable endemic equilibrium (see Figures 2.17 and 2.18). Under these conditions there are two possibilities depending on how many of the boundary equilibria are locally stable. If only one of them is locally asymptotically stable, our computer simulations indicate that this equilibrium is also globally asymptotically stable. Thus the competitive exclusion of one of the strains occurs.

Our computer simulations also indicate that even though the initial outbreak of primary infection is driven by the strain with the highest reproduction number, it is precisely this strain that can be competitively excluded. This occurs if the primary infection enhances (increases susceptibility to) secondary infections (see Figure 2.18). Therefore, the strain with the smallest reproduction number may end up persisting in the host.

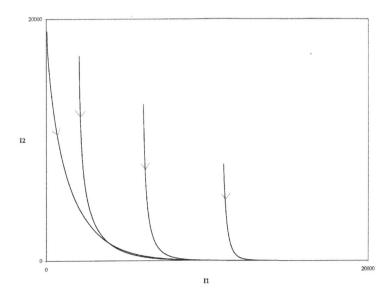

Fig. 2.16 Phase plot in the space (I_1, I_2) for values of the superinfection indices outside the shaded area shown in Figure 2.15. The graph was computed with the same parameter values shown in Figure 2.15 but with $\sigma_1 = 5$, $\sigma_2 = 0.05$, and positive disease-induced death rates $e_1 = 0.0001/$day, $e_2 = 0.0005/$day. These parameter values give $\sigma_2 < g^{-1}(\sigma_1) = 0.1$. In this case strain 1 competitively excludes strain 2. The final outcome of the disease (which strains wins) is independent of initial conditions. The unit of measurement of I_1 and I_2 is number of cases.

The second possibility occurs when both boundary equilibria exist and both are locally asymptotically stable. In this case the outcome of the interaction–competitive exclusion of one strain–depends on the initial conditions (Figure 2.17).

Gupta, Swinton and Anderson [Gupta *et al.* (1994)] show in a model for malaria that coexistence is a likely outcome when cross-immunity is taken into account. Although malaria is a parasitic, not a viral disease, the mathematical structure of the model allows some comparisons with ours because both deal with a vector transmitted disease. Gupta *et al.* generalize directly the Ross-Macdonald model for malaria studied by Aron and May [Aron and May (1982)] introducing cross-immunity and two infected host subtypes: those that are infected and infectious, and those that are infected but uninfectious. Thus, essentially there is a reduction in the net number of infected individuals that can transmit the disease. However all infected individuals can hold the parasite. In particular, the rate at which parasites become ineffective to transmission, i.e., the host becomes infected but not infectious, is exponential, guaranteeing the presence of positive densities (however small) of each type of infected hosts for all time. Thus, the Gupta *et al.* model effectively creates a refuge for each parasite strain. Moreover, the total host population is considered constant. The assumption of constant host population size is achieved by defining the recruitment rate in such a way as to balance the output from all

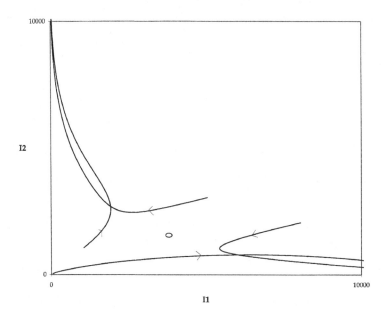

Fig. 2.17   Similar to Figure 2.16 but for values of the superinfection indices outside the shaded area shown in Figure 2.15. Other parameter values are $\sigma_1 = 1$, $\sigma_2 = 4.2$, and positive disease-induced death rates $e_1 = 0.0001$/day, $e_2 = 0.0005$/day. The presence of a saddle point in the interior of the region and the existence of a separatrix may be conjectured. Note that the final outcome of the disease (which strains wins) depends on initial conditions. The circle indicates the endemic equilibrium point.

system compartments. This factor alone when associated with cross-immunity is enough to enhance coexistence in models for directly transmitted diseases [Nowak and May (1994)], [May and Nowak (1994)]. In the case of our general model all infected individuals are infectious; thus there are no refuges. Also, by definition, we take virulence as extra mortality induced by the disease. This prevents the existence of a constant population size for the host. We do not define the recruitment rate for the total population so as to balance disease mortality losses (and therefore achieve a constant size in the host population). This would be equivalent, in our case, to require that the extra-mortality rate is compensated exactly by the cure rate of the disease and, therefore, population variability would be independent of disease dynamics. However, even in this case (no virulence) our model predicts competitive exclusion of one of the strains.

The main reason that explains why coexistence in our model is an improbable outcome resides, we believe, in the coupling of two populations, each with a different pattern of disease progression. The structure of the equations that describe the transmission dynamics in the host population is that of an SIR model with super-infection and variable population size. In a directly transmitted disease with this structure and no virulence one would expect analogous results to those of Nowak

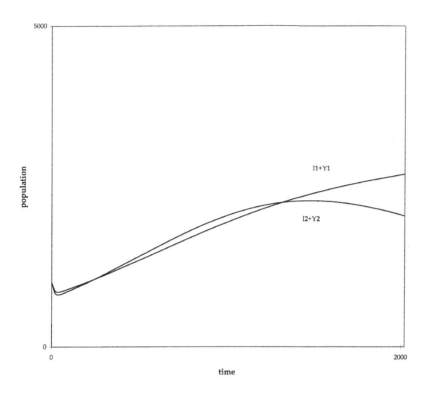

Fig. 2.18 Time plot of model (2.49-2.50) for a period of 5 years. The graph shows the total numbers of infected individuals for each strain $I_1 + Y_1$ and $I_2 + Y_2$. Parameter values are the same as in Figure 2.15 except for the following: $\alpha_1 = 0.005$, $\alpha_2 = 0.005$, $\beta_1 = 0.005$, $\beta_2 = 0.007$, $\sigma_1 = 4$, $\sigma_2 = 1.2$. There are two curves, one for each strain. For about 3 years both strains seem to increase and coexist. Only in the fourth year strain 1 clearly wins over strain 2. Note that strain 2 increases faster at the beginning of the epidemic but it is this strain that goes extinct.

and May [Nowak and May (1994)]: coexistence of both strains as a rule. Our model also incorporates an SI model without superinfection in the vector population. In a directly transmitted disease this structure would predict competitive exclusion of the strain with lower basic reproduction number [Bremermann and Thieme (1989)].

When we couple both of these types of epidemics into one, our host-vector model (1-2), the outcome is competitive exclusion of one of the strains if at least one of the basic reproductive numbers is greater than one. In a sense, the vector dynamics dominates the dynamics of the coupled system. The reason for this is that the vector-host relationship is asymmetric. The vector chooses the host. In this case we have modeled the contact rates according to a generalization of the Ross-Macdonald model: the contact rate is frequency-dependent [Castillo-Chavez *et al.* (1994)] (depends on the ratios of vector numbers to host numbers for both types of strains). Thus, what our results show is that *coexistence promoted by superinfection in the host population is 'broken' by frequency-dependent dynamics in the biting (contact) rates, thus resulting in the competitive exclusion of one strain even when an interior steady-state exists.*

Other models that incorporate cross-immunity and multiple strains have been studied (see, for example, [Levin and Pimentel (1981); Lipsitch and Nowak (1995); Mena-Lorca *et al.* (1999)]). We compare our results with the original one that introduced this idea of competition of multiple strains in epidemic models, namely the Levin and Pimentel paper [Levin and Pimentel (1981)]. In summary, the conclusions of [Levin and Pimentel (1981)] are that in a variable host population system coexistence is possible in a bounded region of parameter space. Outside this region, depending on the relative magnitudes of parameter values one of the two strains wins and competitively excludes the other. This model was originally designed for the theoretical study of myxomatosis as a control factor of an exponentially growing population. The fact that virulence is the growth regulatory factor in this model determines the existence of a coexistence region in parameter space. In the dengue model that we analyze here, the disease is not the unique factor that regulates growth. Permanent immunity is also explicitly introduced into the model. Even in the case when virulence is negligible competitive exclusion is the rule. The existence of frequency-dependent contact rates closes the window of coexistence.

2.4 Age-structured models

Most of the models considered in previous sections do not explicitly include age-dependent factors, either the chronological age or the infection age (time since infection). There are two commonly used approaches for incorporating an age-struction of the population. One approach uses finite discrete age groups, which leads to a system of ordinary differential equations. This requires the assumption that the parameter values are independent of age within each age group, and the aging process may complecate the analysis (see [Hethcote (2000)]). The other approach is to consider age as a continuous variable, which leads to a system of partial differential equations (often fewer equations). Some examples are presented in this section.

2.4.1 *Optimal ages for vaccination*

The model and results presented in this section are taken from [Castillo-Chavez and Feng (1998)]. It concerns age-dependent vaccination strategies for TB. It follows the framework developed in [Hadeler and Müller (1996)]. Divided the population into susceptible, vaccinated, latent, infectious, and treated classes, where $s(t,a), v(t,a), l(t,a), i(t,a)$, and $j(t,a)$ denote the associated density functions with these respective epidemiological age-structure classes, and $n(t,a) = s(t,a) + v(t,a) + l(t,a) + i(t,a) + j(t,a)$. Assume that all newborns are susceptible and that the mixing between individuals is proportional to their age-dependent activity level. Assume also that an individual may become infected only through

contact with infectious individuals, that vaccination is partially effective (i.e., vaccinated individuals can become infected again but with a reduced susceptibility), that
only susceptibles will be vaccinated. The dynamics are governed by the following
initial boundary value problem:

$$\left(\frac{\partial}{\partial t} + \frac{\partial}{\partial a}\right) s(t,a) = -\beta(a)c(a)B(t)s(t,a) - \mu(a)s(t,a) - \psi(a)s(t,a),$$

$$\left(\frac{\partial}{\partial t} + \frac{\partial}{\partial a}\right) v(t,a) = \psi(a)s(t,a) - \mu(a)v(t,a) - \delta\beta(a)c(a)B(t)v(t,a),$$

$$\left(\frac{\partial}{\partial t} + \frac{\partial}{\partial a}\right) l(t,a) = \beta(a)c(a)B(t)(s(t,a) + \sigma j(t,a) + \delta v(t,a))$$

$$\qquad\qquad\qquad -(k+\mu(a))l(t,a), \tag{2.52}$$

$$\left(\frac{\partial}{\partial t} + \frac{\partial}{\partial a}\right) i(t,a) = kl(t,a) - (r+\mu(a))i(t,a),$$

$$\left(\frac{\partial}{\partial t} + \frac{\partial}{\partial a}\right) j(t,a) = ri(t,a) - \sigma\beta(a)c(a)B(t)j(t,a) - \mu(a)j(t,a)$$

with initial and boundary conditions:

$$B(t) = \int_0^\infty \frac{i(t,a')}{n(t,a')}p(t,a,a')da',$$

$$s(t,0) = \Lambda, \quad v(t,0) = l(t,0) = i(t,0) = j(t,0) = 0,$$

$$s(0,a) = s_0(a), \ v(0,a) = v_0(a), \ l(0,a) = l_0(a), \ i(0,a) = i_0(a), \ j(0,a) = j_0(a).$$

The function p represents the mixing between susceptible individuals of age a and
infectious individual of age a', and is assumed to have the form of proportionate
mixing:

$$p(t,a,a') = \frac{c(a')n(t,a')}{\int_0^\infty c(u)n(t,u)du}.$$

Λ is the recruitment/birth rate (assumed constant); $\beta(a)$ is the age-specific (average)
probability of becoming infected through contact with infectious individuals; $c(a)$ is
the age-specific per-capita contact/activity rate; $\mu(a)$ is the age-specific per-capita
natural death rate (all of these functions are assumed to be continuous and to be
zero beyond some maximum age); k is the per-capita rate at which individuals leave
the latent class by becoming infectious; r is the per-capita treatment rate; σ and δ
are the reductions in risk due to prior exposure to TB and vaccination, respectively,
$0 \le \sigma \le 1, 0 \le \delta \le 1$; and $p(t,a,a')$ gives the probability that an individual of age a
has contact with an individual of age a' given that it has a contact with a member
of the population. Here we assume proportionate mixing as introduced earlier by
many authors including Hethcote and Yorke [Hethcote and Yorke (1984)], Dietz

and Schenzle [Dietz and Schenzle (1985b)], Anderson and May [Anderson and May (1984)], and Castillo-Chavez *et al.* [Castillo-Chavez (1987); Castillo-Chavez *et al.* (1989a)]. Hence, using the approach of Busenberg and Castillo-Chavez [Busenberg and Castillo-Chavez (1991)], we assume that $p(t, a, a') = p(t, a')$ as explicitly described above. The initial age distributions are assumed to be known and to be zero beyond some maximum age. The model (2.52) is well-posed and the proof is similar to that found in Castillo-Chavez *et al.* [Castillo-Chavez *et al.* (1989a)].

Notice that $n(t, a)$ satisfies the following equations:

$$\left(\frac{\partial}{\partial t} + \frac{\partial}{\partial a}\right) n(t, a) = -\mu(a) n(t, a),$$

$$n(t, 0) = \Lambda, \quad n(0, a) = n_0(a) = s_0(a) + v_0(a) + l_0(a) + i_0(a) + j_0(a).$$

Using the method of characteristic curves we can solve for n explicitly:

$$n(t, a) = n_0(a) \frac{\mathcal{F}(a)}{\mathcal{F}(a - t)} H(a - t) + \Lambda \mathcal{F}(a) H(t - a),$$

where

$$\mathcal{F}(a) = e^{-\int_0^a \mu(s) ds},$$

$$H(s) = 1, \ s \ge 0; \quad H(s) = 0, \ s < 0.$$

Hence,

$$n(t, a) \to \Lambda \mathcal{F}(a),$$

$$p(t, a) \to \frac{c(a)\mathcal{F}(a)}{\int_0^\infty c(b)\mathcal{F}(b) db} =: p_\infty(a), \quad t \to \infty. \tag{2.53}$$

Introducing the fractions

$$u(t, a) = \frac{s(t, a)}{n(t, a)}, \ w(t, a) = \frac{v(t, a)}{n(t, a)}, \ x(t, a) = \frac{l(t, a)}{n(t, a)}, \ y(t, a) = \frac{i(t, a)}{n(t, a)}, \ z(t, a) = \frac{j(t, a)}{n(t, a)},$$

we get a simplified system of (2.52):

$$\left(\frac{\partial}{\partial t} + \frac{\partial}{\partial a}\right) u(t, a) = -\beta(a) c(a) B(t) u(t, a) - \psi(a) u(t, a),$$

$$\left(\frac{\partial}{\partial t} + \frac{\partial}{\partial a}\right) w(t, a) = \psi(a) u(t, a) - \delta \beta(a) c(a) B(t) w(t, a),$$

$$\left(\frac{\partial}{\partial t} + \frac{\partial}{\partial a}\right) x(t, a) = \beta(a) c(a) B(t)(u(t, a) + \delta w(t, a) + \sigma z(t, a)) - k x(t, a),$$

$$\left(\frac{\partial}{\partial t} + \frac{\partial}{\partial a}\right) y(t, a) = k x(t, a) - r y(t, a), \tag{2.54}$$

$$\left(\frac{\partial}{\partial t} + \frac{\partial}{\partial a}\right) z(t, a) = r y(t, a) - \sigma \beta(a) c(a) B(t) z(t, a),$$

$$B(t) = \int_0^\infty y(t, a) p(t, a) da,$$

$$p(t, a) = \frac{c(a) n(t, a)}{\int_0^\infty c(u) n(t, u) du},$$

$$u(t, 0) = 1, \quad w(t, 0) = x(t, 0) = y(t, 0) = z(t, 0) = 0.$$

Let $\mathcal{F}_\psi(a)$ denote the probability that a susceptible individual has not been vaccinated at age a. Then

$$\mathcal{F}_\psi(a) = e^{-\int_0^a \psi(b)db}.$$

The system (2.54) has the infection-free steady state

$$u(a) = \mathcal{F}_\psi(a), \ w(a) = 1 - \mathcal{F}_\psi(a), \ x(a) = y(a) = z(a) = 0, \ n(a) = \Lambda F(a). \quad (2.55)$$

To study the local stability of the infection-free equilibrium, we linearize (2.54) about (2.55) and consider exponential solutions of the form

$$x(t,a) = X(a)e^{\lambda t}, \ y(t,a) = Y(a)e^{\lambda t}, \ B(t) = B_0 e^{\lambda t} + O(e^{2\lambda t}),$$

where

$$B_0 = \int_0^\infty Y(a)p_\infty(a)da \quad (2.56)$$

is a constant and $p_\infty(a)$ is as in (2.53). Then the linear parts of the x and y equations in (2.54) are of the form

$$\lambda X(a) + \frac{d}{da}X(a) = \beta(a)c(a)B_0\mathcal{V}_\psi(a) - kX(a),$$

$$\lambda Y(a) + \frac{d}{da}Y(a) = kX(a) - rY(a),$$

where

$$\mathcal{V}_\psi(a) = \mathcal{F}_\psi(a) + \delta(1 - \mathcal{F}_\psi(a)). \quad (2.57)$$

An expression for $Y(a)$ can be obtained by solving the above system:

$$Y(a) = B_0 \int_0^a \frac{k}{r-k}\beta(\alpha)c(\alpha)\left(e^{(\lambda+k)(\alpha-a)} - e^{(\lambda+r)(\alpha-a)}\right)\mathcal{V}_\psi(\alpha)d\alpha. \quad (2.58)$$

From (2.56) and (2.58) we get

$$B_0 = B_0 \int_0^\infty \int_0^a \frac{k}{r-k}p_\infty(a)\beta(\alpha)c(\alpha)\left(e^{(\lambda+k)(\alpha-a)} - e^{(\lambda+r)(\alpha-a)}\right)\mathcal{V}_\psi(\alpha)d\alpha da. \quad (2.59)$$

By changing the order of integration, introducing $\tau = a - \alpha$, and dividing both sides by B_0 (because $B_0 \neq 0$) in (2.59) we get the characteristic equation

$$1 = \int_0^\infty \int_0^\infty \frac{k}{r-k}p_\infty(\alpha+\tau)\beta(\alpha)c(\alpha)\left(e^{-(\lambda+k)\tau} - e^{-(\lambda+r)\tau}\right)\mathcal{V}_\psi(\alpha)d\tau d\alpha \quad (2.60)$$

$$:= G(\lambda).$$

Now we are ready to define the net reproduction number as $\mathcal{R}(\psi) = G(0)$; i.e.,

$$\mathcal{R}(\psi) = \int_0^\infty \int_0^\infty \frac{k}{r-k}p_\infty(\alpha+\tau)\beta(\alpha)c(\alpha)\left(e^{-k\tau} - e^{-r\tau}\right)\mathcal{V}_\psi(\alpha)d\tau d\alpha. \quad (2.61)$$

Noticing that

$$G'(\lambda) < 0, \ \lim_{\lambda\to\infty} G(\lambda) = 0, \ \lim_{\lambda\to-\infty} G(\lambda) = \infty,$$

we know that (2.60) has a unique negative real solution λ^* if, and only if, $G(0) < 1$, or $\mathcal{R}(\psi) < 1$. Also, (2.60) has a unique positive (zero) real solution if $\mathcal{R}(\psi) > 1$ ($\mathcal{R}(\psi) = 1$). To show that λ^* is the dominant real part of roots of $G(\lambda)$, we let $\lambda = x + iy$ be an arbitrary complex solution to (2.60). Note that

$$1 = G(\lambda) = |G(x + iy)| \le G(x),$$

indicating that $\Re\lambda \le \lambda^*$. It follows that the infection-free steady state is l.a.s. if $\mathcal{R}(\psi) < 1$, and unstable if $\mathcal{R}(\psi) > 1$. This establishes the following result.

Theorem 2.9. *Let $\mathcal{R}(\phi)$ be the reproduction number defined in (2.61).*

(a) *The infection-free steady-state (2.55) is locally asymptotically stable (l.a.s.) if $\mathcal{R}(\psi) < 1$ and unstable if $\mathcal{R}(\psi) > 1$.*

(b) *There exists an endemic steady state of (2.54) when $\mathcal{R}(\psi) > 1$.*

Note that the basic reproduction number is $\mathcal{R}_0 = \mathcal{R}(0)$, i.e.,

$$\mathcal{R}_0 = \int_0^\infty \int_0^\infty \frac{k}{r-k} p_\infty(\alpha + \tau)\beta(\alpha)c(\alpha)\left(e^{-k\tau} - e^{-r\tau}\right)d\tau d\alpha.$$

The quantities $\mathcal{R}(\psi)$ and $\mathcal{R}_0$ can be used to construct the optimization problems to identify optimal vaccination strategies. Generally speaking, the effect of subjecting a population to a vaccination program is to reduce its reproduction number and to increase the average age of first infection. Ideally, in a vaccination program, one would like to eliminate or eradicate a disease, but vaccinations often can only prevent major epidemic outbreaks. Because elimination is usually highly unlikely, we often try to find ways of reducing the prevalence or incidence of a particular disease. By lowering the reproduction number, we reduce the prevalence and incidence of a disease. In Dietz and Schenzle [Dietz and Schenzle (1985b)], a simpler formula was derived that can be used to determine the function $\psi(a)$ needed to reduce $\mathcal{R}(\psi)$ below 1. Their approximation was constructed for diseases where the length of the infectious period is short. TB has a long and variable period of infectiousness. Therefore, we consider instead the approach used by Hadeler and Müller [Hadeler and Müller (1996)] in the case of HIV, and look at the effectiveness of vaccination policies that are driven by reductions of the reproduction number.

Consider the functional defined by

$$F(\psi) = \mathcal{R}_0 - \mathcal{R}(\psi).$$

Then

$$F(\psi) = \int_0^\infty \int_0^\infty \beta(\alpha)c(\alpha)p(\alpha + \tau)\frac{k}{r-k}(e^{-k\tau} - e^{-r\tau})(1 - \mathcal{V}_\psi(\alpha))d\tau d\alpha, \quad (2.62)$$

where $\mathcal{V}_\psi(\alpha)$ is as in (2.57). Note that $F(\psi)$ gives a measure of the reduction in the reproduction number by the vaccination strategy ψ.

We next formulate two optimization problems by considering costs that are associated with the vaccination strategies. Let $s_\psi(a)$ denote the density function of susceptibles describing the steady demographic state in the absence of disease, then

$$s_\psi(a) = \Lambda \mathcal{F}(a)\mathcal{F}_\psi(a).$$

Let $C(\psi)$ be the total cost associated with the vaccination strategy ψ, and assume that $C(\psi)$ depends linearly on the number of vaccinations (see [Hadeler and Müller (1996)]). Then we can write

$$C(\psi) = \int_0^\infty \kappa(a)\psi(a)s_\psi(a)da,$$

where $\kappa(a)$ is a positive function representing the costs associated with one vaccination at age a. For future use, we note that

$$C(\psi) = \int_0^\infty \Lambda\kappa(a)\psi(a)\mathcal{F}(a)e^{-\int_0^a \psi(s)ds}da. \tag{2.63}$$

Two optimization problems can be defined as follows. Let $\mathcal{R}_*$ and C_* be two constants.

(I) Find a vaccination strategy $\psi(a)$ that minimizes $C(\psi)$ constrained by $\mathcal{R}(\psi) \le \mathcal{R}_*$.
(II) Find a vaccination strategy $\psi(a)$ that minimizes $\mathcal{R}(\psi)$ constrained by $C(\psi) \le C_*$.

The difficulty associated with these optimization problems is due to the fact that $C(\psi)$ and $F(\psi)$ are nonlinear functionals of ψ. Hadeler and Müller [Hadeler and Müller (1996)] showed how to overcome this difficulty. In order to make both $C(\psi)$ and $F(\psi)$ linear functionals we apply the transformation

$$\phi(a) = -\frac{d}{da}e^{-\int_0^a \psi(s)ds} = \psi(a)e^{-\int_0^a \psi(s)ds}.$$

Denote $\bar{F}(\phi) = F(\psi), \bar{C}(\phi) = C(\psi)$. Then, seeing that

$$1 - \mathcal{V}_\psi(a) = (1-\delta)(1 - \mathcal{F}_\psi(a)) = (1-\delta)\int_0^a \phi(s)ds,$$

and also noting that

$$\bar{F}(\phi) = \int_0^\infty \int_0^\infty \beta(\alpha)c(\alpha)p(\alpha+\tau)\frac{k}{r-k}(e^{-k\tau} - e^{-r\tau})(1-\delta)\int_0^\alpha \phi(s)dsd\tau d\alpha$$

$$= \int_0^\infty \left\{\int_a^\infty \int_0^\infty (1-\delta)\beta(\alpha)c(\alpha)p_\infty(\alpha+\tau)\frac{k}{r-k}(e^{-k\tau} - e^{-r\tau})d\tau d\alpha\right\}\phi(a)da,$$

(see (2.63)), we arrive at

$$\bar{F}(\phi) = \int_0^\infty K(a)\phi(a)da,$$

$$\bar{C}(\phi) = \int_0^\infty B(a)\phi(a)da,$$

(see also (2.62)), where

$$K(a) = \int_a^\infty \int_0^\infty (1-\delta)\beta(\alpha)c(\alpha)p_\infty(\alpha+\tau)\frac{k}{r-k}(e^{-k\tau} - e^{-r\tau})d\tau d\alpha,$$

$$B(a) = \Lambda\kappa(a)\mathcal{F}(a). \tag{2.64}$$

Hence, we have replaced two nonlinear functionals with the linear functionals given by $\bar{F}(\phi)$ and $\bar{C}(\phi)$. If we let

$$Q(\phi) = \int_0^\infty \phi(a)da,$$

then it is easy to see that $Q(\phi) \leq 1$.

Letting $\rho = \mathcal{R}_0 - \mathcal{R}_*$ we are able to replace (I) by the following linear optimization problem:

$$\text{minimize} \quad \bar{C}(\phi)$$

$$\text{subject to} \quad f(\phi) \leq 0, \tag{2.65}$$

$$\phi \geq 0,$$

where

$$f(\phi) = \begin{pmatrix} f_1(\phi) \\ f_2(\phi) \end{pmatrix} = \begin{pmatrix} \rho - \bar{F}(\phi) \\ Q(\phi) - 1 \end{pmatrix},$$

and $f(\phi) \leq 0$ is equivalent to $f_i(\phi) \leq 0$ $(i = 1, 2)$. After using (formally) the Saddle Point Theorem of Kuhn and Tucker for the convex optimization problem (see [Stoer *et al.* (1970)]) we can show that (2.65) is mathematically equivalent to the (P1) in [Stoer *et al.* (1970)]. Hence, using the same arguments we arrive at the following conclusion.

The optimization problem (I) can be analyzed using the same methods as the problem (P1) in [Hadeler and Müller (1996)], which leads to the following conclusion.

Theorem 2.10. *There are two possible optimal vaccination strategies in (I):*

(i) *one-age strategy: vaccinate the susceptible population at exactly age A;*
(ii) *two-age strategy: vaccinate part of the susceptible population at age A_1 and the remaining susceptibles at a later age A_2.*

For the two vaccination strategies, the optimal ages can be calculated in the following way: Note that $K(a)$ (see (2.64)) is a strictly decreasing function with $K(0) = \mathcal{R}_0 > \rho$ and $K(a) \to 0$ as $a \to \infty$. Hence, we can find $A_* > 0$ such that $K(A_*) = \rho$. Let A be the minimum of the quotient $B(a)/K(a)$. (See [Hadeler and Müller (1996)] for discussions about the existence of A.) If $A \in [0, A_*]$, then it gives an optimal age for the one-age strategy. If $A \in (A_*, \infty)$, then the optimal two-age strategy is found by minimizing the expression $C(A_1, A_2)$ on $A_1 \in [0, A_*]$ and $A_2 \in [A_*, \infty)$, where

$$C(A_1, A_2) = \frac{\rho - K(A_2)}{K(A_1) - K(A_2)} B(A_1) + \frac{K(A_1) - \rho}{K(A_1) - K(A_2)} B(A_2).$$

For (II), a similar conclusion to Theorem 2.10 can be obtained; i.e., the optimal vaccination strategy is either one- or two-age, and the optimal ages can be determined.

A similar approach is used in [Shim *et al.* (2006)] to explore age-dependent optimal vaccination strategies for rotavirus.

2.4.2 *Infection-age-dependent transmission and progression*

In this section we present two examples that take into consideration a more realistic distribution for the latent period of TB. One is a single strain model studied in [Feng *et al.* (2001)] and the other includes both drug-sensitive and drug-resistant strains as well as a distributed latent period considered in [Feng *et al.* (2002)]. The TB models considered in [Castillo-Chavez and Feng (1997)] assumed that the latent stage is exponentially distributed, which leads to systems of ODEs. This assumption is not realistic for TB. As pointed out by Comstock, "tuberculosis is an infectious disease with an incubation period from weeks to a lifetime". Figure 2.19 shows that TB has a long and variable period of latency.

Figure 2.20 shows the data from observation in adolescents who had developed clinical tuberculosis following primary infection [Styblo (1991)]. It suggests that among the 10% of latent individuals who eventually develop active TB, around 60% will do so during the first year post infection. The rest will develop active TB in either 2 years (20%), 5 years (15%), 20 years (5%) or even longer.

A one-strain TB model with distributed latency

To examine whether or not the incorporation of a more realistic distribution of latent period will alter model outcomes, in [Feng *et al.* (2001)] we extended the one-strain ODE model in [Castillo-Chavez and Feng (1997)] by incorporating a distributed delay to model the long and variable latent period. As the focus is on the distributed latency, the model assumes that treated individuals will return to the susceptible class. The one-strain ODE model reads:

$$S' = \Lambda - \beta S \frac{I}{N} - \mu S + r_1 E + r_2 I,$$

$$E' = \beta S \frac{I}{N} - (\mu + k + r_1)E,$$

$$I' = kE - (\mu + d + r_2)I,$$

$$N = S + E + I.$$

(2.66)

Λ is the constant recruitment rate; β is the average number of susceptible individuals infected by one infectious individual per unit of time, μ is the per-capita natural death rate; k is the rate at which an individual leaves the latent class by becoming infectious; d is the per-capita disease-induced death rate; r_1 and r_2 are the per-capita treatment rates of latent and infectious individuals, respectively.

Let $p(s)$ be a function representing the proportion of those individuals latent at time t and who, if alive, are still infected (but not infectious) at time $t + s$. Then $-\dot{p}(\tau)$ is the rate of removal of individuals from E class into I class τ units of time after becoming latent. Assume that

$$p(s) \geq 0, \quad \dot{p}(s) \leq 0, \quad p(0) = 1,$$

and

$$\int_0^\infty p(s)ds < \infty.$$

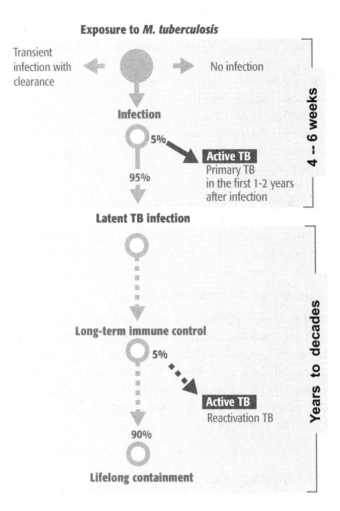

Fig. 2.19   A diagram showing the progression from latent to active TB during the period of infection (adopted from www.biomerieux-diagnos6cs.com). It shows that only about 10% of latently infected will develop active TB and 5% of those will stay in the latent stage for long time.

Then the number of individuals who have been latent from time 0 to t and are still in class E is given by

$$\int_0^t \beta S(s)\frac{I(s)}{N(s)}p(t-s)e^{-(\mu+r_1)(t-s)}ds.$$

Thus the number of individuals who become infectious from time 0 to t and are still alive and in class I is

$$\int_0^t \int_0^\tau \beta S(s)\frac{I(s)}{N(s)}e^{-(\mu+r_1)(\tau-s)}[-\dot{p}(\tau-s)e^{-(\mu+r_2+d)(t-\tau)}]dsd\tau.$$

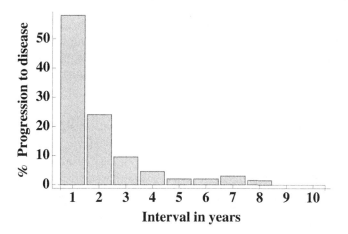

Fig. 2.20 An example of distribution of progression from latent to active TB [Styblo (1991)].

Then we have the following model:

$$S' = \Lambda - \beta S \frac{I}{N} - \mu S + r_1 E + r_2 I,$$

$$E(t) = E_0(t) + \int_0^t \beta S(s) \frac{I(s)}{N(s)} p(t-s) e^{-(\mu+r_1)(t-s)} ds,$$

$$I(t) = \int_0^t \int_0^\tau \sigma S(s) \frac{I(s)}{N(s)} e^{-(\mu+r_1)(\tau-s)} [-\dot{p}(\tau-s) e^{-(\mu+r_2+d)(t-\tau)}] ds\, d\tau, \qquad (2.67)$$

$$+ I_0 e^{-(\mu+r_2+d)t} + I_0(t),$$

$$N = S + E + I,$$

where $E_0(t)$ denotes those individuals in E class at time $t = 0$ and still in the latent class, $I_0(t)$ denotes those initially in class E who have moved into class I and are still alive at time t, and $I_0 e^{-(\mu+r_2+d)t}$ with $I_0 = I(0)$ represents those who are infectious at time 0 and are still alive and in I class. $E_0(t)$ and $I_0(t)$ are assumed to have compact support (that is they vanish for large enough t).

The reproduction number $\mathcal{R}_0$ derived from model (2.66) is

$$\mathcal{R}_0 = \frac{\beta k}{(\mu + d + r_2)(\mu + k + r_1)},$$

which determines if the disease will die out (when $\mathcal{R}_0 < 1$) or persist (when $\mathcal{R}_0 > 1$). For the model with arbitrary delay (2.67), the reproduction number is given by

$$\mathcal{R}_0 = \beta \int_0^\infty a(\tau) d\tau, \qquad (2.68)$$

where $a(u)$ is defined by

$$a(t-s) = \int_s^t e^{-(\mu+r_1)(\tau-s)} [-\dot{p}(\tau-s) e^{-(\mu+r_2+d)(t-\tau)}] d\tau,$$

and represents the "effective" infectious period. It is shown in [Feng *et al.* (2001)] that the qualitative behaviors of the delay model (2.67) are the same as that of the ODE model (2.66). That is, the disease will die out if $\mathcal{R}_0 < 1$ and persist if $\mathcal{R}_0 > 1$, although the expression for $\mathcal{R}_0$ given in (2.68) provides a more detailed description on how each factor may affect $\mathcal{R}_0$ and thus the threshold condition $\mathcal{R}_0 = 1$.

A two-strain TB model with distributed latency

For the case of two strains, we developed a model with distributed latency in [Feng *et al.* (2002)] based on the ODE model in [Castillo-Chavez and Feng (1997)]. The model vin [Feng *et al.* (2002)] is structured by the age-of-infection. The host population is divided into three epidemiological classes or subgroups: Susceptibles, Infected with sensitive TB, and Infected with resistant TB. Introduce the following notation:

$S(t)$ = *number of susceptibles at time t,*

$i(\theta,t)$ = *density of infected individuals with the drug-sensitive strain at time t,*

$J(t)$ = *number of infected individuals with a drug-resistant strain at time t.*

Here the variable θ denotes the *age of the infection with drug-sensitive TB*, i.e., the time that has lapsed since the individual became infected. We note that this class of infected individuals includes both latent and infectious individuals. In fact, as already pointed out in the introduction, the majority of infected individuals remains latent while, as experimentally observed, only a small proportion of them develop and exhibit the disease, becoming infective. Thus we introduce the function $p(\theta)$ $(0 \leq p(\theta) \leq 1)$ as the *proportion of sensitive-strain-infected individuals that become active at class age θ*. This function is assumed constant in time and is based on experimental data. Thus,

$$p(\theta)i(\theta,t) = \text{age density of infectious individuals.} \tag{2.69}$$

The transitions between classed is described in Figure 2.21. It is assumed that the population involved is a closed population undergoing a per capita mortality rate μ and a density dependent growth rate with the per capita birth rate

$$b = b(N), \quad b'(N) < 0. \tag{2.70}$$

In (2.70) $N = S(t) + J(t) + I(t)$ denotes the total number of individuals, where

$$I(t) = \int_0^\infty i(\theta, t)d\theta. \tag{2.71}$$

Concerning the mechanism of infection, we give the following constitutive form to the *force of infection* relative to the sensitive strain:

$$\lambda_1(t) = \frac{\beta_1}{N(t)} \int_0^\infty p(\theta)i(\theta,t)d\theta, \tag{2.72}$$

with β_1 being the per-capita transmission rate of the sensitive strain.

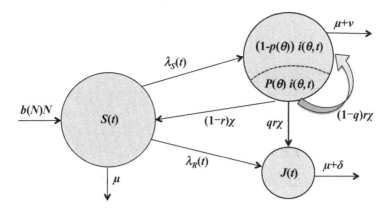

Fig. 2.21   Transition diagram for the two-strain TB model structured by age-of-infection.

The same function $p(\theta)$ is used to give a shape to the removal rate $\gamma(\theta)$, i.e., the rate at which sensitive-strain-infected individuals leave the i class due to treatment. In fact we assume

$$\gamma(\theta) = (1 - r + qr)\chi p(\theta), \qquad (2.73)$$

where χ denotes the *treatment rate*, that is the fraction of infectious people detected and treated per unit of time (for individuals with drug-sensitive TB). The factor $(1-r+qr)$ in (2.73) introduces the effect of incomplete treatment: in fact we assume that a fraction r of the treated individuals with sensitive TB do not recover due to incomplete treatment, and that the remaining fraction $1 - r$ is actually cured and become susceptible again. Moreover, among the individuals who do not finish their treatment, a fraction q of them will develop drug-resistant TB and the remaining fraction will remain latent. Therefore $\gamma(\theta)$ is the sum of the two terms

$$
\begin{aligned}
(1 - r)\chi p(\theta) &= \quad \textit{recovery rate for the treated individuals,} \\
qr\chi p(\theta) &= \quad \textit{rate of developing drug-resistant TB.}
\end{aligned}
\qquad (2.74)
$$

Concerning the resistant class, J, we do not consider its age-structure because these individuals die quickly after acquiring drug-resistant TB. Thus we assume the following constitutive form for the force of infection:

$$\lambda_2(t) = \beta_2 \frac{J(t)}{N(t)}, \qquad (2.75)$$

where β_2 is the transmission rate of resistant strain. We also introduce an additional mortality rate δ for infected individuals of the drug-resistant class with the condition $b(0) > \mu + \delta$, which is necessary to have a sustained population, i.e., when the population size is small, the birth rate needs to exceed the total death rate.

Based on Figure 2.21 the model reads:

$$\frac{dS}{dt} = b(N)N - (\mu + \lambda_1(t) + \lambda_2(t))\,S + (1-r)\chi \int_0^\infty p(\theta)i(\theta, t)d\theta,$$

$$\frac{\partial i}{\partial t} + \frac{\partial i}{\partial \theta} + \mu\, i(\theta, t) + \gamma(\theta)i(\theta, t) = 0,$$

$$\frac{dJ}{dt} = \lambda_2(t)S - (\mu + \delta)J + qr\chi \int_0^\infty p(\theta)i(\theta, t)d\theta,$$

$$i(0,t) = \lambda_1(t)S(t), \quad S(0) = S_0 > 0, \quad i(\theta, 0) = i_0(\theta) \ge 0, \quad J(0) = J_0 > 0.$$

(2.76)

System (2.76) is equivalent to the following system, which is easier to analyze:

$$v(t) = \frac{N(t) - J(t) - \int_0^t K_0(\theta)v(t-\theta)d\theta}{N(t)} - \int_0^t K_1(\theta)v(t-\theta)d\theta + \tilde{F}_1(t),$$

$$\frac{dJ}{dt} = \beta_2 \left(N(t) - J(t) - \int_0^t K_0(\theta)v(t-\theta)d\theta \right) \frac{J(t)}{N(t)}$$

$$\qquad\qquad - mJ(t) + \int_0^t K_2(\theta)v(t-\theta)d\theta + \tilde{F}_2(t),$$

$$\frac{dN}{dt} = b(N)N(t) - \mu N(t) - \delta J(t),$$

(2.77)

where $\tilde{F}_i(t)$ involve parameters and initial condition with $\lim_{t\to\infty} \tilde{F}_i(t) = 0$, $i = 1, 2$.
System (2.77) is the main tool for the stability analysis. Introduce the notation:

$$K_0(\theta) = e^{-\mu\theta - \int_0^\theta \gamma(s)ds},$$

$$K_1(\theta) = \beta_1 p(\theta)K_0(\theta) = -\frac{\beta_1}{\chi(1 - r + qr)} \left(\frac{d}{d\theta}K_0(\theta) + \mu K_0(\theta) \right),$$

$$K_2(\theta) = qr\chi p(\theta)K_0(\theta) = -\frac{qr}{(1 - r + qr)} \left(\frac{d}{d\theta}K_0(\theta) + \mu K_0(\theta) \right),$$

$$\mathcal{K}_i = \int_0^\infty K_i(\theta)d\theta, \quad i = 0, 1, 2.$$

(2.78)

Note that $\mathcal{K}_0 > 0$ and $\mathcal{K}_1 > 0$ while $\mathcal{K}_2$ is positive for $q > 0$ and vanishes at $q = 0$.
The existence and stability of steady states depend on the reproduction numbers

$$\mathcal{R}_1 = \mathcal{K}_1 = \beta_1 \int_0^\infty p(\theta)e^{-\mu\theta - \int_0^\theta \gamma(s)ds}d\theta,$$

and

$$\mathcal{R}_2 = \frac{\beta_2}{\mu + \delta}$$

for the sensitive and resistant strains, respectively. These reproduction number determines the existence and stability of possible steady states.

According to Miller [Miller (1971)], any equilibrium of the system (2.77), if it exists, must be a constant solution of the limiting system associated with (2.77), which is given by the following set of equations:

$$v(t) = \frac{(N(t) - J(t) - K_0 * v)}{N(t)} (K_1 * v),$$

$$\frac{dJ(t)}{dt} = \beta_2 (N(t) - J(t) - K_0 * v) \frac{J(t)}{N(t)} - mJ(t) + K_2 * v, \qquad (2.79)$$

$$\frac{dN(t)}{dt} = b(N)N(t) - \mu N(t) - \delta J(t),$$

where

$$K_i * v = \int_0^\infty K_i(\theta) v(t - \theta) d\theta, \quad i = 0, 1, 2.$$

Thus we look for solutions (v^*, S^*, N^*) of the system

$$v^* = \frac{(N^* - J^* - K_0 v^*)}{N^*} K_1 v^*,$$

$$\beta_2 (N^* - J^* - K_0 v^*) \frac{J^*}{N^*} - mJ^* + K_2 v^* = 0, \qquad (2.80)$$

$$b(N^*)N^* - \mu N^* - \delta J^* = 0.$$

Any solution of this system corresponds to the following steady state for the distribution of infected:

$$i^*(a) = v^* K_0(\theta).$$

The system (2.80) always has the disease-free equilibrium $E_0 = (0, 0, b^{-1}(\mu))$, while existence of non-trivial equilibria will depend on the values of $\mathcal{R}_1$ and $\mathcal{R}_2$. Solving system (2.80) we see that, besides E_0, the following nontrivial equilibria are feasible:

(1) If $\mathcal{R}_1 > 1$ and $q = 0$, then the strain 1-only equilibrium $E_1 = (v_1^*, 0, N_1^*)$ exists where

$$N_1^* = b^{-1}(\mu), \quad v_1^* = \left(1 - \frac{1}{\mathcal{R}_1}\right) \frac{N_1^*}{K_0}.$$

(2) If $\mathcal{R}_2 > 1$, then the strain 2-only equilibrium $E_2 = (0, J_2^*, N_2^*)$ exists where

$$N_2^* = b^{-1}\left(\mu + \delta\left(1 - \frac{1}{\mathcal{R}_2}\right)\right), \quad J_2^* = \left(1 - \frac{1}{\mathcal{R}_2}\right) N_2^*.$$

(3) If $\mathcal{R}_1 > 1$, $q > 0$ and $\mathcal{R}_2 < \mathcal{R}_1$, then the coexistence equilibrium $E_* = (v_*, J_*, N_*)$ exists where

$$N_* = b^{-1}(\mu + \delta \xi K_2), \quad J_* = \xi K_2 N_*, \quad v_* = \xi(\mu + \delta)\left(1 - \frac{\mathcal{R}_2}{\mathcal{R}_1}\right) N_*,$$

with

$$\xi = \frac{(1 - \frac{1}{\mathcal{R}_1})}{K_2 + K_0(\mu + \delta)(1 - \frac{\mathcal{R}_2}{\mathcal{R}_1})}.$$

Note that for the coexistence equilibrium, $\xi \mathcal{K}_2 \leq 1 - \frac{1}{\mathcal{R}_1} \leq 1$, whence $b(0) > \mu + \delta > \mu + \delta \xi \mathcal{K}_2$, and thus $N_* > 0$. For the case of $q = 0$, the boundary equilibria E_i ($i = 1, 2$) exist if and only if $\mathcal{R}_i > 1$. E_* does not exist. For the case of $q > 0$, E_1 exists only if $\mathcal{R}_2 < 1$. When $\mathcal{R}_2 > 1$, E_2 always exists while E_* exists if and only if $\mathcal{R}_1 < \mathcal{R}_2$. The stability properties are summarized below.

Theorem 2.11. *Let E_0, E_1, E_2, and E_* be the equilibria of the system (2.77).*

(a) *if $\mathcal{R}_i < 1$ for $i = 1, 2$, then E_0 is a global attractor, i.e.,*
$$\lim_{t \to \infty} (v(t), J(t), N(t)) \to (0, 0, b^{-1}(\mu)),$$
whereas E_0 is unstable is $\mathcal{R}_1 > 1$ and/or $\mathcal{R}_2 > 1$.

(b) *Assume that $q = 0$. For $i, j = 1, 2$ and $i \neq j$, the boundary equilibrium E_i is stable if $\mathcal{R}_i > 1$ $\mathcal{R}_i > \mathcal{R}_j$ and unstable if $\mathcal{R}_i < \mathcal{R}_j$.*

(c) *Assume that $q > 0$ and $\mathcal{R}_2 > 1$. If $\mathcal{R}_2 > \mathcal{R}_1$ then E_2 is stable, whereas if $\mathcal{R}_2 < \mathcal{R}_1$ then E_2 is unstable, at which time the interior equilibrium E_* is stable.*

The analytic proof of the stability results for the coexistence equilibrium E_* provided in [Feng *et al.* (2002)] required a certain condition of parameters. However, numerical simulations confirmed that the results hold for other cases as well. For the birth function $b(N)$ we choose the logistic form
$$b(N) = b_0 \left(1 - \frac{N}{L}\right)$$
where $b_0 = 2$ and $L = 6 \cdot 10^6$. For the numerical computations, we consider an explicit discretization of problem (2.76), based on backward Euler finite differences for the ODEs, a linearized finite difference method of characteristics for the PDE, and Simpson's rule for the quadratures. Let T be the final time of simulation and let h be the discretization step. Define $M_1 = \frac{\sup\{\theta : i_0(\theta) > 0\}}{h}$ and $M_2 = \frac{T}{h}$. It will be assumed, without loss of generality, that M_1 and M_2 are positive integers. Our numerical method, for $1 \leq n \leq M_2$, $0 \leq j \leq M_1 + n$, is given by

$$
\begin{cases}
\frac{i_j^n - i_{j-1}^{n-1}}{h} + [\mu + \gamma(jh)]\, i_j^n = 0, \\[2mm]
i_0^n = \lambda^{n-1}\left(N^{n-1} - I^{n-1} - J^{n-1}\right), \\[2mm]
\frac{N^n - N^{n-1}}{h} = b_0\left(1 - \frac{N^{n-1}}{l}\right) N^n - \mu N^n - \delta J^{n-1} - \nu A^{n-1}, \\[2mm]
I^n = \frac{h}{3} \sum (i_j^n + 4 i_{j+1}^n + i_{j+2}^n), \\[2mm]
A^n = \frac{h}{3} \sum \left[p(jh) i_j^n + 4 p(jh + h) i_{j+1}^n + p(jh + 2h) i_{j+2}^n \right], \\[2mm]
\lambda^n = \frac{h}{3} \frac{\beta_1}{N^n} \sum \left[p(jh) i_j^n + 4 p(jh + h) i_{j+1}^n + p(jh + 2h) i_{j+2}^n \right], \\[2mm]
\frac{J^n - J^{n-1}}{h} = \beta_2 \frac{J^{n-1}}{N^n}(N^n - I^n - J^n) - (\mu + \delta) J^n + q r \chi A^n,
\end{cases}
$$

where we have used the symbols i_j^n, N^n, J^n, I^n, λ^n to denote, respectively, the approximations of $i(jh, nh)$, $N(nh)$, $J(nh)$, $I(nh)$, $\lambda_1(nh)$, for $j, n \geq 0$.

We explore the behavior of the solution for a fairly realistic set of parameter values including: $\mu = 0.014, \chi = 2, r = 0.5, \beta_1 = 7, \beta_2 = 7, \nu = 0.14, \delta = 1.8$. Furthermore, following Figure 2.19, we take the function $p(\theta)$ as piecewise constant, in the following specific form

$$
p(\theta) = \begin{cases}
0.06, & \theta \in [0, 1), \\
0.084, & \theta \in [1, 2), \\
0.093, & \theta \in [2, 3), \\
0.097, & \theta \in [3, 4), \\
0.099, & \theta \in [5, 6), \\
0.1, & \theta \in [6, \infty),
\end{cases} \tag{2.81}
$$

where θ is measured in *years*.

The outcomes of the numerical simulations are consistent with the results described in Theorem 2.11.

2.5 Models with non-exponentially distributed stages durations

The results presented in this section are adopted from [Feng *et al.* (2007)]. Many of the continuous-time SEIR types of models have taken the approach of using simple ODE models to draw conclusions regarding the effectiveness of various disease control programs. The simplicity of these models is often achieved by making the assumption that the disease stages are exponentially distributed. When the models do not include quarantine and/or isolation, or when the isolation is assumed to be perfect (i.e., isolated individuals do not transmit the disease), the exponential distribution assumption (EDA) and the models that use this assumption have been shown to provide valuable information and important insights into the disease dynamics. However, as demonstrated in this study, the EDA may not be appropriate in models for diseases with relatively long incubation and/or infectious periods when isolation is not completely effective.

Here, we discuss the problem by considering a simple ODE model that is a commonly used *SEIR* type model when quarantine and isolation are included. Assume again that latent individuals, who have not been quarantined yet, progress to the infectious stage at a constant rate α_1 and infectious individuals recover at a constant rate δ_1.

To incorporate control measures such as quarantine and isolation, additional sub-classes can be included. For example, quarantine and isolation can be modeled as follows. Let $\lambda(t)$ denote the force of infection (a specific form is given below). Assume that a fraction b of contacts (susceptible individuals who have had contacts with an infectious person) is actually infected, and that the other fraction $(1-b)$ of contacts remains susceptible who will be quarantined (S_Q) and will return to the S class at a rate r (see, [Lipsitch *et al.* (2003)]). A fraction of the exposed (and infected) individuals (E) who are not quarantined at the time of infection will be quarantined at a constant rate χ throughout the incubation period. (We remark that although for most diseases quarantine is not considered as the exposed individuals show no disease symptoms, the situation for SARS is different in which quarantine was implemented in several places including Hong Kong, Taiwan, and China.) The non-quarantined and quarantined (exposed) individuals will progress to the infectious stage at constant rates α_1 and α_2 respectively (the relationship between α_1 and α_2 will be discussed later). Infectious individuals will be isolated (H) at a rate ϕ and individuals in the H class will recover at a rate δ_2 (the relationship between δ_1 and δ_2 will be discussed later). Because we are considering the case of imperfect isolation, the new infections are now produced at the rate $\lambda(t)S(t)$ with

$$\lambda(t) = \beta \left[\frac{I(t) + (1-\rho)H(t)}{N} \right], \tag{2.82}$$

where $\rho \in [0,1]$ is the fraction of reduction in the transmission rate of isolated individuals with $\rho = 1$, $\rho = 0$, and $0 < \rho < 1$ representing a completely effective, completely ineffective, and partially effective isolation, respectively. The corresponding transmission diagram is shown in Figure 2.22.

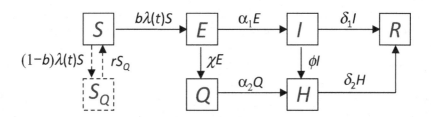

Fig. 2.22   Disease transmission diagrams (birth and death omitted) with quarantine and isolation.

In this study, we consider only the case that most infected people are quarantined during the latency period. In addition, we assume a large population size N (in comparison with the size of the infected population), in which case the quarantine of susceptibles is unlikely to have a significant impact on the disease transmission dynamics and hence will be ignored. This is equivalent to assuming that $b = 1$. (It may not be appropriate to ignore the S_Q class if one is concerned with the cost associated with quarantine, which is not the case in this study.) For simplicity the disease-induced death is ignored and the per-capita birth rate and the natural death rate are assumed equal. Hence, the total population size N remains constant. Then the corresponding ODE model is given by the following system

$$S' = \mu N - \beta S \frac{I + (1 - \rho)H}{N} - \mu S,$$

$$E' = \beta S \frac{I + (1 - \rho)H}{N} - (\chi + \alpha_1 + \mu)E,$$

$$Q' = \chi E - (\alpha_2 + \mu)Q, \tag{2.83}$$

$$I' = \alpha_1 E - (\phi + \delta_1 + \mu)I,$$

$$H' = \alpha_2 Q + \phi I - (\delta_2 + \mu)H,$$

$$R' = \delta_1 I + \delta_2 H - \mu R.$$

" $'$ " denotes the derivative with respect to time t.

Here we point out one of the drawbacks of the simple model (2.83) when used to evaluate intervention policies. We argue that the main reason for these problems is due to the simplifying assumption of exponential distributions for the disease stages. This provides a motivation of using more realistic stage distributions. Non-exponential distributions have been considered in epidemiological models (see, for example, [Hethcote and Tudor (1980); Hethcote *et al.* (1981); Lloyd (2001b,a); Plant and Wilson (1986); Taylor and Karlin (1998); Feng *et al.* (2001); Feng and Thieme (2000a,b)]). However, none of these studies focuses on the evaluation of intervention policies.

In this study, we develop a general model with arbitrarily distributed disease stages. The general setting allows us to identify new models that are improvements to the simple model (2.83) while keeping the improved models as simple as possible. We show that in the case of exponential distributions the general model reduces to the simple model (2.83) with appropriate constraints on model parameters. We also consider a particular non-exponential stage distribution, the gamma distribution, in which case the general model reduces to another ODE model. Analysis for both the general model and the model with the gamma distribution assumption (GDA) are provided. We demonstrate that the model under GDA is indeed an improvement on the model under EDA.

In the model (2.83), relevant parameters that represent disease intervention are χ, ϕ, ρ, α_2 and δ_2. It is very important that these parameters have well-defined meanings to connect them with epidemiological data and to determine their appropriate values, and to ensure that the model predictions are reasonable regarding the effect of various control strategies. Otherwise, the results obtained from the model might be misleading, as demonstrated below.

One of the quantities that can be used to assess the impact of various control measures is the cumulative number, $C(t)$, of infections determined by the equation

$$C'(t) = \lambda(t)S(t).$$

Let $C(0) = 0$ so that $C(t)$ is the cumulative number of new infections at the end of an epidemic (in the case that the disease is driven into extinction). One would expect that the C value will be reduced if we increase the value of any of the control parameters. However, Figure 2.23 (see (a) and (b)) shows that C increases with increasing rates of quarantine (χ) and isolation (ϕ). The parameter values used in Figure 2.23 are the following: $\alpha_1 = 0.2$ and $\delta_1 = 0.15$, which correspond to a latency period of $1/\alpha_1 = 5$ days and an infectious period of $1/\delta_1 \approx 7$ days, respectively. These values are in the realistic range of many infectious diseases. The transmission coefficient is chosen to be $\beta = 0.13$, which corresponds to a reproduction number equal to approximately 0.9 (so that the disease will die out). The isolation efficiency is $\rho = 0.3$ and other parameters have different values in Figure 2.23 (a)-(d) depending on the assumptions. In Figure 2.23 (a) (b), no additional constraints are imposed on α_2 and δ_2, which have values 0.17 and 0.1 respectively.

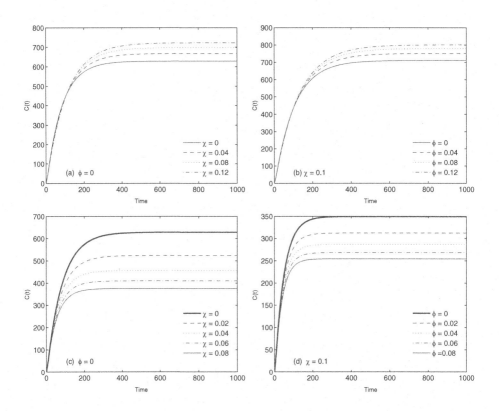

Fig. 2.23   Numerical simulations of the model (2.83). The number of cumulative new infections $C(t)$ is plotted for various values of the control parameters χ and ϕ. (a) and (b) are for the case of no constraints on the parameter values. (c) and (d) are for the case when constraint (2.84) is used.

The lack of constraints on α_2 and δ_2 may be responsible for the problem shown in Figure 2.23 (see (a) and (b)). Our simulation results show that the problem can be avoided if the following constraints are imposed

$$\alpha_2 = \alpha_1, \quad \delta_2 = \delta_1. \tag{2.84}$$

A more rigorous argument for constraint (2.84) will be provided in section 2.5.2. Here, we give only a heuristic argument. The ordinary differential equation model (2.83) implicitly assumes the exponential distribution for the incubation and infectious stages. More precisely, the exponential functions

$$p_E(s) = e^{-\alpha_1 s} \quad \text{and} \quad p_I(s) = e^{-\delta_1 s}$$

have been used to describe the probability of remaining in the incubation stage and the infectious stage, respectively, and the mean durations of incubation and infectious stages are

$$T_E = \int_0^\infty p_E(s)ds = \frac{1}{\alpha_1}$$
$$\tag{2.85}$$
$$T_I = \int_0^\infty p_I(s)ds = \frac{1}{\delta_1}.$$

Similarly, the mean sojourn times in the Q and H classes are respectively

$$T_Q = \frac{1}{\alpha_2} \quad \text{and} \quad T_H = \frac{1}{\delta_2}. \tag{2.86}$$

A fundamental property of the exponential distribution is the memory-less property, which requires that the remaining expected sojourn in the H (or Q) class is independent of the time already elapsed before entering it. This property implies that

$$T_H = T_I \quad \text{and} \quad T_Q = T_E,$$

which is equivalent to the condition given in (2.84) (see (2.85) and (2.86)).

Another argument for the use of (2.84) is the following. The average time individuals (both isolated and non-isolated individuals) stay in the I class is equal to $1/(\delta_1+\phi)$, and the average time an isolated individual stays in the H class is $1/\delta_2$. Notice that $\phi/(\delta_1 + \phi)$ and $\delta_1/(\delta_1 + \phi)$ are fractions of isolated and non-isolated individuals, respectively. Then the weighted average time an individual stays in the infectious stage is

$$\frac{\phi}{\delta_1 + \phi}\left(\frac{1}{\delta_1 + \phi} + \frac{1}{\delta_2}\right) + \frac{\delta_1}{\delta_1 + \phi}\left(\frac{1}{\delta_1 + \phi}\right), \tag{2.87}$$

which is equal to $\frac{1}{\delta_1}$ (i.e., the infectious period) only if we set $\delta_2 = \delta_1$ in accordance with constraint (2.84). It follows from a similar argument that the weighted average time an individual stays in the exposed stage (when quarantine is present) is equal to the latency period only if $\alpha_2 = \alpha_1$. Figures 2.23(c) and 2.23(d) illustrate that the value of $C(t)$ reduces as the values of control parameters increase, showing the improvement compared to Figures 2.23(a) and 2.23(b). For Figures 2.23(c) and (d) all parameter values are the same as in Figures 2.23(a) and (b) except that the constraint (2.84) holds.

Constraint (2.84) seems to provide a partial solution to the problem exhibited in Figures 2.23(a) and 2.23(b). However, it creates a different problem. Because the isolated or quarantined individuals have already spent some time in the infectious or incubation stage before entering the H or the Q class, the condition (2.84) amounts to allowing for a prolonged period of infectiousness for isolated individuals and a prolonged period of latency for quarantined individuals. In reality, if an infectious person already spent some time in the I class before being isolated, then the expected remaining sojourn in the H class should be shorter than the infectious period. Therefore, the model assumption (EDA) conflicts with biological constraints. A similar argument applies to quarantined individuals. In section 2.5.2, we demonstrate how the predictions of the model (2.83) constrained by (2.84) may be in disagreement with that of models using more realistic stage distributions.

It should be pointed out that the purpose of this study is not to argue which assumptions/constraints are more appropriate than others, or whether or not they are correct. The goal is to illustrate the weakness of models that assume exponentially distributed disease stages and to demonstrate possible problems with constraints of the type (2.84), or no constraint at all on parameters. Therefore, models with more realistic stage distributions may need be considered.

Epidemiological models with non-exponential distributions such as the gamma distribution have been previously studied (see, for example, [Hethcote and Tudor (1980); Plant and Wilson (1986); Taylor and Karlin (1998); Lloyd (2001b,a)]). In these studies the authors discussed other objections to the EDA. For example, it is pointed out that constant recovery is a poor description of real-world infections, and they show that in models with more realistic distributions of disease stages less stable behavior may be expected and disease persistence may be diminished [Hethcote *et al.* (1981); Lloyd (2001b,a)]). However, these studies do not focus on the impact of imperfect isolation. In the rest of this section we consider models with more realistic disease stage distributions and study the properties of these models.

2.5.1 *A model with general distributions for stage durations*

Consider an infectious disease for which quarantine and isolation can be effective control measures. As an extension of the SEIR model (2.4), introduce two additional classes, quarantined (Q) and isolated or hospitalized (H). Let $P_E, P_I : [0, \infty) \rightarrow [0, 1]$ describe the durations of the incubation and infectious stages, respectively. More precisely, $P_i(s)$ ($i = E, I$) gives the probability that the disease stage i lasts longer than s time units (or the probability of being still in the same stage at stage age s). Then, the derivative $-P_i'(s)$ ($i = E, I$) gives the rate of removal from the stage i at stage age s by the natural progression of the disease. These duration functions have the following properties

$$P_i(0) = 1, \quad P_i'(s) \le 0, \quad \int_0^\infty P_i(s)ds < \infty, \quad i = E, I.$$

For example, in the special case of exponentially distributed incubation and infectious periods with mean stage durations α and δ, $\tilde{P}_E(s) = e^{-\alpha s}$ and $\tilde{P}_I(s) = e^{-\delta s}$. Let $k(s), l(s) : [0, \infty) \to [0, 1]$ denote respectively the probabilities that exposed, infectious individuals have not been quarantined, isolated at stage age s. Hence, $1 - k(s) =: \bar{k}(s)$, $1 - l(s) =: \bar{l}(s)$ give the respective probabilities of being quarantined, isolated before reaching stage age s. Assume that $k(0) = l(0) = 1$, $k'(s) \le 0$ and $l'(s) \le 0$.

When the survivals from quarantine and isolation are described by the exponential functions $k(s) = e^{-\chi s}$ and $l(s) = e^{-\phi s}$ with χ and ϕ being constants, the model can be described by the following equations:

$$S(t) = \int_0^t \mu N e^{-\mu(t-s)} ds - \int_0^t \lambda(s) S(s) e^{-\mu(t-s)} ds + S_0 e^{-\mu t},$$

$$E(t) = \int_0^t \lambda(s) S(s) P_E(t-s) k(t-s) e^{-\mu(t-s)} ds + E_0(t) e^{-\mu t},$$

$$Q(t) = \int_0^t \int_0^\tau \lambda(s) S(s) [-P_E(\tau-s) k'(\tau-s)] P_E(t-\tau|\tau-s) e^{-\mu(t-s)} ds d\tau + \tilde{Q}(t),$$

$$I(t) = \int_0^t \int_0^\tau \lambda(s) S(s) [-P_E'(\tau-s) k(\tau-s)] P_I(t-\tau) l(t-\tau) e^{-\mu(t-s)} ds d\tau + \tilde{I}(t),$$

$$H(t) = \int_0^t \int_0^u \int_0^\tau \lambda(s) S(s) [-P_E'(\tau-s) k(\tau-s)][-P_I(u-\tau) l'(u-\tau)]$$
$$\times P_I(t-u|u-\tau) e^{-\mu(t-s)} ds d\tau du$$
$$+ \int_0^t \int_0^\tau \lambda(s) S(s) [-P_E'(\tau-s) \bar{k}(\tau-s)] P_I(t-\tau) e^{-\mu(t-s)} ds d\tau + \tilde{H}(t),$$

$$R(t) = \int_0^t \int_0^\tau \lambda(s) S(s) [-P_E'(\tau-s)][1 - P_I(t-\tau)] e^{-\mu(t-s)} ds d\tau + \tilde{R}(t),$$

with $\lambda(t) = \beta \dfrac{I(t) + (1-\rho) H(t)}{N}$,

(2.88)

where $\tilde{X}(t) = X_0(t) e^{-\mu t} + \tilde{X}_0(t)$ ($X = Q, I, H, R$) denote initial conditions with $\tilde{X}(t) \to 0$ as $t \to \infty$ (see [Feng et al. (2007)] for more detailed description of the equations as well as the existence and uniqueness of the solutions). Because we are not focusing on vital dynamics, we use the simplest function $e^{-\mu t}$ for the probability of survival from the natural death. Let the numbers of initial susceptible and removed individuals be $S_0 > 0$ and $R_0 > 0$ respectively. Let $E_0(t) e^{-\mu t}$, $I_0(t) e^{-\mu t}$, $Q_0(t) e^{-\mu t}$, and $H_0(t) e^{-\mu t}$ be the non-increasing functions that represent the numbers of individuals that were initially exposed, infectious, quarantined, isolated, respectively, and are still alive and in the respective classes at time t. $P_E(w|a) = \frac{P_E(w+a)}{P_E(a)}$ denote the conditional remaining function that gives the probability that an exposed individual remains non-infectious for w time units longer given that the person was already exposed for a units of time. Individuals in the H class are from two sources, one from the I class via isolation at the rate $-P_I l'(t)$, and the other from the Q class via disease progression at the rate $-P_E' \bar{k}(t)$. $P_I(w|a) = \frac{P_I(w+a)}{P_I(a)}$ is the conditional remaining function as $P_E(w|a)$. All variables and parameters are listed in Table 2.4.

Table 2.4   Definitions of symbols related to the continuous-time model (2.88).

Symbol	Definition
$S(t)$	Number of susceptible individuals at time t
$E(t)$	Number of exposed (not yet infectious) individuals at time t
$Q(t)$	Number of quarantined (exposed) individuals at time t
$I(t)$	Number of susceptible individuals at time t
$H(t)$	Number of isolated (infectious) individuals at time t
$R(t)$	Number of recovered individuals at time t
$C(t)$	Number of cumulative new infections at time t
$\lambda(t)$	Force of infection at time t
β	Transmission coefficient
$\alpha_1,\ \alpha_2$	Rate at which non-quarantined, quarantined individuals become infectious
α	Same as α_1
$\delta_1,\ \delta_2$	Rate at which non-isolated, isolated individuals become recovered
δ	Same as δ_1
$\chi,\ \phi$	Rate of quarantine, isolation
ρ	Isolation efficiency ($0 \leq \rho \leq 1$)
$p_i(s),\ P_i(s)$	Probability that disease stage i lasts longer than s time units $(i = E, I)$
$k(s),\ l(s)$	Probability of not being quarantined, isolated at stage age s
$T_E,\ T_I$	Mean of $p_E(s) = e^{-\alpha s}$, $p_I(s) = e^{-\delta s}$ $(T_E = 1/\alpha,\ T_I = 1/\delta)$
$\mathcal{M}_i(s),\ M$	Expected remaining sojourn at age s: $\displaystyle\int_0^\infty P_i(t\lvert s)dt\ (i = E, I),\ M = \mathcal{M}(0)$
$\mathcal{T}_E$	Probability of surviving and becoming infectious: $\displaystyle\int_0^\infty [-P_E'(s)]e^{-\mu s}dt$
$\mathcal{T}_{E_k}$	"Quarantine-adjusted" probability (similar to $\mathcal{T}_E$): $\displaystyle\int_0^\infty [-P_E'(s)k(s)]e^{-\mu s}dt$
$\mathcal{T}_I$	Probability an infectious person survives and recovers: $\displaystyle\int_0^\infty [-P_I'(s)]e^{-\mu s}dt$
$\mathcal{T}_{I_l}$	"Isolation-adjusted" probability (similar to $\mathcal{T}_I$): $\displaystyle\int_0^\infty [-P_I'(s)l(s)]e^{-\mu s}dt$
$\mathcal{D}_E$	Mean time in exposed stage (adjusted by death): $\displaystyle\int_0^\infty P_E(s)e^{-\mu s}dt$
$\mathcal{D}_{E_k}$	"Quarantine-adjusted" mean time in exposed stage: $\displaystyle\int_0^\infty P_E(s)k(s)e^{-\mu s}dt$
$\mathcal{D}_I$	Mean time in infectious stage (also adjusted by death): $\displaystyle\int_0^\infty P_I(s)e^{-\mu s}dt$
$\mathcal{D}_{I_l}$	"Isolation-adjusted" mean time in infectious stage: $\displaystyle\int_0^\infty P_I(s)l(s)e^{-\mu s}dt$
$\mathcal{R}_0$	The basic reproduction number
$\mathcal{R}_c$	The reproduction number under control measures
EDA, GDA	Exponential distribution assumption, Gamma distribution assumption
EDM, GDM	Exponential distribution model, Gamma distribution model

The mathematical analysis of the system of integral equations (2.88) is much more difficult than that of systems such as (2.1) and (2.4). Nonetheless, the qualitative behaviors of (2.88) are the same as those of (2.1) and (2.4). That is, the dynamics are completely determined by the reproduction numbers.

Let $\mathcal{R}_c$ denote the control reproduction number. For ease of biological interpretations, introduce the following quantities:

$$\mathcal{T}_E = \int_0^\infty [-P_E'(s)]e^{-\mu s}ds, \quad \mathcal{T}_{E_k} = \int_0^\infty [-P_E'(s)k(s)]e^{-\mu s}ds,$$

$$\mathcal{T}_I = \int_0^\infty [-P_I'(s)]e^{-\mu s}ds, \quad \mathcal{T}_{I_l} = \int_0^\infty [-P_I'(s)l(s)]e^{-\mu s}ds,$$

$$\mathcal{D}_E = \int_0^\infty P_E(s)e^{-\mu s}ds, \quad \mathcal{D}_{E_k} = \int_0^\infty P_E(s)k(s)e^{-\mu s}ds,$$

$$\mathcal{D}_I = \int_0^\infty P_I(s)e^{-\mu s}ds, \quad \mathcal{D}_{I_l} = \int_0^\infty P_I(s)l(s)e^{-\mu s}ds.$$

(2.89)

$\mathcal{T}_E$ and $\mathcal{T}_{E_k}$ represent respectively the probability and the "quarantine-adjusted" probability that exposed individuals survive and become infectious. $\mathcal{T}_I$ and $\mathcal{T}_{I_l}$ represent respectively the probability and the "isolation-adjusted" probability that infectious individuals survive and become recovered. $\mathcal{D}_E$ and $\mathcal{D}_{E_k}$ represent respectively the mean sojourn time (death-adjusted) and the "quarantine-adjusted" mean sojourn time (death-adjusted as well) in the exposed stage. $\mathcal{D}_I$ and $\mathcal{D}_{I_l}$ represent respectively the mean sojourn time (death-adjusted) and the "isolation-adjusted" mean sojourn time (death-adjusted as well) in the infectious stage. The quantities in (2.89) can be used to express the formula for $\mathcal{R}_c$:

$$\mathcal{R}_c = \mathcal{R}_I + \mathcal{R}_{IH} + \mathcal{R}_{QH},$$

(2.90)

where

$$\mathcal{R}_I = \beta \mathcal{T}_{E_k} \mathcal{D}_{I_l},$$

$$\mathcal{R}_{IH} = (1-\rho)\beta \mathcal{T}_{E_k}(\mathcal{D}_I - \mathcal{D}_{I_l}),$$

(2.91)

$$\mathcal{R}_{QH} = (1-\rho)\beta(\mathcal{T}_E - \mathcal{T}_{E_k})\mathcal{D}_I.$$

The three components, $\mathcal{R}_I, \mathcal{R}_{IH}, \mathcal{R}_{QH}$ in $\mathcal{R}_c$ represent contributions from the I class and from the H class through isolation and quarantine, respectively, From $0 \leq k, l \leq 1$ we know that

$$\mathcal{T}_{E_k} \leq \mathcal{T}_E, \qquad \mathcal{D}_{I_l} \leq \mathcal{D}_I.$$

(2.92)

Hence, $\mathcal{R}_{IH}$ and $\mathcal{R}_{QH}$ are both positive. Clearly, each $\mathcal{R}_i$ $(i = I, IH, QH)$ is a product of the transmission rate β (or $(1-\rho)\beta$), the probability of surviving the exposed stage and entering the infectious stage, and the average sojourn time being infectious in the corresponding class (adjusted by natural death).

In the absence of control, i.e., $k(s) = l(s) \equiv 1$, $\mathcal{R}_c$ gives the basic reproductive number:

$$\mathcal{R}_0 = \beta \int_0^\infty [-P_E'(s)]e^{-\mu s}ds \int_0^\infty P_I(s)e^{-\mu s}ds = \beta \mathcal{T}_E \mathcal{D}_I. \tag{2.93}$$

We can express $\mathcal{R}_c$ in terms of $\mathcal{R}_0$ as follows. Notice from (2.91) that $\mathcal{R}_c$ can be simplified to $(1 - \rho)\beta \mathcal{T}_E \mathcal{D}_I + \rho \beta \mathcal{T}_{E_k} \mathcal{D}_{I_l}$. Hence, from (2.90) we get

$$\mathcal{R}_c = \mathcal{R}_0 \left[1 - \rho\left(1 - \frac{\mathcal{T}_{E_k}\mathcal{D}_{I_l}}{\mathcal{T}_E\mathcal{D}_I}\right)\right]. \tag{2.94}$$

From (2.92) it is easy to see that $\mathcal{R}_c < \mathcal{R}_0$. The impact of various (single or combined) control measures represented by ρ, χ, or ϕ on the reduction of $\mathcal{R}_0$ can be evaluated using (2.94).

System (2.88) always has DFE $U_1 = (S_1, E_1, Q_1, I_1, H_1, R_1) = (N, 0, 0, 0, 0, 0)$. If $\mathcal{R}_c > 1$, then there is a unique endemic equilibrium

$$U^* = (S^*, E^*, Q^*, I^*, H^*, R^*) \tag{2.95}$$

with

$$S^* = \frac{N}{\mathcal{R}_c}, \quad E^* = \mathcal{D}_{E_k}\lambda^* S^*, \quad Q^* = (\mathcal{D}_E - \mathcal{D}_{E_k})\lambda^* S^*, \quad I^* = \mathcal{T}_{E_k}\mathcal{D}_{I_k}\lambda^* S^*,$$

$$H^* = (\mathcal{T}_E\mathcal{D}_I - \mathcal{T}_{E_k}\mathcal{D}_{I_l})\lambda^* S^*, \quad R^* = \frac{1}{\mu}\mathcal{T}_E\mathcal{T}_I\lambda^* S^*, \quad \lambda^* = \mu(\mathcal{R}_c - 1).$$

As in many disease models, the DFE U_1 is a global attractor if $\mathcal{R}_c < 1$.

Let

$$A(\tau) = a_1(\tau) + (1 - \rho)\left[a_2(\tau) + a_3(\tau)\right],$$

where

$$a_1(\tau) = e^{-\mu\tau} \int_0^\tau [-P_E'(\tau - u)k(\tau - u)]P_I(u)l(u)du,$$

$$a_2(\tau) = e^{-\mu\tau} \int_0^\tau [-P_E'(\tau - u)k(\tau - u)]P_I(u)\bar{l}(u)du, \tag{2.96}$$

$$a_3(\tau) = e^{-\mu\tau} \int_0^\tau [-P_E'(\tau - u)\bar{k}(\tau - u)]P_I(u)du,$$

with $\bar{k}(s) = 1 - k(s)$, $\bar{l}(s) = 1 - l(s)$. Then the reproductive number $\mathcal{R}_c$ can be written as

$$\mathcal{R}_c = \beta \int_0^\infty A(\tau)d\tau, \tag{2.97}$$

and the I and H equations (for large t) can be rewritten as

$$I = \int_0^t \lambda(s)S(s) \int_0^{t-s} [-P_E'(t - s - u)k(t - s - u)]P_I(u)l(u)e^{-\mu(t-s)}duds$$

$$\tag{2.98}$$

$$= \int_0^t \lambda(s)S(s)a_1(t - s)ds,$$

and

$$H = \int_0^t \lambda(s)S(s) \int_0^{t-s} [-P_E'(t-s-u)k(t-s-u)]P_I(u)\bar{l}(u)e^{-\mu(t-s)}duds$$

$$+ \int_0^t \lambda(s)S(s) \int_0^{t-s} [-P_E'(t-s-u)\bar{k}(t-s-u)]P_I(u)e^{-\mu(t-s)}duds \qquad (2.99)$$

$$= \int_0^t \lambda(s)S(s) [a_2(t-s) + a_3(t-s)] ds.$$

The following observation allows us to study the disease persistence by considering only two equations. Again, to study the large time behavior of solutions we ignore those individuals who are initially in the population. From equations (2.96), (2.98) and (2.99) we have

$$\lambda(t) = \frac{\beta}{N} \int_0^t \lambda(s)S(s)A(t-s)ds. \qquad (2.100)$$

Integrating the S equation in (2.88) we obtain

$$S(t) = N - \int_0^t \lambda(s)S(s)e^{-\mu(t-s)}ds. \qquad (2.101)$$

We observe that equations (2.100) and (2.101) can be studied independently of other variables, and that the behavior of $\lambda(t)$ determines whether or not the disease will die out. Let (S^*, λ^*) denote an equilibrium of (2.100) and (2.101). Then S^* and λ^* are constant solutions of the equations

$$S(t) = N - \int_{-\infty}^t \lambda(t)S(t)e^{-\mu(t-s)}ds,$$

$$\lambda(t) = \frac{\beta}{N} \int_{-\infty}^t \lambda(t)S(t)A(t-s)ds,$$

or satisfy the equations

$$S^* = N - \int_{-\infty}^0 \lambda^* S^* e^{\mu u} du,$$

$$\qquad (2.102)$$

$$\lambda^* = \frac{\beta}{N} \int_{-\infty}^0 \lambda^* S^* A(-u)du.$$

Equations (2.102) have two solutions (equilibria)

$$(S_1^*, \lambda_1^*) = (N, 0), \qquad (S_2^*, \lambda_2^*) = (N/\mathcal{R}_c, \mu(\mathcal{R}_c - 1)).$$

Clearly, the non-trivial equilibrium (S_2^*, λ_2^*) is feasible only if $\mathcal{R}_c > 1$. It is shown in [Feng et al. (2007)] that if $\mathcal{R}_c > 1$, then the DFE (S_1^*, λ_1^*) is unstable and the endemic equilibrium (S_2^*, λ_2^*) is l.a.s.

The stability results for the system (2.88) are summarized in the following theorem.

Theorem 2.12. *The DFE U_1 is a global attractor if $\mathcal{R}_c < 1$. If $\mathcal{R}_c > 1$ then U_1 is unstable and the endemic equilibrium U^* given by (2.95) is l.a.s.*

2.5.2   *The cases of gamma and exponential distributions*

The integral formulation of the model (2.88) enables the comparison between models that differ in their distributions of disease stages, as well as the examination of their consequent epidemiological implications. In this section we apply the results obtained for the general model (2.88) to simplified models that use specific, non-exponential disease stage distributions such as the gamma distribution.

One possibility to replace the exponential stage duration function $p(s) = e^{-\theta s}$, or equivalently the probability density function (PDF) $f(s) = \theta e^{-\theta s}$, is to consider the gamma distribution with parameter θ for which the stage duration function is $p_n(s, \theta) = e^{-n\theta s} \sum_{k=0}^{n-1} \frac{(n\theta s)^k}{k!}$ where $n \geq 1$. We remark that the exponential distribution is a special case of the gamma distribution when $n = 1$. To see the role of n, the graph of the PDF, $f_n(s)$, is shown in Figure 2.24 for $n = 1, 3, 20$. Notice that when $n \to \infty$ it corresponds to a fixed duration. The appropriate value of n may be determined by epidemiological data.

Under the gamma distribution $p_n(s, \theta)$ (or simply denoted by $p_n(s)$) with $n \geq 2$, the expected remaining sojourn at stage age s is

$$\mathcal{M}_n(s) = \int_0^\infty \frac{p_n(t+s)}{p_n(s)} dt = \frac{1}{p_n(s)} \int_s^\infty p_n(t) dt = \frac{1}{n\theta} \frac{\sum_{k=0}^{n-1} \sum_{j=0}^{k} \frac{(n\theta s)^j}{j!}}{\sum_{k=0}^{n-1} \frac{(n\theta s)^k}{k!}}.$$

After checking $\mathcal{M}_n'(s) < 0$ and $\lim_{s \to \infty} \mathcal{M}_n(s) \to T/n$ where $T = 1/\theta$, we know that $\mathcal{M}_n(s)$ strictly decreases with stage age s, and that when s is large the expected remaining sojourn can be as small as T/n. Hence, the expected remaining sojourn in a stage is indeed dependent on the time already spent in the stage. Therefore, the gamma distribution $p_n(s)$ for $n \geq 2$ provides a more realistic description than the exponential distribution $p_1(s)$ for which $\mathcal{M}_1(s) = T$ for all s.

In the special case when P_E and P_I are both exponential with means $1/\alpha$ and $1/\delta$, respectively, the system (2.88) reduces to the following EDM

$$\begin{aligned}
S' &= \mu N - \beta S \frac{I + (1-\rho)H}{N} - \mu S, \\
E' &= \beta S \frac{I + (1-\rho)H}{N} - (\chi + \alpha + \mu)E, \\
Q' &= \chi E - (\alpha + \mu)Q, \\
I' &= \alpha E - (\phi + \delta + \mu)I, \\
H' &= \alpha Q + \phi I - (\delta + \mu)H, \\
R' &= \delta I + \delta H - \mu R.
\end{aligned} \tag{2.103}$$

For the EDM (2.103), the expressions for the control reproductive number ($\mathcal{R}_{c,e}$) and basic reproductive number ($\mathcal{R}_{0,e}$) can be obtained by applying the formulas for $\mathcal{R}_c$ (2.90) and $\mathcal{R}_0$ (2.93), respectively:

$$\mathcal{R}_{c,e} = \frac{\alpha}{\mu + \alpha} \frac{\beta}{\mu + \delta} \left[1 - \rho \left(1 - \frac{\mu + \alpha}{\mu + \alpha + \chi} \frac{\mu + \delta}{\mu + \delta + \phi} \right) \right] \tag{2.104}$$

and

$$\mathcal{R}_{0,e} = \frac{\alpha}{\mu + \alpha} \frac{\beta}{\mu + \delta}. \tag{2.105}$$

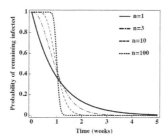 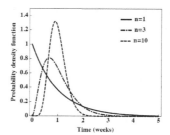

Fig. 2.24 Probability survival and density functions of Gamma distributions.

If P_E and P_I are both gamma distributions with the duration functions for the incubation and infectious stages being $P_E(s) = p_m(s, \alpha)$ and $P_I(s) = p_n(s, \delta)$, respectively, where

$$p_j(s, \theta) = e^{-j\theta s} \sum_{k=0}^{j-1} \frac{(j\theta s)^k}{k!}, \tag{2.106}$$

then with the functions $k(s)$ and $l(s)$ given as before (i.e., $k(s) = e^{-\chi s}$ and $l(s) = e^{-\phi s}$), the system (2.88) simplifies to the following GDM:

$$S' = \mu N - \beta S \frac{I + (1-\rho)H}{N} - \mu S,$$

$$E_1' = \beta S \frac{I + (1-\rho)H}{N} - (\chi + m\alpha + \mu)E_1,$$

$$E_j' = m\alpha E_{j-1} - (\chi + m\alpha + \mu)E_j, \qquad j = 2, \cdots, m,$$

$$Q_1' = \chi E_1 - (m\alpha + \mu)Q_1,$$

$$Q_j' = \chi E_j + m\alpha Q_{j-1} - (m\alpha + \mu)Q_j, \qquad j = 2, \cdots, m,$$

$$I_1' = m\alpha E_m - (\phi + n\delta + \mu)I_1, \tag{2.107}$$

$$I_j' = n\delta I_{j-1} - (\phi + n\delta + \mu)I_j, \qquad j = 2, \cdots, n,$$

$$H_1' = m\alpha Q_m + \phi I_1 - (n\delta + \mu)H_1,$$

$$H_j' = n\delta H_{j-1} + \phi I_j - (n\delta + \mu)H_j, \qquad j = 2, \cdots, n,$$

$$R' = n\delta I_n + n\delta H_n - \mu R,$$

with $I = \sum_{j=1}^{n} I_j$, $H = \sum_{j=1}^{n} H_j$.

This approach of converting a gamma distribution to a sequence of exponential distributions is known as the "linear chain trick". It has been noted that the use of the gamma distribution $p_n(s, \theta)$ for a disease stage, e.g., the exposed stage, is equivalent to assuming that the entire stage is replaced by a series of n sub-stages, and each of the sub-stages is exponentially distributed with the removal rate $n\theta$ and the mean sojourn time T/n, where $T = 1/\theta$ is the mean sojourn time of the entire stage (see, for example, Hethcote and Tudor, 1980; Lloyd, 2001a; MacDonald, 1978).

Again, from the formula (2.90) the control reproductive number for the GDM, $\mathcal{R}_{c,g}$, can be written as

$$
\mathcal{R}_{c,g} = \frac{(m\alpha)^m}{(\mu + m\alpha)^m} \frac{\beta}{\mu + n\delta} \sum_{j=0}^{n-1} \frac{(n\delta)^j}{(\mu + n\delta)^j}
$$

$$
\times \left[1 - \rho \left(1 - \frac{(\mu + m\alpha)^m}{(\mu + m\alpha + \chi)^m} \frac{\mu + n\delta}{\mu + n\delta + \phi} \frac{\sum_{j=0}^{n-1} \frac{(n\delta)^j}{(\mu + n\delta + \phi)^j}}{\sum_{j=0}^{n-1} \frac{(n\delta)^j}{(\mu + n\delta)^j}} \right) \right], \tag{2.108}
$$

and from the formula (2.93) the basic reproductive number for the GDM, $\mathcal{R}_{0,g}$, is

$$
\mathcal{R}_{0,g} = \frac{(m\alpha)^m}{(\mu + m\alpha)^m} \frac{\beta}{\mu + n\delta} \sum_{j=0}^{n-1} \frac{(n\delta)^j}{(\mu + n\delta)^j}. \tag{2.109}
$$

From the partial derivatives

$$
\frac{\partial \mathcal{R}_{c,g}}{\partial \chi} = -\beta\rho \frac{m(m\alpha)^m}{(\mu + m\alpha + \chi)^{m+1}} \sum_{j=0}^{n-1} \frac{(n\delta)^j}{(\mu + n\delta + \phi)^{j+1}} < 0, \tag{2.110}
$$

$$
\frac{\partial \mathcal{R}_{c,g}}{\partial \phi} = -\beta\rho \frac{(m\alpha)^m}{(\mu + m\alpha + \chi)^m} \sum_{j=0}^{n-1} \frac{(j+1)(n\delta)^j}{(\mu + n\delta + \phi)^{j+2}} < 0 \tag{2.111}
$$

it is clear that both quarantine (χ) and isolation (ϕ) will reduce the magnitude of $\mathcal{R}_{c,g}$. Similarly, it is easy to check that $\mathcal{R}_{c,e}$ is also a decreasing function of χ and ϕ. When the EDM and GDM are used to compare model predictions of combined control strategies (i.e., the joint effect of quarantine and isolation), however, inconsistent predictions by the two models are observed. For example, in Figures 2.25(a) and 2.25(b), $\mathcal{R}_c$ for both models is plotted either as a function of ϕ for a fixed value of $\chi = 0.1$, or as a function of χ for a fixed value $\phi = 0.1$, or as a function of both χ and ϕ with $\chi = \phi$. For any vertical line except the one at 0.1, the three curves intersect the vertical line at three points that represent three control strategies. The order of these points (from top to bottom) determines the order of effectiveness (from low to high) of the corresponding control strategies because a larger $\mathcal{R}_c$ value will most likely lead to a higher disease prevalence. The order of these three points (labeled by a circle, a triangle and a square) predicted by the EDM and the GDM is clearly different for the selected parameter sets, suggesting conflict assessments of interventions between the two models. These conflict assessments are also shown when we compare the C values. For example, Figures 2.25(c) and 2.25(d) plot the function $C(t)$ at the end of epidemics (at which time the number of new infection is zero). Obviously they show the same problem as that shown in Figures 2.25(a) and 2.25(b) in which the $\mathcal{R}_c$ values are compared. The parameter values used in Figure 2.25 are $\beta = 0.2$, $\rho = 0.8$, $\alpha = 1/7$, and $\delta = 1/10$, corresponding to a disease with a latency period of $1/\alpha = 7$ days and an infectious period of $1/\delta = 10$ days (e.g., SARS).

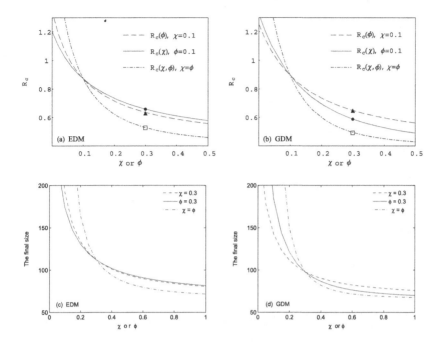

Fig. 2.25   Comparision of the EDM and the GDM on the impact of various control measures. (a) and (b) are plots of the reproductive number $\mathcal{R}_c$. (c) and (d) are plots of the number of cumulative infections C.

To examine in more details the quantitative differences between the two models we conducted intensive simulations of the EDM and the GDM for various control measures, some of which are illustrated in Figure 2.26. For demonstration purposes we have used a different α value, $\alpha = 1/10$. Other parameter values are the same as before. Figures 2.26(a) and 2.26(b) are for the case in which there is no control ($\chi = \phi = 0$). We observe that both models predict the same value of C (see the C curve). Figures 2.26(c) and 2.26(d) are for Strategy I, which implements quarantine alone with $\chi = 0.08$, and Figures 2.26(e) and 2.26(f) are for Strategy II, which implements isolation alone with $\phi = 0.1$. The effectiveness of these control measures is reflected by the corresponding $C(t)$ values. According to Figures 2.26(c) and 2.26(e), the EDM predicts that Strategy II is *more* effective than Strategy I as the number C of cumulative infections under Strategy II is 25% lower than the C value under Strategy I (notice that $C = 2095$ and $C = 1570$ under strategies I and II, respectively). However, according to Figures 2.26(d) and 2.26(f), the GDM predicts that Strategy II is *less* effective than Strategy II as the number C of cumulative infections under Strategy I is 30% lower than the C value under Strategy II (notice that $C = 1540$ and $C = 2270$ under strategies I and II, respectively). Obviously, in this example the predictions by the EDM and by the GDM are inconsistent.

The results in this study suggest that the standard SEIR type of ODE model

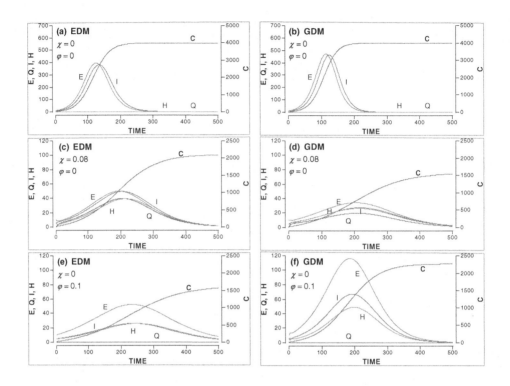

Fig. 2.26   Numerical simulations of the EDM and the GDM under no control (see (a) and (b)), Strategy I (see (c) and (d)) & Strategy II (see (e) and (f)). The number of cumulative infections $C(t)$ as well as other disease variables $E(t)$, $I(t)$, $Q(t)$, and $H(t)$ are plotted.

(such as the model (2.103)), while capable of capturing many essential features of the disease transmission dynamics in the absence of quarantine and isolation, may produce results that are inconsistent with those from models with non-exponentially distributed disease stages. We considered one such model (2.107) by using the gamma distribution with $n = 3$. Obviously, other types of more realistic stage distributions (P_E and P_I) can be used to derive different improved models. In addition to the choice of P_E and P_I, modifications on the model (2.103) can be obtained by making different assumptions on $k(s)$ and $l(s)$ as well. These functions obviously play a very important role for the study of disease intervention as these are the parameters (or parameter functions) that reflect the control measures. Clearly, the biology of the infection ultimately determines the most appropriate forms for the incubation and infectious periods. It is critical to have a good understanding of the advantages and limitations of the models, as well as the biology of the disease.

In another study ([Yan and Feng (2010)]) we also analyzed a model concerning the variability order of the incubation and the infectious periods in a deterministic SEIR epidemic model and evaluation of control effectiveness. Specifically, we use distribution theory and ordering of non-negative random variables to study the SEIR model including quarantine and isolation to reduce the spread of an infectious disease. We identify that the probability distributions of the latent period and the infectious period are primary features of the SEIR model to formulate the epidemic threshold and to evaluate the effectiveness of the intervention measures. If the primary features are changed, the conclusions will be altered in an importantly different way. For the latent and infectious periods with known mean values, it is the dilation, a generalization of variance, of their distributions that ranks the effectiveness of these control measures. We further propose ways to set quarantine and isolation targets to reduce the controlled reproduction number below the threshold using observed initial growth rate from outbreak data. If both quarantine and isolation are 100% effective, one can directly use the observed growth rate for setting control targets. If they are not 100% effective, some further knowledge of the distributions is required.

2.6 Oscillatory dynamic created by isolation

This section includes results from [Feng (1994); Feng and Thieme (1995); Wu and Feng (2000); Feng and Thieme (2000a,b); Nuño *et al.* (2005)].

2.6.1 *An SIQR model with isolation-adjusted incidence*

[Hethcote (1976)] and [Anderson and May (1982b)] consider simple models involving three or four epidemic classes that display dynamics that converge to an epidemic equilibrium in damped oscillations with the quasi-periods being surprisingly close (considering the simplicity of the model) to observed values in some of the childhood diseases ([Anderson and May (1991)], Table 6.1; [Anderson and May (1982b)]). It has been argued that these damped oscillations can be excited to undamped oscillations either by stochastic or periodic deterministic forcing. The impact of stochastic forcing was studied by Bartlett [Bartlett (1957, 1960)] and London and Yorke [London and Yorke (1973)]. The influence of periodic forcing leading to periodic solutions, with the periods being multiples of the period of the forcing, was studied numerically by London and Yorke [London and Yorke (1973)] and Dietz [Dietz (1976)], formally by Grossman *et al.* [Grossman (1980)], and analytically by Smith [Smith (1983a,b)] and Schwartz and Smith [Schwartz and Smith (1983)]. Periodic forcing can even lead to a sequence of period-doubling subharmonic bifurcations and finally chaos [Aron and Schwartz (1984a,b); Schaffer and Kot (1985)]. See [Dietz and Schenzle (1985a); Hethcote and Levin (1989)] for more detailed reviews.

Schenzle [Schenzle (1984)] (see also [Dietz and Schenzle (1985a)]) convincingly argues that a one-year periodic forcing is provided by the school system where long summer vacations interrupt or weaken the chain of infections and new susceptibles are recruited at the beginning of every school year (with the second being apparently more important than the first). Schenzle's work [Schenzle (1984)], which also incorporates age structure and mainly consists of numerical simulations, has motivated several authors to study the question of whether the introduction of age structure alone can be responsible for undamped oscillations in endemic models that have a stable endemic equilibrium without age structure. It has been shown in [Andreasen (1995)] and [Thieme (1991)] that this can indeed be the case, but the conditions they have found so far are rather extreme.

Dietz and Schenzle [Dietz and Schenzle (1985a)] assess that "up to the present day the problem of recurrent epidemics has not been definitely settled" (p. 185) and that there is "reason to think of still another mechanism causing endemic incidence fluctuation to be sustained" (p. 190).

In this study we (partially) rehabilitate the opinion of Hamer [Hamer (1906)] and Soper [Soper (1929)] that autonomous internal forces may be responsible for undamped oscillations in childhood diseases. Standard epidemic models have neglected that infected children who become infective at the end of the latency period get severe symptoms at the end of the incubation period that cause them to stay at home. Though they may still infect their relatives when they are at home, their infectious impact is largely reduced because they are kept from making contacts outside their families. Isolation alone would not create oscillations; traditional models simply merge isolated and immune individuals into one class, and the model behavior is, of course, the same as without discriminating isolated individuals. The point is that immune individuals go back into public life, whereas isolated individuals do not. This may have an effect if, as we suspect, the per capita rate of contacts is basically independent of the number of children present unless this number is quite small.

Adding these two features to standard childhood disease models amounts to adding a new class-isolated individuals (those who stay at home because they are too sick to go out)-and modifying the standard bilinear mass action infection term by dividing it by the number of active (i.e., nonisolated) individuals. The importance of this modification has become apparent before in the study of sexually transmitted diseases, and its destabilizing potential was already discovered in an HIV/AIDS model with infection-age-dependent infectivity (see [Thieme and Castillo-Chavez (1993)] and references therein). Other infection laws that deviate from the usual bilinear mass action term and can lead to undamped oscillations have been considered in [Cunningham (1979); Liu *et al.* (1986, 1987)]. It is not clear, however, what kind of mechanisms they represent in childhood diseases. In infectious diseases that are not of the type considered here, undamped oscillations can be excited by other mechanisms; see [Hethcote and Levin (1989); Thieme and Castillo-Chavez (1993)] for surveys and references.

We split the population (the size of which we denote by N) into individuals that are susceptible to the disease, S, into infective non-isolated individuals, I, into isolated individuals Q (Q like quarantine), and into recovered and immune individuals, R. We neglect a latency period and assume that immunity is permanent. By $A = S + I + R$ we denote the active, i.e., non-isolated individuals. The basic idea consists in assuming that sick children stay at home and so undergo some kind of isolation (or quarantine) that reduces their ability to infect others. For simplicity we assume that they do not infect anybody. The model takes the following form:

$$\frac{dS}{dt} = \Lambda - \mu S - \sigma S \frac{I}{A},$$

$$\frac{dI}{dt} = -(\mu + \gamma)I + \sigma S \frac{I}{A},$$

$$\frac{dQ}{dt} = -(\mu + \xi)Q + \gamma I, \tag{2.112}$$

$$\frac{dR}{dt} = -\mu R + \xi Q,$$

$$A = S + I + R.$$

Λ is the recruitment rate (all newborns are assumed to be susceptible); μ is the per capita mortality rate; σ is the per capita infection rate of an average susceptible individual provided that everybody else is infected (note that we used σ in this model for infection rate instead of β); γ is the rate at which individuals leave the infective class and ξ is the rate at which individuals leave the isolated class; they are all positive constants. $1/\gamma$ gives the length of the infective period and $1/\xi$ gives the length of the isolation period. I/A gives the probability that a given contact actually occurs with an infective individual.

Notice that we have assumed that the per capita number of contacts is basically independent of the number of active individuals. This assumption is widely accepted for sexually transmitted diseases, but there is growing evidence that this is not a bad assumption for otherwise transmitted diseases either unless the number of active individuals is very small (see section 1.1 in [Zhou and Hethcote (1994)] or the Introduction in [Gao *et al.* (1995)]). A possible explanation consists in the number of contacts rather depending on the density of active individuals than their absolute numbers [De Jong *et al.* (1995)]. When there are a few children on a playground or school yard, e.g., they will gather in one part of it, whereas they will spread out over the playground when there are many; so the per capita number of contacts essentially remains the same unless the number of children becomes very small.

More generally one could replace σ by a contact function $C(A)$ (see [Thieme and Castillo-Chavez (1993); Zhou and Hethcote (1994); Gao *et al.* (1995); Thieme (1991)], and the literature cited therein). As far as applications are concerned, this amounts to introducing at least one, but typically several more parameters that have to be estimated. We know of one attempt only that has been taken in this direction, namely for

$$C(A) = \sigma A^\alpha$$

(see [Anderson and May (1991)] and the reference therein). Estimates for five childhood diseases in communities of various size provide that α is between 0.03 and 0.07, and our choice $\alpha = 0$ turns out to be a much better approximation than the traditional $\alpha = 1$.

Adding the differential equations in (2.112) we find for the population size $N = A + Q = S + I + Q + R$ that

$$\frac{dN}{dt} = \Lambda - \mu N.$$

Hence $N(t) \to \Lambda/\mu$ as $t \to \infty$. We assume that the size of the school population has reached its limiting value, i.e.,

$$N \equiv \Lambda/\mu \equiv S + I + Q + R = A + Q.$$

Using $A = N - Q$ and $S = A - I - R$ in (2.112) we can eliminate S from the equations. Further we can scale time such that $\sigma = 1$ by introducing a new, dimensionless, time $\tau = \sigma t$. This gives us the system

$$I' = -(\nu + \theta)I + (1 - (I + R)/(N - Q))\, I,$$

$$Q' = -(\nu + \zeta)Q + \theta I,$$

$$R' = -\nu R + \zeta Q,$$

where

$$\nu = \mu/\sigma, \qquad \theta = \gamma/\sigma, \qquad \zeta = \xi/\sigma,$$

and " $'$ " denotes the derivative in τ.

Introduce the fractions

$$u = \frac{S}{A}, \qquad y = \frac{I}{A}, \qquad q = \frac{Q}{A}, \qquad z = \frac{R}{A}.$$

Note that

$$A' = -Q' = (\nu + \zeta)Q - \theta I, \qquad \Lambda - \mu S = \mu(I + Q + R).$$

Hence, using (2.112) and noticing that

$$u + y + z = 1$$

we obtain

$$y' = y\left(1 - \nu - \theta - y - z + \theta y - (\nu + \zeta)q\right),$$

$$q' = (1 + q)\left(\theta y - (\nu + \zeta)q\right), \tag{2.113}$$

$$z' = \zeta q - \nu z + z\left(\theta y - (\nu + \zeta)q\right).$$

The simplified system (2.113) and the original system (2.112) have the same dynamical behavior. The analytical results below are described in terms of the system (2.113). It can be shown that for any nonnegative initial conditions the system (2.113) has a unique nonnegative solution for all $t > 0$. Moreover, if $y_0 > 0$, then $u(t), y(t), q(t), z(t)$ are strictly positive for $t > 0$. q is bounded by the maximum of q_0 and $\frac{\theta}{\nu + \zeta}$.

The system (2.113) always has the disease-free equilibrium $E_0 = (y_0, q_0, z_0) = (0, 0, 0)$. There is a possible endemic equilibrium $E^* = (y^*, q^*, z^*)$ given by

$$y^* = \nu(\nu + \zeta)\kappa, \quad q^* = \nu\theta\kappa, \quad z^* = \theta\zeta\kappa \tag{2.114}$$

where

$$\kappa = \frac{1 - \nu - \theta}{\nu(\nu + \zeta) + \theta\zeta}.$$

The basic reproduction number is

$$\mathcal{R}_0 = \frac{1}{\nu + \theta} = \frac{\sigma}{\mu + \gamma},$$

and the result below shows that $\mathcal{R}_0$ determines disease extinction or persistence.

Theorem 2.13.

(a) Let $\mathcal{R}_0 \leq 1$. Then, for any solution of (2.113) with nonnegative initial data satisfying $u_0 + y_0 + z_0 = 1$, we have $u(t) \to 1$, $y(t) \to 0$, $q(t) \to 0$, $z(t) \to 0$ as $t \to \infty$.

(b) Let $\mathcal{R}_0 > 1$. Then there exists some $\epsilon > 0$ such that

$$\liminf_{t \to \infty} y(t) \geq \epsilon, \quad \liminf_{t \to \infty} q(t) \geq \epsilon, \quad \liminf_{t \to \infty} z(t) \geq \epsilon$$

for all solutions to (2.113) with nonnegative initial data satisfying $u_0 + y_0 + z_0 = 1, y_0 > 0$.

Notice that ν is much smaller than θ and ζ, the endemic equilibrium of system (2.113) can be written as

$$y^* = \nu\frac{1 - \theta}{\theta} + O(\nu^2), \quad q^* = \nu\frac{1 - \theta}{\zeta} + O(\nu^2), \quad z^* = 1 - \theta + O(\nu). \tag{2.115}$$

It was shown in [Feng and Thieme (1995)] that the endemic equilibrium swiches stability at two critical points $\zeta_0 < \zeta_1$, as stated in the following results.

Theorem 2.14. *(Theorem 5.1 in [Feng and Thieme (1995)])* *There is a function* $\zeta_0(\nu)$ *defined for small* $\nu > 0$

$$\zeta_0(\nu) = \theta^2(1 - \theta) + O(\nu^{1/2}),$$

with the following properties:

(a) *The endemic equilibrium is locally asymptotically stable if $\zeta > \zeta_0(\nu)$ and unstable if $\zeta < \zeta_0(\nu)$, as long as ζ does not become too small.*

(b) *There is a Hopf bifurcation of periodic solutions at $\zeta = \zeta_0(\nu)$ for small enough $\nu > 0$. The periods are approximately*

$$T = \frac{2\pi}{|\Im w_\pm|} \approx \frac{2\pi}{(1 - \theta)^{1/2}\nu^{1/2}} \approx \frac{2\pi}{\sqrt{\theta y^*}}$$

in the neighborhood of the Hopf bifurcation point.

Theorem 2.15. *(Theorem 5.2 in [Feng and Thieme (1995)]) There is a function* $\zeta_1(\nu)$ *defined for small* $\nu > 0$,

$$\zeta_1(\nu) = \frac{1-\theta}{2\theta^2}(1 + \sqrt{1 + 4\theta^2})\nu + O(\nu^{3/2}),$$

with the following properties:

(a) *The endemic equilibrium is locally asymptotically stable if* $\zeta < \zeta_1(\nu)$ *and unstable if* $\zeta > \zeta_1(\nu)$, *as long as* ζ *does not become too large.*

(b) *There is a Hopf bifurcation of periodic solutions at* $\zeta = \zeta_1(\nu)$ *for small enough* $\nu > 0$. *The periods are approximately*

$$T = \frac{2\pi}{|\Im w_{\pm}|} \approx \frac{2\pi}{\sqrt{\theta y^*}}\left(\frac{1}{\sqrt{1-\theta}}\right)$$

in the neighborhood of the Hopf bifurcation point.

Calculations with Doedel's [Doedel (1981)] program *Auto* show that the Hopf bifurcation is supercritical at both bifurcation points for parameters in a certain biologically reasonable region (see Figure 2.27). We also observed that, for parameters in some other regions, when ζ moves away from ζ_1, the period of the periodic solution increases dramatically until the numerical methods fail to work properly. Can we derive an analytical understanding of the possible complex dynamics arising from this epidemic model?

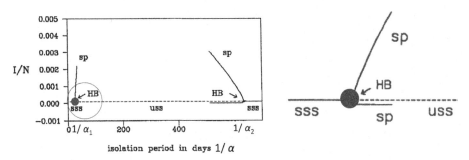

(a) Two hopf bifurcation (HB) points.   (b) Blow-up of the left HB point.

Fig. 2.27 (a) An Auto plot of the lower and upper amplitudes of the periodic solutions (in terms of the fraction of infective (nonisolated) individuals) versus the length of the isolation period. $\mathcal{R}_0$ is chosen at its lower estimate, 6.4. HB, Hopf bifurcation point; sss, (locally asymptotically) stable steady state; uss, unstable steady state; sp, (locally asymptotically orbitally) stable periodic solution. (b) Magnification of the left branch.

We compared the model results to scarlet fever in England and Wales (1897-1978). Anderson and May ([May and Anderson (1983)], Tables 1 and 2; [Anderson and May (1991)], Table 6.1) report an average age of infection (D_S) between 10 and 14 years, a mean interepidemic period (T) of 4.4 years, and an average life expectancy ($D = 1/\mu$) that rises from 60 to 70 years. Hence we assume

$$D = 65 \text{ yr}, \quad D_S = 12 \text{ yr}, \quad T = 4.4 \text{ yr}.$$

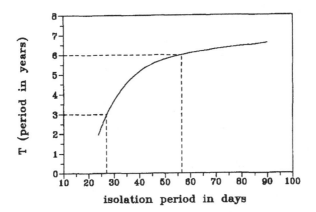

Fig. 2.28 An Auto plot of the periods of the periodic solutions versus the length of the isolation period. $\mathcal{R}_0$ is chosen at its lower estimate, 6.4.

We further identify the temporal mean $\bar{y}$ with the endemic equilibrium value y^* and so obtain that

$$\frac{1}{D_S} < \sigma y^* < -\frac{1}{T} \ln\left(1 - \frac{T}{D_S}\right). \tag{2.116}$$

In numbers,

$$0.0833 < \sigma y^*[\text{yr}^{-1}] < 0.1038.$$

Using the approximate formula $\theta \approx (D\sigma y^* + 1)^{-1}$ we obtain the estimates

$$0.1291 \leq \theta \leq 0.1559.$$

As

$$\mathcal{R}_0 = \frac{1}{\theta + \nu} = \frac{1}{\theta}\left(\frac{1}{1 + \mu/\gamma}\right) \approx \frac{1}{\theta},$$

we obtain the following lower and upper bounds for $\mathcal{R}_0$:

$$6.4 < \mathcal{R}_0 < 7.7.$$

Let $D_Q = 1/\xi = 1/(\sigma\zeta)$ denote the mean isolation period, and let D_Q^0 and D_Q^1 denote the minimum and maximum lengths of the isolation period. Our estimate for θ implies $\theta < 1/2$, that is, the formula

$$\gamma D_Q^0 = \frac{\theta}{\zeta_0} \approx \frac{1}{\theta(1 - \theta)}$$

is decreasing in θ, whereas the formula

$$\frac{D_Q^1}{D} \approx \frac{2\theta^2}{(1 - \theta)(1 + \sqrt{1 + 4\theta^2})}$$

is increasing in θ. This makes it possible to give upper and lower bounds for D_Q^0 and D_Q^1:

$$7.6 < \gamma D_Q^0 < 8.9, \quad 0.0188 < \frac{D_Q^1}{D} < 0.0281.$$

Notice that the lower and upper estimates for $\mathcal{R}_0$ are associated with the lower and upper estimates of γD^0, respectively, whereas the lower estimate for $\mathcal{R}_0$ is associated with the upper estimate for D_Q^1/D, and the upper estimate for $\mathcal{R}_0$ with the lower estimate for D_Q^1/D.

Anderson and May ([Anderson and May (1991)], Table 3.1) report a latent period of 1 or 2 days, an infectious period of 14-21 days, and an incubation period of 2 or 3 days. This suggests an effective infectious period that may be as short as 1 or 2 days, a sum of latent and effective period that is $1/\gamma = 3$ days, and an isolation period between 2 and 3 weeks. Anderson *et al.* [Anderson and Amstein (1948)] report isolation periods of 2 and 3 weeks or even longer. It is possible, however, that there are infections without any symptoms with slight symptoms such that $1/\gamma$ may be larger than 3 days. We stick to $1/\gamma = 3$ days, which, for the lower estimate $\mathcal{R}_0 = 6.4$, gives us a window

$$22.8 \text{ days} < D_Q < 1.83 \text{ yr}$$

for isolation periods that make the endemic equilibrium unstable and a window

$$26.7 \text{ days} < D_Q < 1.22 \text{ yr}$$

for the upper estimate $\mathcal{R}_0 = 7.7$. So the minimum isolation periods that make the endemic equilibrium unstable are of the right order of magnitude.

The formulas in Theorems 2.14 and 2.15 provide the periods of sustained oscillations in case the length of the isolation period is slightly larger than the necessary minimum length D_Q^0 (Theorem 2.14) or slightly smaller than the required maximum length D_Q^1 (Theorem 2.15). Notice that, with

$$1/\gamma \approx D_E + D_l$$

now being the sum of the lengths of the latency period and the effective infectivity period, the formula in Theorem 2.15 is exactly analogous to the formulas given by Dietz [Dietz (1976)] and Anderson and May [Anderson and May (1991)] for a model with latency but without isolation period. In real time the lengths of the interepidemic periods are given by

$$\frac{T_0}{\sigma} = 2\pi \left(\frac{1/\gamma}{\sigma y^*} \right)^{1/2}$$

if the length of the isolation period is slightly larger than the minimum period and by

$$\frac{T_1}{\sigma} = 2\pi \left(\frac{1/\gamma}{\sigma y^*} \right)^{1/2} \left(\frac{1}{\sqrt{1-\theta}} \right)$$

if the lengths of the isolation period is slightly smaller than the maximum period. These formulas provide interepidemic periods that are slightly less than 2 yr and so are much shorter than the reported average length of 4.4 yr but still of the right order of magnitude. We emphasize that the above formulas hold only at the bifurcation points, and we refer to the numerical results in the next section for isolation periods that are not close at the bifurcation points.

We have also conducted numerical simulations for the parameters that we believe mimic the scarlet fever situation in England and Wales from 1897 to 1978. We assume a life expectation of 65 yr and a mean age at infection of 12 yr. We further assume that the sum of the length of the latency period and the effective infectious period (which should be equal to or slightly larger than the length of the incubation period) is 3 days. From these parameters we can calculate ν, once we have estimated θ. Indeed,

$$\frac{\nu}{\theta} = \frac{\mu}{\gamma} = \frac{3 \text{ days}}{65 \text{ yr}} = 0.000126.$$

In the previous section we estimated the basic reproductive number $\mathcal{R}_0$ to be between 6.4 and 7.7 and $\theta \approx 1/\mathcal{R}_0$ to be between 0.1291 and 0.1559. These estimates were based on the possibly unrealistic assumption that the temporal mean of a periodic solution equals the endemic equilibrium value.

The calculations with *Auto* provide that the endemic equilibrium is unstable for isolation periods whose lengths are between 23.8 and 629.5 days if $\mathcal{R}_0$ is at its lower estimate, and for isolation periods where lengths are between 28.7 and 419.8 days if $\mathcal{R}_0 = 7.7$ is at its larger estimate. Recall that the approximate formulas in section 5 suggest an unstable endemic equilibrium when the length of the isolation period is between 22.8 and 667.3 days for $\mathcal{R}_0 = 6.4$ and between 26.7 and 447 days for $\mathcal{R}_0 = 7.7$.

In the following numerical calculations we have chosen $\mathcal{R}_0$ to be at its lower estimated value, $\mathcal{R}_0 = 6.4$. Recall that the lower estimate $\mathcal{R}_0 = 6.4$ would be sharp if the endemic were at its (mathematical) equilibrium state. The lower estimate is the one that is typically taken in the literature. Compared to the upper estimate, the lower estimate as the advantage of being independent of the length of the interepidemic period. The Auto calculations show that the Hopf bifurcation (with the length of the isolation period being the bifurcation parameter) is supercritical at both bifurcation points (Figure 2.27). We conjecture that the two bifurcation points are connected by a global branch of periodic solutions, though the *Auto* calculations do not confirm this.

The periods of the periodic solutions depend rather dramatically on the length of the isolation period (Figure 2.28). The observed lengths of the interepidemic periods (3-6 yr) according ([Anderson and May (1991)], Table 6.1) are feasible with isolation periods between 27 and 57 days. These lengths seem to be unrealistically high, but this may be due to the crudeness of our model. With the model without an isolation period, Anderson and May ([Anderson and May (1991)], Table 6.1) can mimic interepidemic periods between 4 and 5 years.

Homoclinic bifurcation in the SIQR model

The results in [Feng and Thieme (1995)] were extended in [Wu and Feng (2000)]. Note that important disease dynamics of the model can be determined by the threshold $\mathcal{R}_0 = 1$, where $\mathcal{R}_0 = 1/(\nu + \theta)$ and $\nu = \mu/\sigma, \theta = \gamma/\sigma$. Also note that $D = 1/\mu$ denote the average life expectation of humans, and $D_I = 1/\gamma$ is the length of the infective period. The life span is on the order of decades whereas the infective period is on the order of days. Hence μ is much smaller than γ, and so ν is much smaller than θ. This leads us to choose the parameter values $\nu = 0$, $\theta = 1$ at the threshold $\mathcal{R}_0 = 1$, and consider the perturbations of the corresponding system.

To analyze the system corresponding to $\nu = 0$, $\theta = 1$, we first introduce some notations and the normal form theorem. Let H_2 denote the linear space of vector fields whose coefficients are homogeneous polynomials of degree 2, $J = \begin{pmatrix} 0 & 1 \\ 0 & 0 \end{pmatrix}$, and

$$L_J\left(\tilde{h}(\mathbf{x}) \right) \equiv -\left(D\tilde{h}(\mathbf{x})J\mathbf{x} - J\tilde{h}(\mathbf{x}) \right),$$

where $\mathbf{x} \in \mathbf{R}^2$, $\tilde{h} : \mathbf{R}^2 \to \mathbf{R}^2$, and $L_J : H_2 \to H_2$. Choose a complementary G_2 for $L_J(H_2)$ in H_2, so that $H_2 = L_J(H_2) \oplus G_2$. The following theorem is taken from [Wiggins (2003)].

Theorem 2.16. *(Normal Form Theorem) By an analytic coordinate change* $\mathbf{x} = \mathbf{y} + \tilde{h}(\mathbf{y})$; *the system*

$$\mathbf{x}' = J\mathbf{x} + F_2(\mathbf{x}) + O(|\mathbf{x}|^3)$$

can be transformed into

$$\mathbf{y}' = J\mathbf{y} + \widetilde{F}_2(\mathbf{y}) + O(|\mathbf{y}|^3),$$

where $\widetilde{F}_2(y) \in G_2$.

From the above theorem, we can choose $\tilde{h}(\mathbf{y})$ such that only $O(|\mathbf{y}|^2)$ terms that are in G_2 remain. In the new coordinate system, only second order terms are in a space complementary to $L_J(H_2)$.

When $\nu = 0$, $\theta = 1$ (2.113) (with $y = I/A$ replaced by p for ease of notation) becomes

$$\dot{p} = p(-z - \zeta q),$$
$$\dot{q} = (1 + q)(p - \zeta q), \qquad (2.117)$$
$$\dot{z} = \zeta q + z(p - \zeta q).$$

The system (2.117) has one negative and double zero eigenvalues at the equilibrium $e_0 = (p_0, q_0, z_0) = (0, 0, 0)$. By the linear transformation

$$\begin{pmatrix} p \\ q \\ z \end{pmatrix} = \begin{pmatrix} 0 & \zeta & 0 \\ 0 & 1 & 1 \\ \zeta & 0 & -1 \end{pmatrix} \begin{pmatrix} u \\ v \\ w \end{pmatrix} \qquad (2.118)$$

the system (2.117) can be transformed into

$$
\begin{pmatrix} \dot{u} \\ \dot{v} \\ \dot{w} \end{pmatrix} = \begin{pmatrix} 0 & 1 & 0 \\ 0 & 0 & 0 \\ 0 & 0 & -\zeta \end{pmatrix} \begin{pmatrix} u \\ v \\ w \end{pmatrix} + \begin{pmatrix} uv + v^2 - vw/\zeta - \zeta uw \\ -\zeta uv - \zeta v^2 + vw - \zeta vw \\ \zeta uv + \zeta v^2 - vw - \zeta w^2 \end{pmatrix}.
\tag{2.119}
$$

Note that $u - v$ plane is associated with a pair of zero eigenvalues, while the w axis corresponds to the eigenvalue $-\zeta$. Thus, there exists a smooth stable manifold $W^s(e_0)$ and center manifold $W^c(e_0)$. A center manifold at e_0 can be locally represented as follows:

$$
W^c(e_0) = \{(u, v, w) | w = g(u, v), |u| < \delta, |v| < \delta, g(0) = Dg(0) = 0\}
$$

for δ sufficiently small, where g can be computed as a Taylor expansion up to a certain order to obtain the desired degree of accuracy, and the dynamics of (2.119) restricted to the center manifold is given by the first two equations in (2.119) with w replaced by $g(u, v)$. Note that g must have neither a constant nor a linear term, so we can compute the quadratic terms of g and get an approximation to the center manifold

$$
w = \left(1 - \frac{1}{\zeta}\right)v^2 + uv + O(3).
\tag{2.120}
$$

Substitution of (2.120) into (2.119) yields the center manifold reduction up to order 3:

$$
\begin{aligned}
\dot{u} &= v + v^2 + uv - \zeta u^2 v + \left(1 - \tfrac{1}{\zeta} - \zeta\right)uv^2 + \left(\tfrac{1}{\zeta^2} - \tfrac{1}{\zeta}\right)v^3 + O(4), \\
\dot{v} &= -\zeta uv - \zeta v^2 + \left(1 - \zeta\right)uv^2 + \left(2 - \tfrac{1}{\zeta} - \zeta\right)v^3 + O(4).
\end{aligned}
\tag{2.121}
$$

The second order terms in (2.121) can be simplified by using the normal form theorem and computing a normal form up to order 2. We know from [Guckenheimer and Holmes (1983)] that

$$
H_2 = span\left\{ \begin{pmatrix} x^2 \\ 0 \end{pmatrix}, \begin{pmatrix} xy \\ 0 \end{pmatrix}, \begin{pmatrix} y^2 \\ 0 \end{pmatrix}, \begin{pmatrix} 0 \\ x^2 \end{pmatrix}, \begin{pmatrix} 0 \\ xy \end{pmatrix}, \begin{pmatrix} 0 \\ y^2 \end{pmatrix} \right\},
$$

$$
L_J(H_2) = span\left\{ \begin{pmatrix} -2xy \\ 0 \end{pmatrix}, \begin{pmatrix} y^2 \\ 0 \end{pmatrix}, \begin{pmatrix} x^2 \\ -2xy \end{pmatrix}, \begin{pmatrix} xy \\ -y^2 \end{pmatrix} \right\},
$$

and one choice of G_2 is given by

$$
G_2 = span\left\{ \begin{pmatrix} 0 \\ x^2 \end{pmatrix}, \begin{pmatrix} 0 \\ xy \end{pmatrix} \right\}.
$$

For the system (2.121), if we choose the near-identity transformation

$$
\begin{pmatrix} u \\ v \end{pmatrix} = \begin{pmatrix} x \\ y \end{pmatrix} + \begin{pmatrix} \frac{1-\zeta}{2}x^2 + xy \\ -\zeta xy \end{pmatrix},
\tag{2.122}
$$

i.e.,

$$
\tilde{h}(x, y) = \begin{pmatrix} \frac{1-\zeta}{2}x^2 + xy \\ -\zeta xy \end{pmatrix},
$$

then (2.121) can be transformed into the normal form

$$\dot{x} = y + O(3),$$
$$\dot{y} = -\zeta xy + O(3). \tag{2.123}$$

We are interested in studying perturbations of (2.123), which then represent the different types of dynamics that can occur for the full system (2.117) near the bifurcation. The following result holds.

Theorem 2.17. *The vector field (2.113) at the point $\nu = 0$, $\theta = 1$ has a singularity of codimension greater than two.*

The two-parameter unfolding of the system (2.123) consiered in [Wu and Feng (2000)] has the form:

$$\dot{x} = \sigma_1 x + y,$$
$$\dot{y} = (\sigma_2 - \sigma_1)y + \alpha x^2 - \zeta xy. \tag{2.124}$$

Here σ_1, σ_2, and α are small parameters, $\alpha \geq 0$. It is easy to verify that $y \geq 0$ remains invariant under the unfolding. Consider first the case when α is a fixed positive constant. Rescale by letting

$$x \to \frac{\alpha}{\zeta^2} x, \quad y \to \frac{\alpha^2}{\zeta^3} y, \quad t \to \frac{\zeta}{\alpha} t, \quad \sigma_1 \to \frac{\alpha}{\zeta} \mu_1, \quad \sigma_2 \to \frac{\alpha}{\zeta} \mu_2$$

to obtain a transformed system of (2.124):

$$\dot{x} = \mu_1 x + y,$$
$$\dot{y} = (\mu_2 - \mu_1)y + x^2 - xy, \tag{2.125}$$

where μ_1 and μ_2 are new parameters. System (2.125) has two equilibria $E_0 = (0,0)$ and $E^* = (x^*, y^*)$, where

$$x^* = \frac{\mu_1(\mu_2 - \mu_1)}{1 + \mu_1}, \quad y^* = \frac{\mu_1^2(\mu_1 - \mu_2)}{1 + \mu_1}.$$

For illustration purposes, we restrict our attention to the parameter region $\mu_2 \leq \mu_1$.

Theorem 2.18. *A Hopf bifurcation occurs along the curve $H = \{(\mu_1, \mu_2) | \mu_2 = -\mu_1^2, \mu_1 > 0\}$, and the bifurcation is supercritical.*

Let $\Omega = \{(\mu_1, \mu_2) | -2\mu_1^2 < \mu_2 < 0, \mu_1 > 0\}$, then $H \subset \Omega$. It can be shown that, for $(\mu_1, \mu_2) \in \Omega$, $(\text{trace}(A))^2 - 4\det(A) < 0$, and hence E^* is a focus. Using arguments as in the analysis of Takens-Bogdanov's equations (see section 7.3 in [Guckenheimer and Holmes (1983)]) we suspect the existence of global bifurcations.

Theorem 2.19. *A homoclinic bifurcation occurs along the curve*

$$HL = \left\{ (\mu_1, \mu_2) \Big| \mu_2 = -\frac{6}{7}\mu_1^2 + O(\mu_1^3), \mu_1 > 0 \right\}.$$

Our knowledge leads to the bifurcation diagram for system (2.125) shown in Figure 2.29. The following result describes the number of periodic orbits for given μ_1 and μ_2.

Theorem 2.20. *For sufficiently small μ_1 and μ_2, (i) there is a unique limit cycle in the region $-\mu_1^2 < \mu_2 < \frac{6}{7}\mu_1^2$, which shrinks to the stable focus as (μ_1, μ_2) tends to H, and it tends to the homoclinic loop as (μ_1, μ_2) tends to HL; (ii) there are no periodic orbits in the regions $\mu_2 < -\mu_1^2$ (near H) and $\mu_2 > -\frac{6}{7}\mu_1^2$ (near HL).*

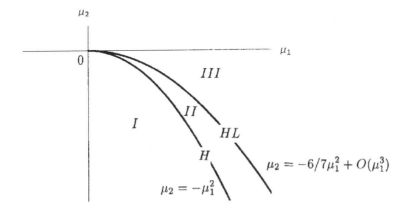

Fig. 2.29   Bifurcation diagram for the system (2.125) in the (μ_1, μ_2) plane. H and HL denote the curves for Hopf bifurcation and homoclinic bifurcation, respectively.

Numerical simulations of the system (2.125) using Ermentrout's program XP-PAUT show that the equilibrium E^* loses its stability for $\mu_2 > -\mu_1^2$ and $(\mu_1, \mu_2) \in \Omega$ (see Figure 2.30). Here we have chosen $\mu_1 = 0.5$ for all plots in Figure 2.30. The plot in Figure 2.30a shows that E^* is a stable focus for $\mu_2 = -0.3$ (so that (μ_1, μ_2) is in region I). Note that the solution with initial data $y(0) > 0$ remains in the upper half plane. Figure 2.30b is for $\mu_2 = -0.23$, and hence $(\mu_1, \mu_2) \in II$. We see that E^* is unstable, and there is a unique ø-limit cycle. We have observed that the limit cycle expands as μ_2 increases until $\mu_2 \approx -0.2005$, at which time the limit cycle hits the saddle point E_0 and a homoclinic orbit occurs (see Figure 2.30c). Hence $(\mu_1, \mu_2) = (0.5, -0.2005) \in HL$. When $\mu_2 = -0.15$; i.e., $(\mu_1, \mu_2) \in III$ (see Figure 2.30d), we see that the stable and unstable manifolds of the saddle point E_0 change their relative positions comparing with the case of $(\mu_1, \mu_2) \in II$. It is clear that our analytical results are confirmed by these numerical computations.

More realistic distributions for the disease stages

We have shown ([Feng (1994); Feng and Thieme (1995)]) that a sufficiently long isolation (or quarantine) period may lead to instability of the endemic equilibrium

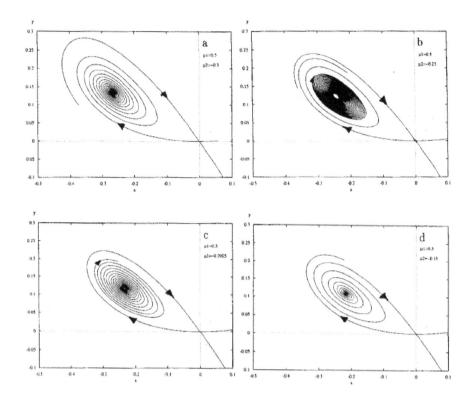

Fig. 2.30   Numerical integration of the system (2.125) according to XPPAUT. μ_1 and μ_2 have been chosen such that (μ_1, μ_2) is in regions I, II, HL and III, respectively. These regions are as in Figure 2.29.

and, via Hopf bifurcation, to periodic oscillations of the disease dynamics. Lumping the latent and infectious period of at least 23 days makes the endemic equilibrium unstable. In this study we separate latent and infectious periods. If both have exponentially distributed durations, an isolation period of at least 18 days average duration will make the endemic equilibrium unstable. If both periods have fixed durations, an isolation period of at least 10 days destabilizes the endemic equilibrium. If we interpolate reported maximum and minimum stage durations by piecewise linear duration functions, an isolation period of at least 11 days is sufficient. We mention that Anderson, Arnstein, and Lester (1962) [Anderson *et al.* (1962)] report isolation periods for scarlet fever that last two or three weeks or even longer.

For most other childhood diseases, our analysis rules out that isolation alone is a cause for unstable endemic equilibria. While long isolation periods can theoretically destabilize the endemic equilibrium, the required lengths turn out not to be in a realistic range. It is conceivable, however, that isolation adds to the other factors that have been found to lead to recurrent epidemic outbreaks like seasonal variation in per capita infection rates, stochastic effects, Allee-type effects in the infection rate, and more dynamic demographics than the constant influx of susceptibles that we will assume in this study (see [Dietz and Schenzle (1985a); Hethcote and Levin (1989); Feng and Thieme (1995); Gao *et al.* (1995)]).

In this context, our main point consists of illustrating that the required minimum length for an isolation period to destabilize the endemic equilibrium is cut into almost half, when exponential length distributions of the latent and infectious periods are replaced by realistic distributions with the same average lengths. This may not be of practical relevance for most childhood diseases, because the required minimum length is still not in a realistic range, but should be reason enough to look out for similar phenomena in other situations.

Assume that the change of the susceptible population obeys the following law:

$$\dot{S}(t) = \Lambda - \mu S(t) - B_0(t) + B_n(t),$$

$$I(t) = (I_1(t), \cdots, I_n(t)),$$

$$B_0(t) = f(S(t), I(t)).$$

Here $\Lambda > 0$ is the recruitment rate of new individuals into the epidemiologically relevant part of the population, all freshly entering individuals are assumed to be susceptible. If the whole population is epidemiologically relevant, Λ is the population birth rate. $\mu > 0$ is the per capita mortality rate not due to disease-related causes. $B_0(t)$ is the incidence, i.e., the infection rate at time t, which is a function of the number of susceptibles and the number of individuals in the various infected classes. $B_n(t)$ is the rate of individuals who have recovered from the disease but lose their immunity and return to the susceptible class.

A convenient concept to model arbitrary distributions of stage durations is stage (or class) age ([Hoppensteadt (1974, 1975)]). Stage age, here usually denoted by a, is the time that has elapsed since entering the stage. We stratify the individuals in the jth stage of infection as

$$I_j(t) = \int_0^{a_j} u_j(t,a)\,da,$$

where $u_j(t,\cdot)$ denotes the stage age density at time t and $a_j > 0$ is the maximum sojourn time in stage j (a_j may be finite or infinite), and

$$u_j(t,a) = \begin{cases} B_{j-1}(t-a)\mathcal{F}_j(a)P_j(a)e^{-\mu a}; & 0 \le a < t, \\[2mm] \tilde{u}_j(a-t)\dfrac{\mathcal{F}_j(a)P_j(a)}{\mathcal{F}_j(a-t)P_j(a-t)}e^{-\mu a}; & 0 \le t \le a \le a_j, \\[2mm] \tilde{u}_j(a); & t = 0 \le a \le a_j. \end{cases}$$

We introduce functions $P_j, \mathcal{F}_j : [0,\infty) \to [0,1]$ that describe the duration of the jth stage and the disease-related mortality in the jth stage. More precisely $P_j(a)$ is the probability that the jth stage lasts longer than a time units. Further $1 - \mathcal{F}_j(a)$ gives the probability to die from disease related causes during the jth stage before reaching stage age a. $\mathcal{F}_j$ and P_j are nonnegative, nonincreasing functions on $[0,\infty)$,

$$\mathcal{F}_j(0) = 1 = P_j(0).$$

Recalling that a_j denotes the maximum sojourn in the jth stage,

$$\mathcal{F}_j(a)P_j(a) > 0, \quad 0 \le a < a_j,$$

$$\mathcal{F}_j(a)P_j(a) = 0, \quad a > a_j, \quad \text{whenever } a_j < \infty.$$

The average duration of the jth stage, D_j, is given by

$$D_j = \int_0^\infty P_j(a)da,$$

while the average sojourn time in the jth stage is given by

$$T_j = \int_0^\infty e^{-\mu a} \mathcal{F}_j(a)P_j(a)da.$$

We assume that $D_j < \infty$ for all stages j except possibly for the last stage, $j = n$.

Notice that P_j and $\mathcal{F}_j$ are not necessarily continuous or absolutely continuous. This allows us to include the case that a stage has a fixed duration. We assume, however, that P_j and $\mathcal{F}_j$ have no joint discontinuities.

To close our model we must still decide $B_j(t)$, the rate at which individuals leave stage j and enter stage $j + 1$ if $j < n$ or return to the susceptible class if $j = n$,

$$B_j(t) = -\int_0^{a_j} \frac{u_j(t, a)}{P_j(a)} P_j(da).$$

At this point we formally interpret the integral in the sense of a Stieltjes integral, though we will later need to reinterpret it as a Lebesgue-Stieltjes integral. This formula can be most readily understood by assuming for a moment that P is differentiable and realizing that $-P_j'(a)/P_j(a)$ is the instantaneous per capita rate of leaving the jth stage at stage age a other than dying. The equation for I_j is

$$I_j'(t) = B_{j-1}(t) - \mu I_j(t) + \int_0^{a_j} \frac{u_j(t, a)}{\mathcal{F}_j(a)} \mathcal{F}_j(da) - B_j(t) \text{ (a.e.)} \tag{2.126}$$

with B_{j-1} and B_j given as above and the following interpretation: The rate of change of the number of individuals in the jth stage is the influx into the stage minus the rate of deaths due to disease-unrelated causes, the rate of deaths from the disease, and the outflux of the stage. The symbols and the model equations are collected in Table 2.5.

Substituting the expressions for u_j into the expressions for I_j and B_j yields the following system of one differential and several Volterra integral and Volterra Stieltjes integral equations:

$$\dot{S}(t) = \Lambda - \mu S(t) - B_0(t) + B_n(t), \quad S(0) = \tilde{S},$$

$$I_j(t) = \int_0^t B_{j-1}(t-a)\mathcal{F}_j(a)P_j(a)e^{-\mu a}da + e^{\mu t}\tilde{I}_j(t),$$

$$B_j(t) = -\int_0^t B_{j-1}(t-a)\mathcal{F}_j(a)e^{-\mu a}P_j(da) + e^{\mu t}\tilde{B}_j(t), \quad j = 1, \cdots, n,$$

$$B_0(t) = f(S(t), I_1(t), \cdots, I_n(t)),$$

Table 2.5 Notations and definitions used in the models in this section.

Symbol	Definition
Dependent variables	
$S(t)$	number of susceptibles at time t
$u_j(t, \cdot)$	stage age density of infected individuals in the jth stage at time t
$\tilde{u}_j(\cdot)$	stage age density of infected individuals in the jth stage at time 0
$I_j(t)$	number of infected individuals in the jth stage at time t
$B_0(t)$	incidence (infection rate) at time t
$B_j(t)$	flux from the jth stage of infection into the $(j+1)$st stage at time t, $j = 1, \cdots, n-1$
$B_n(t)$	return rate from the nth (the last) stage of infection into the susceptible class
Parameters and parameter functions	
n	number of stages of infection
Λ	influx rate of new susceptibles
μ	basic per capita mortality rate
$1 - \mathcal{F}_j(a)$	probability to die from disease-related causes in the jth stage before stage a
$P_j(a)$	probability that the jth stage lasts longer than a time units
a_j	maximum sojourn time in the jth stage

where the forcing functions $\tilde{I}_j, \tilde{B}_j$ are given by

$$\tilde{I}_j(t) = \int_t^\infty \frac{\tilde{\mu}_j(A-t)}{\mathcal{F}_j(a-t)P_j(a-t)} \mathcal{F}_j(a)P_j(a)da,$$

$$\tilde{B}_j(t) = -\int_t^\infty \frac{\tilde{\mu}_j(A-t)}{\mathcal{F}_j(a-t)P_j(a-t)} \mathcal{F}_j(a)P_j(da).$$

The relevant dependent variables in this model formulation are $S, I_1, \cdots, I_n$, $B_0, \cdots, B_n$, but $B_0, \cdots, B_n$ can be eliminated. Note that

$$p_j = (-1) \int_0^\infty \mathcal{F}_j(a)e^{\mu a} P_j(da) \tag{2.127}$$

is the probability of getting through the jth stage alive;

$$q_1 = 1, \quad q_j = p_1 \cdots p_{j=1}, \quad j = 2, \cdots, n+1, \tag{2.128}$$

give the probabilities that an infected individual will survive all the stages 1 to $j-1$.

The reproductive number is given by

$$\mathcal{R}_0 = \sum_{j=1}^n \frac{\delta f}{\delta I_j}(\Lambda/\mu, 0, \cdots, 0)T_j q_j. \tag{2.129}$$

Theorem 2.21. *The disease-free equilibrium is locally asymptotically stable if $\mathcal{R}_0 < 1$. It is unstable if $\mathcal{R}_0 > 1$.*

At an endemic equilibrium we have

$$B_0^* = f(S^*, I_1^*, \cdots, I_n^*), \tag{2.130}$$

where

$$S^* - \frac{\Lambda - B_0^*(1 - q_{n+1})}{\mu}, \quad I_j^* = B_0^* T_j q_j, \quad j = 1, \cdots, n. \tag{2.131}$$

Note that the exponential solution is nontrivial if and only if $\bar{B}_0 \neq 0$. Substitution into the equation for $\bar{B}_0$ and division by $\bar{B}_0$ yields the characteristic equation

$$
\begin{aligned}
1 = & -\frac{1 - \Pi_{k=1}^n K_k(\lambda + \mu)}{\lambda + \mu} \frac{\delta f}{\delta S}(S^*, I^*) L_1(\lambda + \mu) \\
& + \Sigma_{j=2}^n \frac{\delta f}{\delta I_j}(S^*, I^*), L_j(\lambda + \mu) \Pi_{k=1}^{j-1} K_k(\lambda + \mu),
\end{aligned}
\tag{2.132}
$$

where

$$K_j(z) = -\int_0^\infty e^{-za} \mathcal{F}_j(a) P_j(da), \quad L_j(z) = \int_0^\infty e^{-za} \mathcal{F}_j(a) P_j(a) da.$$

Then, the following results were shown in [Feng and Thieme (2000a)].

Theorem 2.22. *An equilibrium solution with S^* and I_j^* giving the equilibrium numbers of individuals in the susceptible and the various infected stages is locally asymptotically stable if all roots of the characteristic equation (2.132) have strictly negative real part. The equilibrium solution is unstable if there exists at least one root with strictly positive real part.*

Theorem 2.23. *Let all partial derivatives $\frac{\delta f}{\delta I_j}(S^*, I)$ be nonnegative and monotone nonincreasing in $I \in (0, \infty)^n$ (with the canonical order). Further let $P_n \equiv 1$. Then the endemic equilibrium is locally asymptotically stable whenever it exists.*

This implies local asymptotic stability for mass action incidence and, if there are no disease fatalities, for standard incidence.

The stability properties will change when we introduce a quarantine class and consder a model of $SEIQR$ type with the incidence function

$$f(S, E, I, Q, R) = \kappa \frac{SI}{S + E + I + R}.$$

In this case, under certain conditions, the endemic equilibrium may lose stability when $\mathcal{R}_0 > 1$ and periodic solutions can occur (see Theorms 6.1 and 6.2 in [Feng and Thieme (2000a)]). We can apply the model results to Scarlet fever in England and Wales, as was done in [Feng and Thieme (1995)], and examine if the model with more realistic stage distributions can provide better estimates of the periods for oscillations. As mentioned in section 2.6.1, for scarlet fever in England and Wales (1897-1978), Anderson and May ([May and Anderson (1983)], Tables 1 and 2; [Anderson and May (1991)], Table 6.1) report an average age of infection between 10 and 14 years and an average life expectancy that rises from 60 to 70 years. We assume $D = 65$ years and $D_S = 12$ years that leads to a basic reproductive number of $\mathcal{R}_0^\diamond \approx 6.4$. Anderson and May ([Anderson and May (1991)], Table 3.1) give a latent period of 1 or 2 days, and infectious period of 14-21 days, and an incubation period of 2 or 3 days.

Assuming that children stay at home as soon as symptoms occur, we assume an effective infectious period of 1 or 2 days and an isolation period between 12 and 20 days. To be specific we assume $D_E = D_I = 1.5$ days. If we assume that the latency period and the effective infectious period are exponentially distributed we obtain

$$D_Q^\sharp = \frac{\mathcal{R}_0^\circ(\mathcal{R}_0^\circ + 1)}{2(\mathcal{R}_0^\circ - 1)}3 + \frac{\mathcal{R}_0^\circ}{2}1.5 \approx 18 \text{ days.}$$

If we assume fixed latency and effective infectious periods, then

$$D_Q^\sharp \approx 10 \text{ days.}$$

Finally we use the approach of interpolating between minimum and maximum durations by a piecewise linear duration function,

$$D_E^\circ = 1, \quad D_E^\bullet = 2, \quad D_I^\circ = 1, \quad D_I^\bullet = 2.$$

Then

$$D_E = D_I = 1.5, \quad \bar{D}_E = \bar{D}_I = \frac{7}{9}, \quad V_E = D_E(2\bar{D}_E - D_E) = \frac{1}{12}.$$

Because the length of the infectious period is symmetric about its mean, we can use the first inequality in (6.5) as equality

$$a_{22} = \frac{1}{3}D_I^2 - (D_I - \hat{D}_I)^2 = \frac{37}{162} \approx 0.228,$$

so

$$\frac{a_{22} + V_E}{(\bar{D}_I + D_E)^2} = \frac{0.228 + 0.083}{5.188} \approx 0.06,$$

which confirms our previous impression that this term is of minor practical importance in the evaluation of $D_Q^\sharp$. Finally we find

$$D_Q^\sharp \approx 10.4 \text{ days.}$$

We inferred from the data in [Anderson and May (1991)] that the isolation period should lie between 12 and 20 days. This agrees with [Anderson et al. (1962)] who report isolation periods for scarlet fever that last 2 or 3 weeks or even longer. The interepidemic period

$$\tau = 2\pi\sqrt{A(D_E + \bar{D}_I)} \tag{2.133}$$

gives an interepidemic period of 1.7 years, which is much lower than the observed 3-6 years [Anderson and May (1991)], Table 6.1). However, it must be noticed that, for this model, the formula (2.133) that has been obtained by linearization gives the interepidemic period only as long as the endemic equilibrium is stable and the oscillations are damped) or has just lost its stability such that the oscillations have small amplitudes, i.e., for isolation periods smaller than 11 days. Numerical calculations for the ODE model considered in Feng and Thieme [Feng and Thieme (1995)] have shown that the interepidemic period very sensitively increases with the length of the isolation period and that (2.133) can no longer be used for its determination.

2.6.2 *A two-strain influenza model with cross immunity*

Several studies have focused on the identification of mechanisms capable of support-ing multiple-strain coexistence for diseases that provide permanent or temporary immunity [Hethcote and Levin (1989); Hethcote (2000)]. Although there is still lim-ited understanding on the role of cross-immunity (form of interference competition) between strains of a given virus, host variability (behavioral and immunological) is known to play a key role in maintaining virus diversity. Influenza epidemics and pandemics are closely linked to two types of mechanisms that maintain viral genetic diversity: antigenic drift, the driver of strain heterogeneity, and antigenic shift, the generator of subtype variability [Thacker (1986)].

In 1918, the Spanish Flu pandemic caused the largest number of flu-related deaths worldwide in a single season [Thacker (1986)]. More than 500,000 people died in the United States with 20–50 million deaths worldwide. The Asian Flu, a result of an antigenic shift in the hemmaglutinin and neuraminidase surface proteins, was responsible for about 70,000 deaths in the United States in 1969 [CDC (2013)]. The most recent and least lethal pandemic, the Hong Kong pandemic, is attributed to the appearance of the H3N2 subtype [CDC (2013)].

The main focus of this study is on the identification of competitive outcomes (mediated by cross-immunity) that result from the interactions between two strains of influenza A in a population where sick individuals may be isolated. Single-strain susceptible-infected-quarantined-recovered (SIQR) models with vital dynamics can generate sustained oscillations [Feng and Thieme (1995); Hethcote *et al.* (2002)]. The introduction of a second strain increases the competition for susceptibles, a process mediated by cross-immunity in our setting. Will such competition preclude the possibility of sustained multistrain oscillations? We show that coexistence of both strains in the oscillatory regime is not uncommon and that oscillatory dynamics are possible for reasonable values of influenza parameters [Couch and Kasel (1983); Fox *et al.* (1982); Thacker (1986)].

Divide the population into ten different classes: susceptibles (S), infected with strain i $(I_i,$ primary infection), isolated with strain i (Q_i), recovered from strain i $(R_i,$ as a result of primary infection), infected with strain i $(V_i,$ secondary infection), given that the population had recovered from strains $j \neq i$, and recovered from both strains (W). Let A denote the population of non-isolated individuals and let $\frac{\beta_i S(I_i + V_i)}{A}$ be the rate at which susceptibles become infected with strain i. That is, the ith $(i \neq j)$ incidence rate is assumed to be proportional to both the number of susceptibles and the available modified proportion of i-infectious individuals, $\frac{(I_i + V_i)}{A}$. Let σ_{ij} denote a measure of the cross-immunity provided by a prior infection with strain i to exposure with strain j $(i \neq j)$. We use data from epidemiological studies conducted in Houston and Seattle [Taber *et al.* (1981); Fox *et al.* (1982)] to generate rough measures of cross-immunity. From these studies it is clear that $\sigma_{ij} \in [0, 1]$. A transition diagram is described in Figure 2.31.

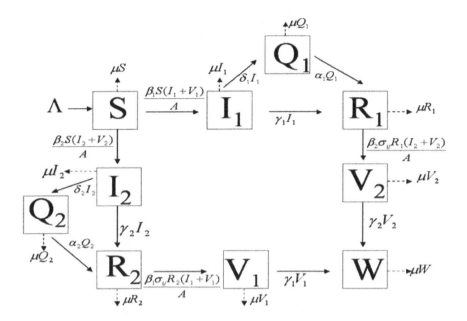

Fig. 2.31 Schematic diagram of the dynamics in host exposed to two co-circulating influenza strains. The cross-immunity parameter takes on values in $(0,1)$ with $\sigma_{ij} = 0$ corresponding to total cross-immunity while $\sigma_{ij} = 1$ indicates no cross-immunity between strains. We assume strong protection if $0 \leq \sigma_{ij} \ll 1$, and weak if $0 \ll \sigma_{ij} \leq 1$.

Using the flow diagram in Figure 2.31, we arrive at the model

$$\frac{dS}{dt} = \Lambda - \sum_{i=1}^{2} \beta_i S \frac{(I_i + V_i)}{A} - \mu S,$$

$$\frac{dI_i}{dt} = \beta_i S \frac{(I_i + V_i)}{A} - (\mu + \gamma_i + \delta_i)I_i,$$

$$\frac{dQ_i}{dt} = \delta_i I_i - (\mu + \alpha_i)Q_i,$$

$$\frac{dR_i}{dt} = \gamma_i I_i + \alpha_i Q_i - \beta_j \sigma_{ij} R_i \frac{(I_j + V_j)}{A} - \mu R_i, \qquad j \neq i \qquad (2.134)$$

$$\frac{dV_i}{dt} = \beta_i \sigma_{ij} R_j \frac{(I_i + V_i)}{A} - (\mu + \gamma_i)V_i, \qquad j \neq i$$

$$\frac{dW}{dt} = \sum_{i=1}^{2} \gamma_i V_i - \mu W,$$

$$A = S + W + \sum_{i=1}^{2}(I_i + V_i + R_i).$$

System (2.134) has at least four equilibria. Analysis of the local stability of the trivial equilibrium (absence of disease) helps identify conditions under which the "flu" can invade. Hence, we first focus on establishing the conditions that make it impossible (at least locally) for both strains to invade a disease-free population, simultaneously. In the analytical results obtained here it is assumed that $\sigma_{12} = \sigma_{21} = \sigma$.

The basic reproductive number for the two strains are

$$\mathcal{R}_i = \frac{\beta_i}{\mu + \gamma_i + \delta_i}.$$

Using the perturbation technique based on the fact that μ is a much smaller parameter than other parameters, we can identify two curves in the $(\mathcal{R}_1, \mathcal{R}_2)$ plane that determine the stability of the non-trivial equilibria. Let $f(\mathcal{R}_1)$ and $g(\mathcal{R}_2)$ be two functions defined by

$$f(\mathcal{R}_1) = \frac{\mathcal{R}_1}{1 + \sigma(\mathcal{R}_1 - 1)\left(1 + \frac{\delta_2}{\mu + \gamma_2}\right)\left(1 - \frac{\mu(\mu + \alpha_1)}{(\mu + \gamma_1)(\mu + \alpha_1) + \alpha_1 \delta_1}\right)} \tag{2.135}$$

and

$$g(\mathcal{R}_2) = \frac{\mathcal{R}_2}{1 + \sigma(\mathcal{R}_2 - 1)\left(1 + \frac{\delta_1}{\mu + \gamma_1}\right)\left(1 - \frac{\mu(\mu + \alpha_2)}{(\mu + \gamma_2)(\mu + \alpha_1) + \alpha_2 \delta_2}\right)}. \tag{2.136}$$

Let $\mathcal{R}_i^* = \mathcal{R}_i$ $(i = 1, 2)$ be evaluated at $\mu = 0$. Then the following result holds.

Theorem 2.24. *There exists a function $\alpha_{1c}(\mu)$ defined for small $\mu > 0$ by*

$$\alpha_{1c}(\mu) = \frac{\delta_1}{\mathcal{R}_1^*}\left(1 - \frac{1}{\mathcal{R}_1^*}\right) + O(\mu^{1/2}),$$

with the following properties: (i) The boundary endemic equilibrium E_1 is locally asymptotically stable if $\mathcal{R}_2 < f(\mathcal{R}_1)$ and $\alpha_1 < \alpha_{1c}(\mu)$, and unstable if $\mathcal{R}_2 > f(\mathcal{R}_1)$ or $\alpha_1 > \alpha_{1c}(\mu)$. (ii) When $\mathcal{R}_2 < f(\mathcal{R}_1)$, periodic solutions arise at $\alpha_1 = \alpha_{1c}(\mu)$ via Hopf-bifurcation for small enough $\mu > 0$. The period can be approximated by

$$T = \frac{2\pi}{|\Im\omega_{2,3}|} \approx \frac{2\pi}{\left((\gamma_1 + \delta_1)(\mathcal{R}_1^* - 1)\right)^{1/2}\mu^{1/2}}.$$

Because we have focused on the symmetric case, an analogous result for the second boundary equilibrium E_2 can be stated immediately. That is, the boundary endemic equilibrium E_2 is locally asymptotically stable if $\mathcal{R}_1 < g(\mathcal{R}_2)$ and $\alpha_2 < \alpha_{2c}(\mu)$. It becomes unstable if $\mathcal{R}_1 > g(\mathcal{R}_2)$ or $\alpha_2 > \alpha_{2c}(\mu)$. A summary of the stability results as presented in Theorem 2.24 for strain 1 is obtained for strain 2 by replacing the parameter indices 1's with 2's and replacing $f(\mathcal{R}_1)$ with $g(\mathcal{R}_2)$. Functions $f(\mathcal{R}_1)$ and $g(\mathcal{R}_2)$ help determine the stability and coexistence regions for strains 1 and 2. In fact, changes in the regions of stability for a single and both strains can be illustrated by varying the coefficient of cross-immunity. For instance, from (2.135) we can compute the value of σ at which

$$f'(\mathcal{R}_1) \equiv \frac{\partial f(\mathcal{R}_1, \sigma)}{\partial \mathcal{R}_1}\bigg|_{\sigma_1^*} = 0, \tag{2.137}$$

namely

$$\sigma_1^* = \frac{1}{\left(1 + \dfrac{\delta_2}{\mu + \gamma_2}\right)\left(1 - \dfrac{\mu(\mu + \alpha_1)}{(\mu + \gamma_1)(\mu + \alpha_1) + \alpha_1 \delta_1}\right)}.$$

Hence, for all $\mathcal{R}_1 > 1$,

$$f'(\mathcal{R}_1) > (<, =) \, 0, \qquad f(\mathcal{R}_1) > (<, =) \, 1 \qquad \text{if } \sigma < (>, =) \, \sigma_1^*.$$

These properties can be easily checked by noticing from (2.135) that

$$f(\mathcal{R}_1) = \frac{\mathcal{R}_1}{1 + \dfrac{\sigma}{\sigma_1^*}(\mathcal{R}_1 - 1)} \qquad \text{and} \qquad f'(\mathcal{R}_1) = \frac{1 - \dfrac{\sigma}{\sigma_1^*}}{\left(1 + \dfrac{\sigma}{\sigma_1^*}(\mathcal{R}_1 - 1)\right)^2}.$$

Using the symmetry between two strains, we can show that similar properties hold for another threshold value σ_2^* (interchanging the subscripts 1 and 2 in the expression of σ_1^*) and a function $\mathcal{R}_1 = g(\mathcal{R}_2)$. The properties of f and g are illustrated in Figure 2.32. The first two plots in Figure 2.32 are for the special case when the two strains have identical parameter values ($\sigma_1^* = \sigma_2^* = \sigma^*$), whereas the last two plots are for the case $\sigma_1^* \neq \sigma_2^*$. $\mathcal{R}_2 < f(\mathcal{R}_1)$ is a necessary condition for the stability of strain 1 (either a stable boundary endemic equilibrium E_1 or the equilibrium associated with strain 1 oscillations). Hence, E_2 is unstable when $\mathcal{R}_2 > f(\mathcal{R}_1)$. Similarly, E_1 is unstable when $\mathcal{R}_1 > g(\mathcal{R}_2)$. Hence, coexistence is expected when $\mathcal{R}_2 > f(\mathcal{R}_1)$ and $\mathcal{R}_1 > g(\mathcal{R}_2)$.

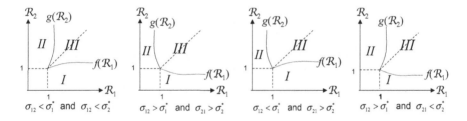

Fig. 2.32 Bifurcation diagrams in the $(\mathcal{R}_1, \mathcal{R}_2)$ plane for various combination of σ_1 and σ_2 values in relation to the threshold levels of σ_1^* and σ_2^* (cross-immunity). Regions I, II, III denote the existence and stability of E_1, E_2, and coexistence, respectively.

Figure 2.33 depicts the stability of the equilibria and periodic solutions when $(\mathcal{R}_1, \mathcal{R}_2)$ lies in Regions I and III. It illustrates how the stabilities of equilibria and periodic orbits change their stability when the parameters α_1 and α_2 change their values crossing the critical points α_{ic} ($i = 1, 2$).

"Flu" epidemic patterns include yearly outbreaks (antigenic drift), the explosive onset of outbreaks, the rapid termination of local epidemics (despite an "abundance" of susceptible individuals), and potentially major pandemics (antigenic shift). The continuous generation (most likely from random mutations in the NS gene) of new

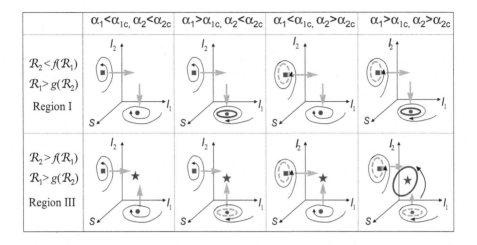

Fig. 2.33 Depiction of the stability properties of equilibria and periodic solutions for parameter values in different regions. A solid circle represents a stable boundary (strain 1 only) equilibrium. A solid square represents an unstable boundary (strain 2 only) equilibrium. A star represents a stable interior (coexistence of both strains) equilibrium. A solid (dashed) closed orbit represents a stable (unstable) period solution.

"flu" strains (minor genetic changes) and the sudden generation of subtypes (radical genetic changes) and their impact on the history of acquired (age-dependent) immunity of host populations make the study of influenza dynamics and its control challenging and fascinating [Fitch *et al.* (1997)].

The focus of this study is on the time evolution of influenza A in a nonfixed landscape driven by tight coevolutionary interactions (that is, interactions where the fate of the host and the parasite are intimately connected) between human hosts and competing strains. The process is mediated by intervention (behavioral changes) and cross-immunity. In other words, the nature of the invading landscape (susceptible host) changes dynamically from behavioral changes (isolation, short time scale) and past immunological experience (cross-immunity, long time scale).

The "partial" herd-immunity generated by past history of invasions on the host population can have a huge impact on the quantitative dynamics of the "flu" at the population level. The assumption that $\sigma_{12} = \sigma_{21} = \sigma$ for $i \neq j$ naturally results in a dynamic landscape that is not too different (in the oscillatory regime) than the one observed on single-strain models with isolation [Feng and Thieme (1995); Hethcote *et al.* (2002)]. That is, a lack of heterogeneity in cross-immunity results in a system "more or less" driven (in the oscillatory regime) by the process of isolation.

In all cases, sustained oscillations with periods that are consistent with influenza epidemics/pandemics are possible [Couch and Kasel (1983); Thacker (1986)]. These results are consistent with those obtained in single-strain models [Feng and Thieme (1995)] (i.e., sustained oscillations are preserved), except that the oscillations are now possible for "realistic" isolation periods. The introduction of a second strain enhances the possibilities. Numerical simulations illustrate various outcomes, including competitive exclusion, coexistence, and subthreshold coexistence. The interepidemic periods range from 2 to 10-13 years, depending on the levels of cross-immunity. Strong intermediate asymmetric cross-immunity leads to interepidemic periods in the range of 10-13 years. Symmetric cross-immunity reduces the range to 1-3 years. The results of intermediate (symmetric) cross-immunity are consistent with those found in [Castillo-Chavez (1987); Castillo-Chavez *et al.* (1989a)]. Documented evidence on the cocirculation of strains belonging to the same subtype [Couch and Kasel (1983); Thacker (1986)] appears to be consistent with these results.

Our results show that multiple strain coexistence is highly likely for antigenically distinct (weak cross-immunity) strains and not for antigenically similar under symmetric cross-immunity ("competitive exclusion" principle [Bremermann and Thieme (1989)]). As the levels of cross-immunity weaken, the likelihood of subthreshold coexistence increases. However, "full" understanding of the evolutionary implications that result from human host and influenza virus interactions may require the study of systems that incorporate additional mechanisms such as seasonality in transmission rates, age-structure, individual differences in susceptibility or infectiousness, and the possibility of coinfections of different strains. Thacker [Thacker (1986)] notes that the observed seasonality of influenza in temperate zones may be the key to observed patterns of recurrent epidemics. Superinfection may also be a mechanism worth consideration, even though studies in [Thacker (1986)] show that it is only moderately possible for young individuals to become infected with two different strains in one "flu" season.

The recent flu epidemic that has invaded all 50 states (3003–2004) and our experiences with the SARS epidemic [Chowell *et al.* (2003)] are a source of concern. While isolation and quarantine [Castillo-Chavez *et al.* (2003)] seem effective [Chowell *et al.* (2003)], they can destabilize flu dynamics (oscillations) and generate some level of uncertainty. The results in this study suggest the need to explore the long-term impact of current U.S. vaccination policies on the levels of cross-immunity generated by herd-immunity in the case of the flu. Whether or not they increase or reduce the likelihood of a future major outbreak is a question worth further modeling studies.

2.7 Coupled dynamics of biological processes

Most mathematical models of biological systems are concered with a single biological process. However, in many cases multiple processes are coupled together (e.g., epidemiology and immunology; epidemiology and genetics; etc.), and the dynamic interactions between these processes may generate outcomes that are different when the interactions are absent. For simplicity, epidemiological models are usually formulated by ignoring the influence from other processes or by assuming static inputs due to other processes. In this section, we consider two examples that explicitly couple multiple biological processes in a single model. One example concerns the epidemiological process of malaria transmission dynamics and the genetic process of sickle cell gene frequency, and the other example deals with the coupling of within-host and between-host dynamics. These results have been presented in [Feng *et al.* (2004b,c, 2012a)].

Clearly, when two coupled biological processes are considered in a model, the dimension of the system will in general increase significantly, which can make the mathematical analysis more difficult. Nonetheless, when the two processes occur on very different time scales, tools from singular perturbation theory may be helpful for reducing the full system to lower dimensional sub-systems that have the same qualitative behavior as the full system.

2.7.1 *Malaria epidemiology and sickle cell genetics*

As for dengue, malaria is also a mosquito-transmitted disease. One of the earliest model for malaria is the Ross-McDonald model. Let x and y denote the fractions of infectious human population and mosquito population, respectively. Let a and b denote the number of bites per unit time and the probability that a bite can produce infection, respectively. Other parameters are the duration of infection in human host, $1/\gamma$, the lifespan of mosquito, $1/\mu$, and the ratio of mosquito to human, $m = M/N$, where M and N denote the numbers of mosquitoes and humans, respectively. Then the RossMacDonald model takes the following form

$$
\begin{aligned}
\frac{dx}{dt} &= abm(1-x)y - \gamma x, \\
\frac{dy}{dt} &= ax(1-y) - \mu y,
\end{aligned}
\tag{2.138}
$$

with appropriate initial conditions. The basic reproductive number for the mdoel (2.138) is

$$
\mathcal{R}_0 = \sqrt{\frac{ma^2 b}{\mu\gamma}}.
$$

The square root reflects the fact that the life cycle of the parasite involves both the host and the vector. It can be shown that the disease will die out if $\mathcal{R}_0 < 1$ and persist if $\mathcal{R}_0 > 1$. Because $\mathcal{R}_0$ is a decreasing function of m and therefore M, the disease can be eliminated if a control measure can reduce the mosquito population M to be below a threshold such that $\mathcal{R}_0 < 1$. Our model in [Feng *et al.* (2004b)] is structured in the similar way except that it includes more epidemiological classes, as described below.

Although the population dynamics of malaria and the population genetics of the sickle-cell genes occur on very different time scales, it is straightforward to develop an appropriate model relating these. The high mortality associated with malaria has led to strong historical selection for resistance, and hence for single major genes conferring resistance in heterozygotes, despite the associated burden borne by homozygotes. We thus build on existing detailed knowledge of the genetics of resistance, focusing on a single locus with two alleles. Our foundation is thus the classical Ross-McDonald model for the spread of malaria, expanded to include the relevant genetic structure of the host.

Let u_1 denote the density of uninfected humans of genotype AA. Similarly, u_2 denotes the population density of genotype AS. Furthermore, let v_1 and v_2 represent the population densities of infected individuals of each genotype. We ignore SS individuals; high mortality rates from sickle-cell disease are typical in countries with high transmission rates of falciparum malaria, so these individuals rarely reach reproductive maturity. An extended model including the SS individuals can be studied using similar methods but it is very difficult to interpret the threshold conditions due to the complexity of the model. Finally, let z be the fraction of mosquitoes that are transmitting malaria. The fraction of the AS individuals in the population is

$$w = \frac{u_2 + v_2}{N}$$

where $N = u_1 + v_1 + u_2 + v_2$ is the total human population density. The frequency of the S gene is denoted by $q = w/2$ and the frequency of the A gene is denoted by $p = 1 - q$.

Let $b(N)$ denote the human per-capita birth rate, possibly density dependent (e.g., the logistic growth), with a constant per-capita natural mortality of m. To couple ecology and evolution, we make two assumptions. First, we assume that the ratio of mosquitoes to humans is a constant, c. This is a standard assumption in the modeling of malaria (ever since the original Ross-McDonald model). Any other assumption about variability in the ratio of mosquitoes to humans (M/H) would need to be justified.

Second, we assume that the fraction of each genotype born into the population P_i is given by

$$P_1 = p^2, \qquad P_2 = 2pq.$$

The transmission of malaria between humans and mosquitoes is governed by some basic epidemiological parameters. The human biting rate is denoted by a, and average life of an infected mosquito is $1/\delta$. The probability that a human develops a parasitemia from a bite is denoted θ_i; we assume that $\theta_1 \geq \theta_2$. The disease induced death rate is denoted by α_i, and we assume that $\alpha_1 \gg \alpha_2$. In addition, we consider that AS individuals may die faster than AA individuals from causes other than malaria, and the excess rate of mortality for AS individuals is ν. The probability that a mosquito acquires plasmodium from biting an individual of type i is denoted by ϕ_i. The average time until a victim of malaria recovers, denoted $1/\gamma_i$, may be different in AA and AS individuals.

The changes in population density of each genotype with each infection status are described by a set of five coupled ordinary differential equations:

$$\dot{u}_i = P_i b(N)N - m_i u_i - a\theta_i c z u_i + \gamma_i v_i,$$

$$\dot{v}_i = a\theta_i c z u_i - (m_i + \gamma_i + \alpha_i)v_i, \tag{2.139}$$

$$\dot{z} = (1 - z)\left(a\phi_1 \frac{v_1}{N} + a\phi_2 \frac{v_2}{N}\right) - \delta z, \qquad i = 1, 2,$$

where $m_1 = m$ and $m_2 = m + \nu$.

It is both mathematically convenient and biologically relevant to introduce new variables for prevalence of malaria infections in each genotype, $x_i = u_i/N$ and $y_i = v_i/N$, as well as the frequency of the S-gene, $w = x_2 + y_2 = 2q$. The equations in the new variables are derived from the original (2.139) using the chain rule. We note for clarification that

$$x_1 + y_1 + x_2 + y_2 = 1 \quad \text{and} \quad x_1 + y_1 = 1 - w.$$

We also introduce notation to reduce the number of parameters, $\beta_{hi} = a\theta_i c$, $\beta_{vi} = a\phi_i$, $i = 1, 2$. Then we obtain the following equivalent system to (2.139) in the terms that describe important epidemiological, demographic, and population genetic quantities, y_1, y_2, z, w, and N:

$$
\begin{cases}
\dot{y}_1 = \beta_{h1} z(1 - w - y_1) - (m_1 + \gamma_1 + \alpha_1)y_1 - y_1 \dot{N}/N, \\[2mm]
\dot{y}_2 = \beta_{h2} z(w - y_2) - (m_2 + \gamma_2 + \alpha_2)y_2 - y_2 \dot{N}/N, \\[2mm]
\dot{z} = (1 - z)(\beta_{v1}y_1 + \beta_{v2}y_2) - \delta z, \\[2mm]
\dot{w} = P_2 b(N) - \alpha_2 y_2 - m_2 w - w\dot{N}/N, \\[2mm]
\dot{N} = N\left((P_1 + P_2)b(N) - m_1(1 - w) - m_2 w - \alpha_1 y_1 - \alpha_2 y_2\right).
\end{cases}
\tag{2.140}
$$

Although most of the equations assume a general birth function $b(N)$, our detailed mathematical analysis for the specific case in which $b(N)$ is a density dependent per-capita birth function, $b(N) = b(1 - N/K)$, where b is a constant (the maximum birth rate when population size is small) and K is approximately the density dependent reduction in birth rate.

The relevant parameters vary across many orders of magnitude. For example, the demographic parameters (b and m_i) and the genetic parameters (α_i) are on the order of 1/decades, and the malaria disease parameters (β_{hi}, γ_i, β_{vi}, and δ) are on the order of 1/days. Hence, although the malaria disease dynamics and the changes in genetic composition are two coupled processes, the former occurs on a much faster time scale than the latter. Let $m_i = \epsilon \tilde{m}_i, \alpha_i = \epsilon \tilde{\alpha}_i$, and $b = \epsilon \tilde{b}$ with $\epsilon > 0$ being small. We can use this fact to simplify the mathematical analysis of the full model with the use of singular perturbation techniques, which allows us to separate the time scales of the different processes. By letting $\epsilon = 0$ we obtain the following system for the fast dynamics:

$$\begin{cases} \dot{y}_1 = \beta_{h1}z(1 - y_1 - w) - \gamma_1 y_1, \\ \\ \dot{y}_2 = \beta_{h2}z(w - y_2) - \gamma_2 y_2, \\ \\ \dot{z} = (1 - z)(\beta_{v1}y_1 + \beta_{v2}y_2) - \delta z, \end{cases} \qquad (2.141)$$

which describes the epidemics of malaria for a given distribution of genotypes determined by w. Here, on the fast time scale, w is considered as a parameter. On the fast time scale, the basic reproductive number of malaria disease can be calculated as the leading eigenvalue of the next generation matrix:

$$\mathcal{R}_0 = \mathcal{R}_1(1 - w) + \mathcal{R}_2 w, \qquad (2.142)$$

where $\mathcal{R}_i = \frac{\beta_{hi}\beta_{vi}}{\gamma_i \delta}$, $i = 1, 2$ involves parameters associated with malaria transmission between mosquitoes and humans of genotype i. In fact, $\mathcal{R}_i$ (or $\sqrt{\mathcal{R}_i}$) is the basic reproductive number when the population consists of entirely humans of genotype i. It can be shown that, when $\mathcal{R}_0 < 1$, the disease-free equilibrium of the system (2.141) is locally asymptotically stable; and when $\mathcal{R}_0 > 1$, the system (2.141) has a unique non-trivial equilibrium $E^* = (y_1^*, y_2^*, z^*)$ given by

$$y_1^* = \frac{T_{h1}z^*}{1 + T_{h1}z^*}(1 - w), \quad y_2^* = \frac{T_{h2}z^*}{1 + T_{h2}z^*}w, \qquad (2.143)$$

where z^* is the unique positive solution of a quadratic equation whose coefficients are functions of w, and

$$T_{hi} = \frac{\beta_{hi}}{\gamma_i}, \qquad i = 1, 2. \qquad (2.144)$$

By using the re-scaled time $\tau = \epsilon t$, we can re-write the full system (2.140) as

$$\begin{cases} \epsilon\dfrac{dy_1}{d\tau} = \beta_{h1}z(1 - y_1 - w) - \gamma_1 y_1 - \epsilon y_1\big((\tilde{m}_1 - \tilde{m}_2)w \\ \\ \qquad\qquad + \tilde{\alpha}_1(1 - y_1) - \tilde{\alpha}_2 y_2 + (P_1 + P_2)\tilde{b}(N)\big), \\ \\ \epsilon\dfrac{dy_2}{d\tau} = \beta_{h2}z(w - y_2) - \gamma_2 y_2 - \epsilon y_2\big((\tilde{m}_1 - \tilde{m}_2)(w - 1) \\ \\ \qquad\qquad - \tilde{\alpha}_1 y_1 + \tilde{\alpha}_2(1 - y_2) + (P_1 + P_2)\tilde{b}(N)\big), \\ \\ \epsilon\dfrac{dz}{d\tau} = (1 - z)(\beta_{v1}y_1 + \beta_{v2}y_2) - \delta z, \\ \\ \dfrac{dw}{d\tau} = \big((1 - w)P_2 - wP_1\big)\tilde{b}(N) + (\tilde{m}_1 - \tilde{m}_2)w(1 - w) \\ \\ \qquad\qquad + \tilde{\alpha}_1 wy_1 - \tilde{\alpha}_2(1 - w)y_2, \\ \\ \dfrac{dN}{d\tau} = N\big((P_1 + P_2)\tilde{b}(N) - \tilde{m}_1(1 - w) - \tilde{m}_2 w - \tilde{\alpha}_1 y_1 - \tilde{\alpha}_2 y_2\big). \end{cases} \qquad (2.145)$$

This system has a two-dimensional slow manifold:

$$M = \{(y_1, y_2, z, w, N) : y_1 = y_1^*(w, N), \ y_2 = y_2^*(w, N), \ z = z^*(w, N)\},$$

which is normally hyperbolically stable as it consists of a set of such equilibria of the fast system (2.141). Here y_1^* and y_2^* are given in (2.143). The slow dynamics on M is described by the equations

$$
\begin{cases}
\dfrac{dw}{d\tau} = \left((1 - w) P_2 - w P_1\right) b(N) + (\tilde{m}_1 - \tilde{m}_2) w(1 - w) \\[2mm]
\qquad\qquad + \tilde{\alpha}_1 w y_1^* - \tilde{\alpha}_2 (1 - w) y_2^*, \\[4mm]
\dfrac{dN}{d\tau} = N\left((P_1 + P_2)\, \tilde{b}(N) - \tilde{m}_1(1 - w) - \tilde{m}_2 w - \tilde{\alpha}_1 y_1^* - \tilde{\alpha}_2 y_2^*\right).
\end{cases}
\tag{2.146}
$$

Define the fitness of the S-gene by

$$\mathcal{F} = \left(\frac{1}{w}\frac{dw}{d\tau}\right),$$

which represents the initial per-capita growth of the S-gene. Then the following formula can be derived:

$$\left(\frac{1}{w}\frac{dw}{d\tau}\right)\bigg|_{w=0} = (\tilde{m}_1 + W_1\tilde{\alpha}_1) - (\tilde{m}_2 + W_2\tilde{\alpha}_2),
\tag{2.147}$$

where

$$W_1 = \frac{T_{h1}(\mathcal{R}_1 - 1)}{(1 + T_{h1})\mathcal{R}_1}, \qquad W_2 = \frac{T_{h2}(\mathcal{R}_1 - 1)}{(1 + T_{h1})\mathcal{R}_1 + T_{h1} - T_{h2}}.
\tag{2.148}$$

Let

$$\sigma_i = \tilde{m}_i + W_i\tilde{\alpha}_i.
\tag{2.149}$$

Then $\sigma_i \geq 0$ is the total per-capita death rate of type i individuals weighted by W_i, which depends only on malaria epidemiological parameters. The biological interpretation of $\mathcal{F}$ suggests that, when the S-gene is initially introduced into a population, it may or may not establish itself depending on whether the fitness is positive or negative, which is equivalent to whether $\sigma_2 < \sigma_1$ or $\sigma_2 > \sigma_1$. This is indeed confirmed by both analytical and numerical studies of the slow system. Figure 2.34(a) shows a bifurcation diagram of the slow dynamics with σ_1 and σ_2 being the bifurcation parameters. In Figure 2.34, $\tilde{b}_1^*$ is a constant larger than the maximum per-capita birth rate $\tilde{b}$ ($=b/\epsilon$); $\sigma_2 = h(\sigma_1)$ is a decreasing function satisfying $h(\tilde{b}) = \tilde{b}$ and $h(\tilde{b}^*) = 0$. Some of the results from this bifurcation diagram are summarized as follows:

Case 1: $\sigma_2 < \sigma_1$ (positive fitness).
  (a) If $\sigma_1 \leq \tilde{b}$, or $\tilde{b} < \sigma_1 < \tilde{b}^*$ and $\sigma_2 < h(\sigma_1)$, then there is a unique interior
      equilibrium $E_* = (w_*, N_*)$ that is globally asymptotically stable (g.a.s.),

(b) If $\tilde{b} < \sigma_1 < \tilde{b}^*$ and $\sigma_2 > h(\sigma_1)$, then the population will be wiped out (due to the death rates being too much higher than the "birth" rate) with the fraction of AS individuals tending to a positive constant as $t \to \infty$.

Case 2: $\sigma_2 > \sigma_1$ (negative fitness). The fraction of AS individuals will tend to zero as $t \to \infty$, whereas the total population size will tend to either K (when σ_2 is small) or zero (when σ_2 is large).

In either case, the system (2.146) has neither periodic solutions nor homoclinic loops.

An analytic proof of these results can be found in [Feng *et al.* (2004c)]. We point out that Case 1(b) is due to the standard incidence form of infection rate used in the z equation. Similar scenarios have been observed in other population models (see for example [Diekmann and Heesterbeek (2000)]), and such scenarios may not be present if the mass action form is used. The standard incidence form is more appropriate if the number of contacts is relatively constant, independent of density. Figures 2.34(b)-(e) demonstrate some numerical calculations of solutions of the system (2.146) for (σ_1, σ_2) in different regions. It shows a couple of possible scenarios when $\sigma_1 < \sigma_2$. It is interesting to notice that it is possible for the system to have two locally asymptotically stable equilibria. One is the boundary equilibrium at which $N > 0$ and $w = 0$, and the other is one of the two interior equilibria. The regions of attraction of the two stable equilibria are divided by the separatrix formed by the stable manifold of the unstable interior equilibrium. This type of bi-stability can occur in several different ways. There are also cases when the system (2.146) has none, or one, or two equilibria on the positive w-axis. It shows that, if $w(0)$ is small, i.e., the initial population size of AS individuals is small, then the S-gene will be extinct due to a negative fitness. However, if for some reason (e.g., immigration of AS individuals) w suddenly becomes large (large enough to be on the right side of the separatrix), then the S-gene will be able to establish itself, even though the fitness is negative.

These results can be used to study questions related to the evolution of associated traits. We see from above that whether or not the S-gene can invade and establish itself in a population is determined by whether the fitness coefficient is positive or negative. Recall that the fitness is given by the difference $\sigma_1 - \sigma_2$, where the total death rate σ_i is a sum of weighted death rates m_i and α_i (see (2.149)) with the weight W_i given by (2.148). The quantity W_i contains all the malaria transmission parameters. Noticing that $\tilde{m}_1 - \tilde{m}_2 = -\tilde{\nu}$ and using (2.148) we have

$$\mathcal{F} = -\tilde{\nu} + \frac{\tilde{\alpha}_1 T_{h1}(\mathcal{R}_1 - 1)}{(1 + T_{h1})\mathcal{R}_1} - \frac{\tilde{\alpha}_2 T_{h2}(\mathcal{R}_1 - 1)}{(1 + T_{h1})\mathcal{R}_1 + T_{h1} - T_{h2}}. \qquad (2.150)$$

Let

$$T_{vi} = \frac{\beta_{vi}}{\delta}, \qquad i = 1, 2. \qquad (2.151)$$

Then $\mathcal{R}_i = T_{hi}T_{vi}$. Notice that T_{hi} involves parameters related to malaria infection of humans of genotype i by mosquitoes, and T_{vi} involves parameters related to

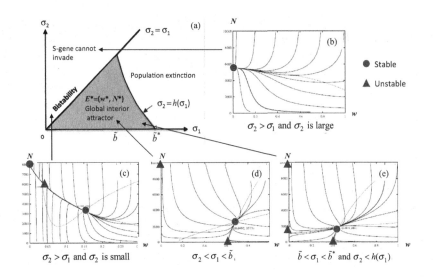

Fig. 2.34   Phase portraits (a) of the slow system in the (σ_1, σ_2) plane. Plots (b)-(e) illustrate the phase portraits of the slow system for (σ_1, σ_2) in different regions.

malaria infection of mosquitoes by humans of genotype i. Clearly, these transmission coefficients affect $\mathcal{F}$ in nonlinear ways.

To assess how the frequency of S-gene may influence the endemic level of malaria, we examine the threshold quantity $\mathcal{R}_0$ given in (2.142). Rewrite (2.142) as $\mathcal{R}_0 = (\mathcal{R}_2 - \mathcal{R}_1)w + \mathcal{R}_1$. It is easy to see that $\mathcal{R}_0$ is either a decreasing function of w if $\mathcal{R}_2 < \mathcal{R}_1$, or an increasing function of w if $\mathcal{R}_2 > \mathcal{R}_1$. Recall that $w = 2q$ and q is the frequency of S-gene in the population. Relative magnitudes of $\mathcal{R}_1$ and $\mathcal{R}_2$ are determined by several epidemiological parameters. Figure 2.35 illustrates an example by changing γ_2 ($1/\gamma_2$ is the duration of malaria infection in AS individuals). It shows that $\mathcal{R}_0$ increases with w for smaller values of γ_2 and decreases with w for larger values of γ_2. In general, the disease prevalence increases with $\mathcal{R}_0$. This is confirmed by numerical simulations of the fast system (2.141), which is shown in Figures 2.36(a,b). $y_1 + y_2$ is the fraction of the population infected with malaria. Notice that in Figure 2.36(b), higher S-gene frequencies correspond to higher endemic levels of malaria at equilibrium.

Thus, increased duration of parasitaemia in heterozygotes (decreasing γ_2) leads to higher endemic prevalence of malaria and increased selection for the S-gene. These changes have very little effect on $\mathcal{F}$. This raises the interesting question of whether traits that affect γ_2 are under selection. Thus, traits associated with disease transmission may coevolve with traits associated with disease resistance.

As discussed in [Feng *et al.* (2004b)], by coupling the dynamics of the epidemiology of malaria and the genetics of sickle cell gene, our model allows for joint investigation of (1) impact of malaria on the selection of S-gene, (2) influence of

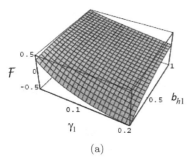

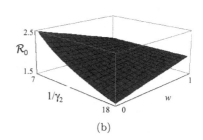

(a)   (b)

Fig. 2.35   Plot of the fitness $\mathcal{F}$ vs. recovery rate γ_1 and transmission rate β_{h1} (a) and the reproductive number $\mathcal{R}_0$ vs. $1/\gamma_2$ and w (b). It illustrates the impact of malaria on the S-gene fitness and the influence of S-gene frequency on $\mathcal{R}_0$.

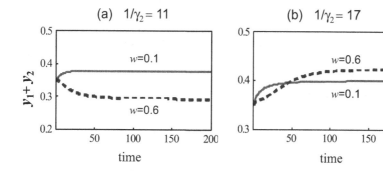

Fig. 2.36   Plots of solutions of the fast system for lower infectious period $1/\gamma_2$ (a) and higher infectious period (b) in type 2 host. The sum $y_1 + y_2$ represents the fraction of the population infected with malaria.

genetic composition of a population on the maintenance of malaria, and (3) evolution of associated traits. Our results are based on threshold conditions derived from our model by separating malaria disease dynamics on the fast time scale and the dynamics of S-gene on the slow time scale and by conducting stability analysis. The epidemic threshold condition $\mathcal{R}_0 > 1$ (under which malaria is endemic) is related to the S-gene frequency through w, and the threshold condition for the fitness of S-gene $\mathcal{F} > 0$ (under which the rare gene is able to invade and maintain itself) depends on epidemiological parameters as well as the endemic level of malaria. We illustrate the uses of these thresholds for the studies of questions related to (1)-(3). These results cannot be obtained from epidemiology models without genetic or genetic models without epidemics.

Standard population genetic models often use discrete-generations, and assume that when both parents are heterozygous, a fourth of all births are S-gene homozygotes. In our models, compensatory reproductive decisions can reduce the fitness costs associated with the S-gene [Hasting (2000)]. For example, S-gene homozygote children may be lost *in utero* or early in infancy. In these cases, the interbirth interval may be shorter following the birth of S-gene homozygote offspring. Alternatively, voluntary decisions to limit family size may reduce the fitness cost of the S-gene; pairings to S-gene heterozygotes will tend to have the same number of children. The net effect in both cases is a marginal delay in the birth rate, and the total fitness cost of the S-gene is somewhat lower than that predicted by the standard models.

The S-gene may affect the expression of several traits associated with malaria transmission dynamics. The S-gene can improve fitness by reducing the probability of becoming parasitaemic, by reducing the duration of a parasitaemia, or by reducing the probability of developing malaria per parasitaemia. If one of the former mechanisms is responsible for the enhanced fitness of S-gene heterozygotes, the population would benefit from increased frequency of the S-gene because heterozygotes would be a sink for Plasmodium, serving the same function as alternative hosts. Alternatively, the S-gene may reduce disease, influencing the expression of traits that increase the selective pressure acting on it. The evidence is mixed, but tends to support the notion that the gene is acting selfishly. One experiment challenged individuals with infectious mosquitoes and showed that 2/15 S-gene heterozygotes developed parasitaemia compared with 14/15 homozygotes lacking the S-gene [Allison (1954)]. On the other hand, the prevalence of parasitaemia is similar in S-gene heterozygotes compared with non-S homozygotes [Allen *et al.* (1992); Le Hesran *et al.* (1999); Lell *et al.* (1999); Stirnadel *et al.* (1999)]. If there is a real difference in the probability of becoming parasitaemic, but no real difference in the prevalence of parasitaemia, parasitaemia may last longer in S-gene heteroygotes. Other evidence suggests that transmission to mosquitoes is higher from S-gene heterozygotes [Trager and Gill (1992); Robert *et al.* (1996); Drakeley *et al.* (1999); Trager *et al.* (1999)]. On balance, it seems that increased frequency of the S-gene leads to enhanced transmission rates for Plasmodium.

There are several intrinsic shortcomings in the Macdonald [Macdonald (1957)] model of malaria transmission, on which our model is based. In particular, it does not account for the complex effects of acquired immunity on transmission [Dietz *et al.* (1974)], for fluctuations in transmission intensity [McKenzie *et al.* (2001a)], or for the existence of multiple parasite genotypes and meiotic recombination among them [McKenzie *et al.* (2001b)]. These shortcomings may have unexpected importance in the current context, in that recent evidence suggests that sickle-cell trait may differentially affect different parasite genotypes, as defined at immunogenic loci, and may influence superinfection frequencies [Ntoumi *et al.* (1997b,a)]. Our model includes several other simplifications with respect to empirical data; for instance, our assumption that all sickle homozygotes (SS) die is only an approximation [Aluoch (1997)]. Furthermore, relationships between malaria prevalence (or incidence) and malaria-induced mortality (or morbidity) are far more complex than assumed here: Plasmodium infection is necessary but by no means sufficient to produce disease in malaria [Marsh and Snow (1999); Smith *et al.* (2001)].

2.7.2 *Coupling within- and between-host dynamics*

In [Feng *et al.* (2012a)], we presented a new model for the linking of within- and between-host dynamics. We use this as a conceptual model for the dynamics of *Toxoplasma gondii*, in which the parasite's life cycle includes interactions with the environment. We postulate the infection process to depend on the size of the infective inoculum that susceptible hosts may acquire by interacting with a contaminated environment. Because the dynamical processes associated with the within- and between-host occur on different time scales, the model behavior can be analyzed by using a singular perturbation argument, which allows us to decouple the full model by separating the fast- and slow-systems. We define new reproductive numbers for the within-host and between-host dynamics for both the isolated systems and the coupled system. Particularly, the reproductive number for the between-host (slow) system dependent on the parameters associated with the within-host (fast) system in a very natural way. We show that these reproductive numbers determine the stability of the infection-free and the endemic equilibrium points. Our model may present a so-called backward bifurcation.

For most infectious diseases there are two key processes in the host-parasite interaction. One is the epidemiological process associated with the disease transmission, and the other is the immunological process at the individual host level. Viral dynamic models (e.g., Anderson and May [Anderson and May (1991)], De Boer and Perelson [De Boer and Perelson (1998)], Nowak and May [Nowak and May (2000)], Wodarz [Wodarz (2007)]) consider the within-host dynamics independent of the interaction at the population level, whereas epidemiological models of population dynamics (e.g., Anderson and May [Anderson and May (1991)], Thieme [Thieme (2003)] and references therein) consider the interaction between susceptible and infected hosts without an explicit link to the viral dynamics within the hosts. There are, however, questions that can only be studied by using models that explicitly link the two processes. Such questions include: i) How does the within-host dynamics influence the transmission of a pathogen from individual to individual? ii) What is the effect of population dynamics of disease transmission on the viral dynamics at the individual level? iii) Will the model predictions in terms of the virulence and basic reproductive number of the pathogen be altered if the two processes are dynamically linked ([Feng *et al.* (2012b)])? Gilchrist and Coombs [Gilchrist and Sasaki (2002)] have used a nested model to evaluate the direction of natural selection (in the study of evolution of virulence) at the within- and between-host levels. In this nested model, the within-host system is independent of the transmission dynamics at the population level.

In this study, we propose a framework that explicitly links the epidemiological and immunological dynamics through an environmental compartment. Our approach is based on the idea of separating biological time scales, a fast time scale associated with the within-host dynamics, an intermediate time scale associated with the epidemiological process, and a slow time scale associated with the environment. In a simpler case the processes associated with epidemiology and environment can be merged. We demonstrate our framework by using as a simplified model system for the infection by *Toxoplasma gondii*. *Toxoplasma* is an infectious disease for which contamination of the environment plays a major and determinant role in transmission. One of the advantages of this modeling approach (relative to other approaches for modeling within-host between-host transmission processes) is that the explicit linkage between the two processes can be established through environmental contamination, and it allows to postulate an 'inoculum' size related to the degree of infectiousness of the contaminated environment. To the best of our knowledge, all existing models that attempt to couple immunological and epidemiological dynamics confront the difficult and controversial questions as to how to model the influence of the epidemiological dynamics on the within-host cellular infection. We propose, as a first step in that direction, the study of infectious diseases where the environmental component is important.

For more information about the complex life cycle of *Toxoplasma gondii*, the recently published studies [Sullivan *et al.* (2012)] provide detailed descriptions relevant to mathematical modeling of *Toxoplasma*. Only a brief description is provided in this section. *Toxoplasma gondii* is an obligate intracellular parasite that can infect all warm-blooded vertebrates, including mammals and birds. Infections in humans reaches 30% and can cause encephalitis in immunocompromised persons such as AIDS patients or recipients of organ transplants. Infection acquired during pregnancy may cause severe damage to the fetus. *T. gondii* has a complex life cycle and a simplified diagram of the cycle is depicted in Figure 2.37.

The parasite reproduces sexually in felines. Once the cat becomes infected, it sheds oocysts, which contaminate the environment. These oocysts can be ingested by mammals and birds that then become infected with the parasite [Sullivan *et al.* (2012)]. Eating another organism that is infected can also infect the secondary hosts. These are the main facts that will be used in our model below, although the parasite's life cycle within a host is much more complicated. Again, following Sullivan *et al.* [Sullivan *et al.* (2012)] and references therein we know that, within a host, *T. gondii* exists in two stages that can be reached intermitently: bradyzoites and tachyzoites. Bradyzoites are the slow-growing and encysted form, whereas tachyzoites are the fast-replicating parasites, which lead to the acute phase of infection. After cysts are ingested by the host, the cysts are digested and bradyzoites, which are resistant to gastric conditions in the stomach, will subsequently invade the host's epithelial cells and convert into tachyzoites. While most of tachyzoites in immunocompetent hosts are eliminated by the innate and adaptive immune responses, some tachyzoites differentiate into the dormant bradyzoite stage inside host cells. The differentiation of tachyzoites into the bradyzoite stage plays an essential role in the

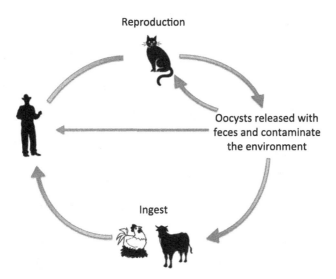

Fig. 2.37   A depiction of the T. gondii life cycle relevant to the model presented here.

development of tissue cysts, which allows life-long persistence of the parasites in the host. Reactivation of bradyzoites back to tachyzoites can lead to life threatening infection.

The between-host dynamics is governed by the SI system

$$\frac{dS}{dt} = \mu N - \lambda E S - \mu S,$$

$$\frac{dI}{dt} = \lambda E S - \mu I, \tag{2.152}$$

$$N = S + I.$$

The variables $S = S(t)$ and $I = I(t)$ represent the numbers on susceptible and infectious individuals, respectively, at time t. The parameters μ is the per-capita host natural mortality rate, which is assumed to be the same as the per-capita birth rate so that the total population size N remains constant for all time t. The parameter λ denotes the per-capita infection rate of hosts in a contaminated environment, and $E = E(t)$ denotes the level of environment contamination at time t, or the concentration of oocysts per unit area or volume of a region being considered ($0 \leq E \leq 1$). This level of contamination is dependent on the number of infected hosts (I) and the average parasite load (V) within the host as described by the equation

$$\frac{dE}{dt} = \theta I V (1 - E) - \gamma E, \tag{2.153}$$

where θ is the rate of contamination and γ is the clearance rate.

For the within-host dynamics, the sub-system reads

$$\frac{dT}{dt} = \Lambda - kVT - mT,$$

$$\frac{dT^*}{dt} = kVT - (m+d)T^*, \tag{2.154}$$

$$\frac{dV}{dt} = g(E) + pT^* - cV.$$

The variables T, T^*, and V represent the density of healthy cells, infected cells, and parasite load, respectively. The parameter k denotes the per-capita infection rate of cells; m and d are the per capita background mortality and infection-induced mortality rates of cells, respectively; p is the parasite reproductive rate by an infected cell; and c is the within-host mortality rate of parasites. The function $g(E)$ represents the rate at which an average host is inoculated.

In order to link the environmental contamination with the infection process at the individual level, we must postulate a mechanism. In this work we assume that environmental contamination is measured by or related to the concentration of pathogenic forms living in the environment, and that hosts acquire infection by ingesting contaminated food (which exists in the environment). The function g expresses the fact that if the environmental contamination is high then the inoculum (the average per capita concentration of pathogen's infectious forms introduced into a given host) is also high. These biological considerations suggest that the function g should have the following properties:

$$g(E) \geq 0, \quad g(0) = 0, \quad g'(E) > 0. \tag{2.155}$$

Although in general the function g may reach a saturation level, we consider here a simpler linear form as our aim is to illustrate the framework of linking within- and between-host dynamics. A saturating "functional response" will be explored with detail in further studies. Thus, we consider the linear form

$$g(E) = aE, \tag{2.156}$$

where a is a positive constant.

We point out that the inclusion of the inoculation $g(E)$ is the key for linking the within-host dynamics to the between-host dynamics. In directly-transmitted diseases, this function is of the form pT and is directly associated with the target cells that the pathogen infects, whereas in the current model the link is established though the external variable E. In the absence of this link, the model has the same form as the cell-virus models usually used for HIV dynamics. We have previously analyzed variations of such models (see [Rong *et al.* (2007a,c,b)]).

Let $\mathcal{R}_v(E)$ denote the within-host reproductive number, which is a function of the environmental contamination level, and let

$$\mathcal{R}_{v0} = \mathcal{R}_v(0).$$

.

The formula for $\mathcal{R}_v(E)$ when $E > 0$ will be specified later. It is easy to derive that

$$\mathcal{R}_{v0} = \frac{T_0 kp}{c(m+d)}. \tag{2.157}$$

The biological meaning of $\mathcal{R}_{v0}$ is clear. T_0 is the total target cells in the absence of parasite; p/c is the net virion production per cell before clearance, and $k/(m+d)$ is the infection rate during an average timespan of an infected cell. Thus, $\mathcal{R}_{v0}$ represents the baseline within-host reproductive number. We have shown that U_0 is l.a.s. if $\mathcal{R}_{v0} < 1$ and unstable if $\mathcal{R}_{v0} > 1$.

Individual- and population-based models have strengths and weaknesses. Individual-based models capture the chance nature of interpersonal contacts and permit concurrent membership in multiple risk groups (e.g., households and schools or workplaces). Results are presented as frequency distributions from multiple realizations of stochastic processes, allowing policymakers to determine the risk of outcomes more extreme than desired under particular conditions. In contrast, the systems of differential equations comprising population-level models can often be analyzed for general insights. Moreover, their fewer parameters can be more easily estimated from observations. And deficiencies are easier to remedy by comparing predictions to observations and determining the cause of any discrepancies. While existing formulae represent contacts within sub-populations (e.g., age classes) and between each such group and all others, recently published empirical studies of encounters by which respiratory diseases might be transmitted indicate that parents and children and co-workers also mix preferentially. We generalize the model of [Jacquez *et al.* (1988)] to include these contacts explicitly, permitting more realistic assessments of the risks.

The subsystem for the within-host dynamics (2.154) can be considered as the fast system in which the variable E is treated as a constant (i.e., it is not changing with time on the fast time scale).

Consider the case when $g(E) > 0$. Let $\tilde{U}(E) = \left(\tilde{T}(E), \tilde{T}^*(E), \tilde{V}(E)\right)$ denote a nontrivial equilibrium (i.e., $\tilde{T}^*(E) > 0, \tilde{V}(E) > 0$). Then

$$\tilde{V}(E) = \frac{1}{c}\left(g(E) + p\tilde{T}^*(E)\right), \quad \tilde{T}^*(E) = \frac{m}{m+d}(T_0 - \tilde{T}(E)), \tag{2.158}$$

and $\tilde{T}(E)$ is a solution of the following quadratic equation:

$$T^2 - a_1 T + a_2 = 0 \tag{2.159}$$

where

$$a_1 = \frac{g(E)[m+d]}{pm} + T_0\left[1 + \frac{1}{\mathcal{R}_{v0}}\right] > 0,$$

$$\tag{2.160}$$

$$a_2 = \frac{T_0^2}{\mathcal{R}_{v0}} > 0.$$

Thus, (2.159) has two positive real solutions given by

$$\tilde{T}_{\pm}(E) = \frac{1}{2}\left(a_1 \pm \sqrt{a_1^2 - 4a_2}\right). \tag{2.161}$$

Note that $a_1'(E) = g'(E)(m+d)/(pm) > 0$ and that

$$\tilde{T}_{\pm}'(E) = \frac{1}{2}a_1'(E)\left(1 \pm \frac{a_1}{\sqrt{a_1^2 - 4a_2}}\right).$$

Let $\tilde{U}_{\pm}(E)$ denote the nontrivial equilibria corresponding to $\tilde{T}_{\pm}(E)$. From $\tilde{T}_{+}'(E) > 0$ and $\tilde{T}_{+}(0) \geq T_0$ we know that $\tilde{T}_{+}(E) > T_0$ for all $E > 0$. But $\tilde{T}_{+}(E) \leq T_0$; and thus, $\tilde{U}_{+}(E)$ does not exist. We know that $\tilde{U}_{-}(E)$ exists when $\mathcal{R}_{v0} > 1$ and it coincides with U_0 when $\mathcal{R}_{v0} \leq 1$.

The existence condition for $\tilde{U}_{-}(E)$ motivates the following definition for $\mathcal{R}_v(E)$ in the case of $E > 0$

$$\mathcal{R}_v(E) =: \frac{T_0}{\tilde{T}_{-}(E)} \tag{2.162}$$

where $\tilde{T}_{-}(E)$ is given in (2.161).

Let $\tilde{U}_{-}(E) = (\tilde{T}_{-}(E), \tilde{T}_{-}^*(E), \tilde{V}_{-}(E))$ where $\tilde{T}_{-}^*(E)$ and $\tilde{V}_{-}(E)$ are given by (2.158) in which $\tilde{T}(E)$ is replaced by $\tilde{T}_{-}(E)$. The existence and stability of equilibria for the fast system are summarized in the following result.

Theorem 2.25. *Let $g(E) > 0$ and let $\mathcal{R}_v(E)$ be defined in (2.162).*

(i) $\mathcal{R}_v(E)$ is an increasing function of E.
(ii) The fast system has a unique nontrivial equilibrium $\tilde{U}_{-}(E)$ if and only if $\mathcal{R}_{v0} > 1$.
(iii) U_0 is l.a.s. when $\mathcal{R}_{v0} < 1$ and unstable when $\mathcal{R}_{v0} > 1$. For all $E > 0$, $\tilde{U}_{-}(E)$ is always l.a.s. whenever it exists.

Because $\mathcal{R}_{v0}$ does not depend on E, the influence of E on $\mathcal{R}_v(E)$ and on $\tilde{T}_{-}(E)$ can be dependent on the magnitude of $\mathcal{R}_{v0}$. Such a dependence is illustrated in Figure 2.38. This figure plots the curves $\mathcal{R}_v(E)$ and $\tilde{T}_{-}(E)$ for two different $\mathcal{R}_{v0}$ values, one with $\mathcal{R}_{v0} > 1$ (the left figure) and the other with $\mathcal{R}_{v0} < 1$ (the right figure). We observe that while $\tilde{T}_{-}(E)$ is more sensitive to E in the case when $\mathcal{R}_{v0} < 1$ than in the case when $\mathcal{R}_v(0) > 1$, $\mathcal{R}_v(E)$ is more sensitive to changes in E when $\mathcal{R}_{v0} > 1$. In this figure, we have used a linear function for $g(E) = aE$ where a is a constant and, for demonstration purposes, the parameter values used are $\Lambda = 5000, m = 0.31, d = 0.1, p = 10^4, c = 20, a = 2 \times 10^7$. The two $\mathcal{R}_{v0}$ values used in the two plots are determined by using different k values (e.g., $k = 10^{-7}$ for $\mathcal{R}_{v0} = 2$).

Numerical simulations of the full sytem for the case of $\mathcal{R}_{v0} > 1$ are illustrated in Figure 2.39. The solution curves for infected T-cells (T^*) and viruses (V) are shown in (a), while the solution curves for the level of environmental contamination (E) and the prevalence at the population level (I/N) are shown in (b). We observe

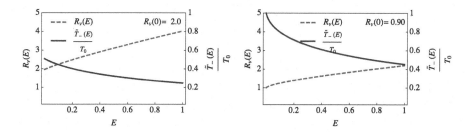

Fig. 2.38 Plots of the within-host reproductive number $\mathcal{R}_v^{\pm}(E)$ as a function of environmental contamination level E for different $\mathcal{R}_{v0}$ values. The left figure is for the case when $\mathcal{R}_{v0} = 2 > 1$, in which case $\lim_{E \to 0} \mathcal{R}_v^+(E) = \mathcal{R}_{v0}$, whereas the figure on the right is for the case of $\mathcal{R}_{v0} = 0.9 < 1$, in which case $\lim_{E \to 0} \mathcal{R}_v^-(E) = \mathcal{R}_{v0}$.

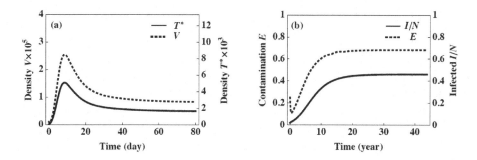

Fig. 2.39 Simulation results of the full system for the case of $g(E) = aE > 0$. The left figure plots the fast variables T^* and V while the right figure plots the slow variables I and E. We observe that the system stabilizes at the nontrivial equilibrium $\tilde{U}_-$ at $t \to \infty$. The parameter values used are: $\mu = 0.0004, \lambda = 0.0005, \gamma = 0.02, a = 3 \times 10^5, \theta = 10^{-10}, \Lambda = 5 \times 10^3, m = 0.3, d = 0.1, k = 5.5 \times 10^{-7}, p = 5 \times 10^3, c = 10^2$.

that the fast dynamics quickly stabilizes at the interior equilibrium $\tilde{U}_-$, and the convergence at the population level is much slower.

We consider the epidemiological and environmental variables to be slow variables that consist of S, I and E. Assume that $\mathcal{R}_v(E) > 1$ so that the fast system is at the stable nontrivial equilibrium $\tilde{U}_-(E) = (\tilde{T}_-(E), \tilde{T}*_-(E), \tilde{V}_-(E))$ with

$$\tilde{T}_- = \frac{T_0}{\mathcal{R}_v(E)}, \quad \tilde{V}_- = \frac{1}{c}\left[g(E) + \frac{p\Lambda}{m+d}\left(1 - \frac{1}{\mathcal{R}_v(E)}\right)\right],$$

$$\tilde{T}^*_- = \frac{\Lambda}{m+d}\left(1 - \frac{1}{\mathcal{R}_v(E)}\right) \tag{2.163}$$

and $\mathcal{R}_v(E)$ is defined in (2.162) or

$$\mathcal{R}_v(E) = \frac{2T_0}{a_1 - \sqrt{a_1^2 - 4a_2}}$$

where a_1 and a_2 are given in (2.160).

Because $N = S + I$ remains constant for all time, we can eliminate the S equation and get the following two-dimensional slow system:

$$\dot{I} = \lambda E(N - I) - \mu I$$
$$\dot{E} = \theta I \tilde{V}_-(E)(1 - E) - \gamma E \qquad (2.164)$$

where $V = \tilde{V}_-(E)$ is given in (2.163).

Notice that the dynamic linkage for the between- and within-host systems is through the enviromental contamitation variable E. The dependence of E on the within-host dynamics is through the term $\theta V I$ in the E equation. The two subsystems will be decoupled if V is a constant, in which case we can obtain an isolated reproductive number (i.e., a reproductive number that does not depend on the within-host system) for the between-host system. Without loss of generality, assume that the constant is $V = 1$ (otherwise the scaling constant can be absorbed in the transmission coefficient θ), which leads to the following baseline reproductive number for the between-host dynamics:

$$\mathcal{R}_{h0} =: \frac{\theta \lambda N}{\gamma \mu}. \qquad (2.165)$$

The quantity $\mathcal{R}_{h0}$ plays the role of the classical basic reproductive number for the between-host infection in the absence of the link to the vithin-host dynamics, as will be shown later in this section.

When the two processes are coupled with the fast system being near the equilibrium $\tilde{U}_-(E)$, the variable V in the E equation will be replaced by $\tilde{V}_-(E)$ given in (2.163). Let $\hat{W} = (\hat{I}, \hat{E})$ denote an interior equilibrium of the slow system, i.e., $\hat{I} > 0$ and $\hat{E} > 0$. Then

$$\hat{I} = \frac{\lambda \hat{E} N}{\lambda \hat{E} + \mu},$$

and $\hat{E}$ satisfies the equation

$$\frac{1 - \hat{E}}{c}\left[g(\hat{E}) + \frac{p\Lambda}{m + d}\left(1 - \frac{1}{\mathcal{R}_v(\hat{E})}\right)\right] = \frac{\gamma \hat{E}}{\theta N} + \frac{1}{\mathcal{R}_{h0}} \qquad (2.166)$$

where N is a constant total population size. Let $F(E)$ and $G(E)$ denote the functions on the left and right hand sides of the equation (2.166), respectively, i.e.,

$$F(E) = \frac{1 - E}{c}\left[g(E) + \frac{p\Lambda}{m + d}\left(1 - \frac{1}{\mathcal{R}_v(E)}\right)\right], \quad G(E) = \frac{\gamma E}{\theta N} + \frac{1}{\mathcal{R}_{h0}}. \qquad (2.167)$$

Then, $\hat{E}$ is a solution of the equation $F(E) = G(E)$ with $0 < \hat{E} < 1$.

By examining the properties of F and G, it suggests that the condition for the existence of $\hat{W} = (\hat{I}, \hat{E})$ depends on the following reproductive number for the slow (between-host) system when it is coupled with the fast (within-host) system:

$$\mathcal{R}_h =: \frac{pmT_0}{c(m + d)}\left(1 - \frac{1}{\mathcal{R}_{v0}}\right)\mathcal{R}_{h0}. \qquad (2.168)$$

The following result is shown in [Feng et al. (2012a)].

Theorem 2.26. *Let $\mathcal{R}_{v0} > 1$ and let $\mathcal{R}_h$ be defined in (2.168). Then the slow system has at least one endemic equilibrium $\hat{W} = (\hat{I}, \hat{E})$ if $\mathcal{R}_h > 1$.*

The analysis presented here cannot exclude the possibility of multiple endemic equilibria for the slow system. For the parameter values used in the numerical simulations presented in this study, the endemic equilibrium is unique as the curves of $F(E)$ and $G(E)$ have only one intersection for $E \in (0,1)$. However, some preliminary results suggest that a backward bifurcation may occur, i.e., endemic equilibria exist when $\mathcal{R}_h < 1$. The analysis for possible existence of multiple endemic equilibria will be presented in future publications.

The Jacobian matrix of the slow system (2.164) at $\hat{W} = (\hat{I}, \hat{E})$ is given by

$$J(\hat{W}) = \begin{pmatrix} -\lambda\hat{E} - \mu & \lambda(N - \hat{I}) \\ \theta\tilde{V}_-(\hat{E})(1 - \hat{E}) & \theta\hat{I}(1 - \hat{E})\tilde{V}'_-(\hat{E}) - \theta\hat{I}\tilde{V}_-(\hat{E}) - \gamma \end{pmatrix},$$

or using the notation of $F(E)$ and $G(E)$

$$J(\hat{W}) = \begin{pmatrix} -\mu F(\hat{E})\mathcal{R}_{h0} & \dfrac{\lambda N}{\mathcal{R}_{h0}G(\hat{E})} \\ \theta G(\hat{E}) & -\dfrac{\gamma}{F(\hat{E})}\left(F(\hat{E}) - \hat{E}F'(\hat{E})\right) \end{pmatrix}.$$

To obtain the above matrix, the following relations have been used:

$$\bar{V}_-(E) = \frac{F(E)}{1-E} = \frac{G(E)}{1-E}, \qquad \theta\hat{I} = \frac{\gamma\hat{E}}{F(\hat{E})},$$

$$\lambda\hat{E} = \frac{\lambda\theta N}{\gamma}\left(F(\hat{E}) - \frac{1}{\mathcal{R}_{h0}}\right) = \mu\left(F(\hat{E})\mathcal{R}_{h0} - 1\right).$$

Note that $F(E) > 0$ as $\mathcal{R}_v(E) > \mathcal{R}_{v0} > 1$, and note that $F'(\hat{E}) < 0$. Thus,

$$K =: F(\hat{E}) - \hat{E}F'(\hat{E}) > 0. \tag{2.169}$$

The trace and determinant of $J(\hat{W})$ are

$$\mathrm{tr}(J(\hat{W})) = -\mu F(\hat{E})\mathcal{R}_{h0} - \frac{\gamma}{F(\hat{E})}K < 0,$$

$$\det(J(\hat{W})) = \gamma\mu\left[\mathcal{R}_{h0}K - 1\right].$$

Therefore, $\hat{W}$ is l.a.s. if and only if $\det(J(\hat{W})) > 0$. Let

$$\mathcal{R}_{h0}^* = \frac{1}{K} \tag{2.170}$$

where K is defined in (2.169). Then the following result holds.

Theorem 2.27. *An endemic equilibrium $\hat{W}$ of the slow system is l.a.s. if and only if $\mathcal{R}_{h0} > \mathcal{R}_{h0}^*$.*

Theorems 2.25-2.27 suggest that the existence and stability of equilibrium points of the fast and slow systems are connected to three reproductive numbers: $\mathcal{R}_{v0}$, $\mathcal{R}_{h0}$ and $\mathcal{R}_h$, which are defined in (2.157), (2.165) and (2.168), respectively. These results can be summarized as follows:

1) For the fast system, the infecction-free equilibrium U_0 is l.a.s. $\mathcal{R}_{v0} < 1$. When $\mathcal{R}_{v0} > 1$, U_0 is unstable and a unique interior equilibrium $\tilde{U}_-(E)$ exists and is l.a.s.

2) For the slow system, under the condition that the fast system has the stable interior equilibrium $\tilde{U}_-(E)$ (i.e., when $\mathcal{R}_{v0} > 1$), there exists at least one endemic equilibrium $\hat{W}$ if $\mathcal{R}_h > 1$.

3) The endemic equilibrium $\hat{W}$ of the slow system is l.a.s. if and only if $\mathcal{R}_{h0} > \mathcal{R}_{h0}^*$.

One of the major findings of this study is the possibility of disease prevalence in the host population even when the isolated between-host reproductive number is less than one. For example, although it is not easy to see from (2.169) whether $K < 1$ or $K > 1$, for the parameter values used in our simulations (e.g., Figures 2.38 and 2.39) we have $K > 1$, in which case $\mathcal{R}_{h0}^* < 1$. Note that $\mathcal{R}_{h0}$ represents the baseline between-host reproductive number. Thus, our result shows that the system can have a stable endemic equilibrium even when $\mathcal{R}_{h0} < 1$. That is, the system may have a backward bifurcation.

The coupled system can be used to determine how the within-host reproductive number ($\mathcal{R}_v(E)$) and the infection level of the within-host system ($\tilde{T}(E)$) depend on the environmental contamination E. This is illustrated in Figure 2.38. Particularly, we observe that T^* is more sensitive to changes in E when $\mathcal{R}_{v0}$ is small, and that T^* is less sensitive to $\mathcal{R}_{v0}$ when the level of environmental contamination is high.

Our model analyses also revealed interesting results about how the within-host dynamics may influence the between-host process. For example, the existence and stability of an interior equilibrium of the full system do not require the baseline between-host reproductive number $\mathcal{R}_{h0}$ to be greater than 1, suggesting the existence of a backward bifurcation. This outcome is not likely from either the within-host model alone or from the between-host model alone. The threshold conditions, $\mathcal{R}_{v0} > 1$, $\mathcal{R}_h > 1$ and $\mathcal{R}_{h0} > \mathcal{R}_{h0}^*$ also provide valuable information on how these processs depend on each other and on the environmental conditions.

Finally, as mentioned earlier, the model considered in this study is more appropriate for an environmentally-driven infectious disease and the parasite has a similar but simpler life cycle than the case of toxoplasma. This model can be modified to include a preydator-prey interaction representing the cats and rats populations that can both be infected.

PART 2

Applications to Public Health Policymaking

Chapter 3

Applications of models to evaluations of disease control strategies

In this chapter we present examples of applying results from modeling studies in Part 1 to address specific biological and public health questions. Some of the examples focus on the evaluation of strategies for disease control and prevention, other examples explore the evolutionary consequences (in hosts and/or pathogens) influenced by disease control programs. Specific diseases considered in this chapter include influenza, SARS, TB, schistosomiasis, malaria, HIV, and HSV-2.

3.1 Influenza

The most commonly used measures for influenza control and prevention are vaccination and drug treatment. The distributions of vaccines and medications may have a significant influence on the outcomes of the disease spread and control. Heterogeneities in mixing patterns bwtween individuals from various groups and in transmission due to seasonality can play a critical role in the effects of control programs. Several questions related to the implementation of control programs are considered in this section. Questions related to the synergy between multiple diseases, such as that between HIV and HSV-2 and that between HIV and TB, are also discussed.

3.1.1 *Age- or group-targeted vaccination*

Compartmental models with non-homogeneous mixing between sub-populations have been used to study disease control and prevention (see, for example, [Glasser *et al.* (2010); Medlock and Galvani (2009); Del Valle *et al.* (2007); Chow *et al.* (2011)], etc.). The examples considered in this section are from Glasser *et al.* [Glasser *et al.* (2010)] and [Chow *et al.* (2011)].

Seasonal influenza causes an estimated 200,000 hospitalizations and 36,000 deaths on average in the United States, most among the elderly [Fiore *et al.* (2009)]. If a 1918-like pandemic occurred today, 10 million hospitalizations and 1.9 million deaths–many among younger adults–are expected. Vaccination affords the best protection, especially for those at risk of pneumonia and other life-threatening complications [Fiore *et al.* (2009)].

Development and production of influenza vaccines is challenging. In the northern hemisphere, the World Health Organization (WHO) collects relevant information every February for review by experts. Based on which viruses they believe will most likely be circulating, the experts select 3 strains for inclusion in the upcoming season's vaccine. Almost every year, at least one vaccine constituent is replaced, because viral strains drift; i.e., undergo constant genetic change. Even small changes can result in novel strains, and mismatch with circulating strains can reduce vaccine effectiveness, as occurred during the 2007–2008 influenza season [Belongia *et al.* (2008)]. Once the experts have identified the strains likely to circulate next season, a vaccine must be manufactured in a slow process that has changed little since its invention. Testing, approval, and distribution also take several months. Problems encountered during production, such as inability to grow sufficient quantities of a viral strain, may cause vaccine shortages or delays in distribution. Such problems have affected vaccine availability in recent influenza seasons in the United States [CDC (2010)].

During influenza pandemics, these challenges are compounded. Pandemic strains may emerge when antigenic shifts – major changes in the genetic makeup of a virus – occur in influenza A, creating new viral subtypes against which populations have little or no immunity [Earn *et al.* (2002)]. Even when effective vaccines are created, acute shortages are possible, especially in areas with limited production capacity that also have little advance warning, making it difficult or impossible to obtain sufficient vaccine in time to protect at-risk populations. During the recent pandemic, even in wealthy countries that developed and produced an H1N1 vaccine as soon as possible, vaccine supplies were inadequate to accommodate all who sought timely vaccination. The prospect of a shortage motivated health authorities in influenza vaccine-producing countries to devise strategies for ensuring that people who were most likely to suffer complications of influenza were vaccinated first. In the United States, the CDC's Advisory Committee on Immunization Practices (ACIP) determined that pregnant women, caregivers of young infants, health care workers, and people too young to have antibodies to H1N1 had first priority. Next were those most vulnerable to complications of influenza, generally the elderly [ACIP (2009)].

In such circumstances, other strategies for using scarce influenza vaccine efficiently also warrant consideration. Among such strategies is indirect protection; that is, immunizing those who might infect vulnerable people. One group whose vaccination might achieve the benefits of indirect protection is school children. The merits of vaccinating school children against influenza, partly to protect others, such as the elderly, have been argued from community-intervention trials [Monto *et al.* (1970)], natural experiments [Reichert *et al.* (2001)], and individual-based models [Longini *et al.* (2004)]. While trials generally are better controlled than natural experiments, they are relatively expensive and time-consuming. Moreover, only models allow examination of alternative vaccination strategies in exactly the same setting. Models should be evaluated against historical observations to check their predictive ability, but identifying and remedying deficiencies of individual-based models can be prohibitively difficult. Population models are simple enough for evaluation before use to inform public policy making. Analytical results, such as the optimal targets for interventions against infectious diseases, can also be derived.

To identify vaccine allocation strategies with the greatest potential to reduce influenza morbidity and mortality, we studied an age-structured population model whose infection rates we estimated from observed proportions infected [Chin *et al.* (1960)] and interpersonal contacts weighted by duration [Del Valle *et al.* (2007)]. Our models disease-induced mortality rates were either quotients of deaths attributed to pneumonia or influenza and populations at risk or products of those rates and ratios of 1918 and average 1913–1917 mortalities [Luk *et al.* (2001)]. We refer to the latter as contemporary 1918-like mortality.

An SIR model with multiple age groups and age-dependent mixing, as discussed in section 2.2, is used in this study. The age groups considered are < 1, 1–4. 5–9, ..., 80–84, 85+ years. For the estimates of c_{ij}, several recently published studies of face-to-face conversations or periods in proximity with others during which respiratory diseases might be transmitted. Particularly, using the values $C_{ij} = a_i c_{ij}$ from Del Valle *et al.* [Del Valle *et al.* (2007)], the values of $a_{ij} = \sum_j C_{ij}$ can be obtained and thus the estimates of c_{ij}. A logistic regression model is used to fit to the $y_i = I_i/N_i$ reported in [Chin *et al.* (1960)], which can provide the risks of infection $\lambda_i = -\ln(1 - y_i)$. Then, from the relation $\lambda_i = a_i \beta_i \sum_j c_{ij} y_j$, the probabilities of infection (β_i) can be estimated.

In experiments, all else should be equal. We simulated our model without vaccination, with 60% of infants < 1 year and adults ≥ 65 years of age or the same percentage of children aged 1–9, adolescents 10–19, or young adults 20–29 years being vaccinated. These groups are roughly the same size, but coverage actually is <60% among persons <65 years of age [Schiller and Euler (2009)]. Our hypothetical annual influenza vaccine protected 70% of people 1–64 years of age, but lower proportions of infants and older adults. Efficacy was 35% among infants and declined linearly with age over 64 years (i.e., was 60% among people aged 65–69 years, 50% among those aged 70–74 years, and so on). The resulting efficacies among elderly adults correspond roughly to those reported by Govaert *et al.* [Govaert *et al.* (1994)]. Annual vaccination occurred November through January; pandemic vaccination began 30 days later and continued for 6 months. Pandemic efficacy was half annual, but as roughly twice as many doses were eventually administered, similar numbers of people were protected.

To assess the impact of these alternative strategies on morbidity and mortality, we averaged daily differences between age-specific cases or deaths with and without vaccination over 365-day periods. Averaging was necessary because our simulation model is stochastic (i.e., we employ Renshaw's discrete event/time method [Renshaw (1993)]). Finally, in age-structured models, the average number of effective contacts, R0, may be calculated as the dominant eigenvalue of the next-generation matrix [Diekmann *et al.* (1990)] whose associated eigenvector describes the age-specific contributions [Wallinga *et al.* (2010, 2006)].

The matrix of effective contacts or infection rates illustrates preferential mixing (see Figure 3.1), not only among contemporaries – which is particularly intense among older children, adolescents, and young adults [Hurford *et al.* (2010)] – but also that between parents and children and among co-workers evident in more recent, higher-resolution observations [Del Valle *et al.* (2007); Mossong *et al.* (2008); Zagheni *et al.* (2008)]. As indirect effects emanate from off-diagonal matrix elements, such observations increase the accuracy of assessments of intervention impacts via transmission modeling. Similarly, our 1918-like mortalities (see Figure 2 in [Glasser *et al.* (2010)]) resemble those of the 2009 pandemic, although this swine H1N1 was much less virulent than that avian strain.

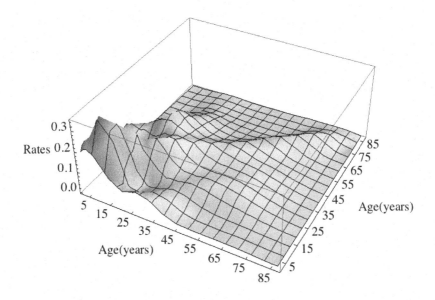

Fig. 3.1 Effective contact or infection rates derived from attack "rates" during the 1957 pandemic [Chin *et al.* (1960)] and daily contacts weighted by duration [Del Valle *et al.* (2007)]. doi:10.1371/journal.pone.0012777.g001

During simulated pandemic as well as annual influenza outbreaks, vaccinating older children, adolescents, and young adults reduced morbidity the most, especially among target age groups (Figure 3.2). Despite a contact matrix with relatively large off-diagonal elements, only 20-25% of cases averted were in groups not targeted. By contrast, vaccinating infants and elderly adults reduced mortality most during simulated annual influenza outbreaks (Figure 3.3(a)), but vaccinating young adults also reduced mortality during simulated pandemics (Figure 3.3(b)).

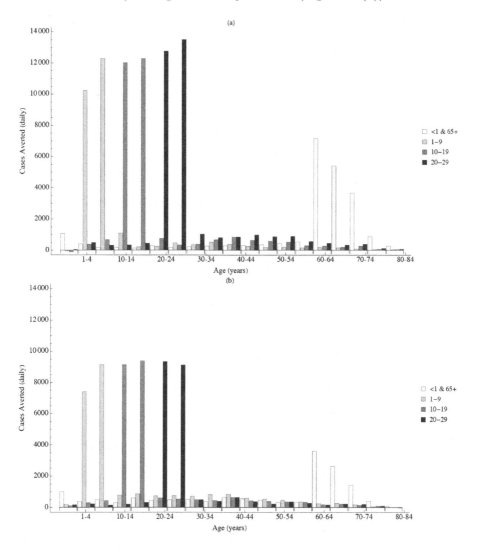

Fig. 3.2 **Cases averted by vaccination.** Similar patterns in cases averted by vaccinating people aged < 1 year and 65+ years, 1-9 years, 10-19 years, and 20-29 years during hypothetical annual (a) and pandemic (b) outbreaks. doi:10.1371/journal.pone.0012777.g003

While target age groups are similar in size, the numbers of cases averted depend on the vaccine efficacies as well as age distribution of the 2005 U.S. population. Cases averted per efficacy adjusted dose correspond to the proportionate contributions to $\mathcal{R}_0$ (Figure 3.4), which identifies the optimal target for interventions to reduce transmission.

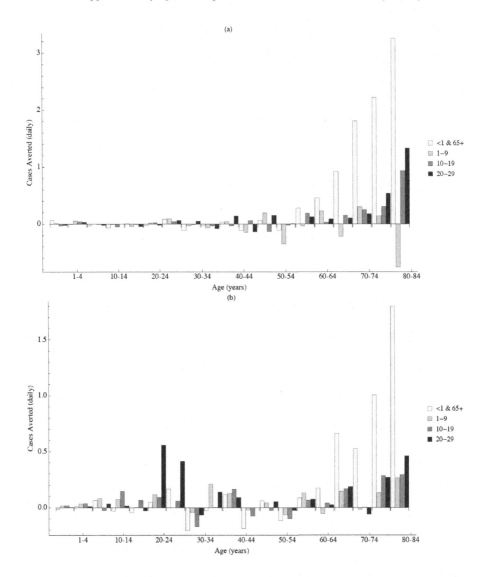

Fig. 3.3 Deaths averted by vaccination. Dissimilar patterns in deaths averted by vaccinating people aged < 1 year and 65+ years, 1-9 years, 10-19 years, and 20-29 years during hypothetical annual (a) and pandemic (b) outbreaks. doi:10.1371/journal.pone.0012777.g004

To summarize the results in this study, we adapted a classic age-structured population model with parameters chosen to maximize indirect effects due to vaccinating older children, adolescents, and young adults, and to accurately assess direct effects due to vaccinating elderly adults. Comparing the impact of vaccinating these age groups against influenza, we found that vaccinating children, adolescents, and young adults would reduce morbidity the most, with 20–25% of the reduction in other age groups. However, while vaccinating infants and older

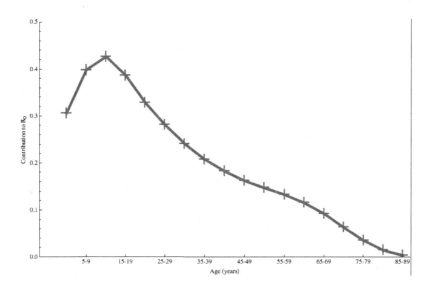

Fig. 3.4   Plot showing the normalized age-specific contributions to the reproduction number. doi:10.1371/journal.pone.0012777.g005

adults would mitigate mortality most during annual outbreaks, vaccinating young adults would also mitigate mortality during contemporary 1918-like pandemics. Evidently, which vaccination strategy is superior depends on the objective: mitigating morbidity or mortality, and if mortality, its age-distribution. For many years, U.S. vaccination policy was designed to mitigate mortality, particularly among elderly adults. Relatively recently, it was redesigned to also mitigate morbidity, initially among young children, but then progressively among older children, adolescents, and adults [http://www.cdc.gov/ media/pressrel/2010/r100224.htm]. Unlike this policy, in which the 6-month lower age of recommended vaccination has not changed as the upper age has increased, our experimental design maintained similar target group sizes by increasing both lower and upper ages of vaccination simultaneously.

Our findings are comparable to those obtained via other methodologies. The observation that mortality attributed to influenza and pneumonia among elderly Japanese was lower when children were vaccinated routinely [Reichert *et al.* (2001)] suggests that susceptible young people pose a risk to elderly ones, but not necessarily directly. While few such studies are unequivocal, numerous U.S. experiences [Glezen (1996)] are consistent with this deduction. Similar conclusions have been reached via community intervention trials [King *et al.* (2005); King Jr *et al.* (2006); Loeb *et al.* (2010)] as well as individual-based modeling [Longini *et al.* (2004)]. As our findings support results of these studies using other methodologies, they make a strong case for using relatively simple population models to examine pressing public health issues, and therefore to arrive relatively quickly at sound conclusions about the effectiveness of alternative interventions.

Influenza vaccination strategies have been compared recently using a variety of modeling approaches and perspectives. In 2007, Dushoff *et al.* [Dushoff *et al.* (2007)] explored the same strategies in a 2-group model, one more effective at transmitting the pathogen and other more vulnerable to its effects. These researchers were reluctant to choose among the many interesting scenarios described by various combinations of their parameters, and urged only caution. In 2006, Bansal *et al.* [Bansal *et al.* (2006)] adopted a more detailed network model with which they also evaluated these strategies, obtaining results qualitatively similar to ours. Three years later, Medlock and Galvani [Medlock and Galvani (2009)] used an age-structured population model with a mixing matrix whose off-diagonal elements are relatively small [Mossong *et al.* (2008)]. Nonetheless, they concluded that vaccinating older children, adolescents, and young adults was the best strategy, regardless of objective.

Impacts of other control measures for pandemic influenza have also been explored recently, by modeling individual members of socially and spatially structured populations [Ferguson *et al.* (2006); Germann *et al.* (2006); Glass *et al.* (2006); Haber *et al.* (2007); Longini *et al.* (2005)]. Our work illustrates several advantages of simpler population models [May (2004)]. Insofar as plausible mixing scenarios are modeled, individual behavior is extraneous. Furthermore, systems of equations can be analyzed, whereas computer programs cannot; for example, Areno *et al.* [Arino *et al.* (2006)] not only reproduced results with a proportionately-mixed, age-structured population model that had been obtained with a relatively complex individual-based model [Longini *et al.* (2004)], but also deduced several analytical results. Finally, population models use observations and make predictions familiar to epidemiologists, who group individuals based on characteristics of interest, both in disease surveillance, and to develop and implement interventions. As recently as 2008, for example, Vynnycky and Edmunds used a population model to investigate the impact of school closures on the spread of influenza during a pandemic [Vynnycky and Edmunds (2008)]. Because people of some ages are more active than others, immunizing those potential "super-spreaders" reduces the average number of secondary infections disproportionately. As Figure 3.4 indicates, adolescents and young adults are the optimal targets for reducing morbidity. Because the main diagonal predominates in all known mixing matrices [Del Valle *et al.* (2007); Wallinga *et al.* (2006); Mossong *et al.* (2008); Zagheni *et al.* (2008)], however, direct effects exceed indirect ones. Unless vaccine efficacy is very low, consequently, the best strategy for reducing mortality will be to vaccinate members of at-risk groups [Nichol *et al.* (2007)]. This analytical result is not limited to vaccination; it may be applied to other interventions that prevent infection or reduce the magnitude or duration of infectiousness. For example, as neuraminidase inhibitors are most effective when administered early [Moscona (2005)], timely medication of ill children, adolescents, and young adults could reduce the number needing treatment and possibly the duration of treatments. Treating optimally would be much less costly than widespread prophylaxis, and reduce the risk of drug-resistant strains emerging [Lipsitch *et al.* (2007)].

Age-specific infection rates are the essence of population models. We calculated risks of infection from Chin *et al.*'s prospective study of household transmission following illnesses among school children [Chin *et al.* (1960)]; households without school-aged children were not represented. Together with clinical observations and individual onset dates, a cross-sectional serological survey would remedy this possible deficiency and might resolve uncertainty about the contribution of asymptomatic infections to transmission. Anderson and May [Anderson and May (1991)] described "who-acquires-infection-from-whom" matrices with as many unique elements as risks of infection, but Nold [Nold (1980)] formulated mixing as a convex combination of age-specific activities (number of contacts per person per day) and constant preference (proportion with others in the same group), and Jacquez *et al.* [Jacquez *et al.* (1988)] allowed preference to vary with age. Recent empirical observations enabled us to include contacts between parents and children and among co-workers [Glasser *et al.* (2012)]. Insofar as mixing differs from society to society, if not between rural and urban sub-populations, more diverse subjects would permit continued refinement of methods to permit rapid, robust analysis and interpretation of alternative actions to address public health priorities.

3.1.2 *Preferential mixing and group-targeted vaccination*

In section 2.2.1, we presented models with preferential mixing and the derivation of the reproduction number. The next example demonstrate that incorporation of preferential mixing in the model can provide valuable information about control programs, although it will make the mathematical analysis more challenging. The model is used to examine how the effects of vaccination programs might be affected by various factors including the degree of mixing preference within subgroups, group activity levels, and population sizes of subgroups. The evaluations will be based on the control reproduction numbers and the epidemic sizes.

To demonstrate how mixing may affect disease dynamics, consider an extension of the classical SIR model (2.1) by incorporating mutiple groups. This example is chosen from [Chow *et al.* (2011)]. Consider a network of n populations whose sizes are denoted by N_i for $i = 1, 2, \cdots, n$. These population sizes remain constant for all time by assuming equal per-capita birth and death rates (μ). Each of the sub-groups is divided into three epidemiological classes: susceptible (S_i), infectious (I_i), and recovered (R_i). The recovery rate (γ) is assumed to be the same for all sub-groups. All individuals are born susceptible. For each sub-group i, a fraction p_i is vaccinated and immune. The multi-group model is a system consisting of the following ordinary differential equations:

$$
\begin{cases}
\dfrac{dS_i}{dt} = \mu N_i (1 - p_i) - (\lambda_i(t) + \mu)S_i, \\[2mm]
\dfrac{dI_i}{dt} = \lambda_i(t)S_i - (\gamma + \mu)I_i, \\[2mm]
\dfrac{dR_i}{dt} = \mu N_i p_i + \gamma I_i - \mu R_i, \\[2mm]
N_i = S_i + I_i + R_i, \qquad i = 1, 2, \cdots, n.
\end{cases}
\tag{3.1}
$$

Here, λ_i represents the force of infection for susceptibles in group i given by

$$
\lambda_i = a_i \beta \sum_{j=1}^{n} c_{ij} \frac{I_j}{N_j},
\tag{3.2}
$$

where a_i denotes the average number of contacts an individual in sub-population i has per unit of time (which represents the activity level of group i), and β is the probability of infection per contact when the contact is with an infectious individual. The fraction I_j/N_j gives the probability that a contact is with an infectious individual in sub-population j. The contact matrix (c_{ij}) has the same form as the preferential mixing considered in [Jacquez *et al.* (1988)] with

$$
c_{ij} = \varepsilon_i \delta_{ij} + (1 - \varepsilon_i) f_j, \quad i, j = 1, 2, \cdots, n.
\tag{3.3}
$$

The parameter ε_i is the fraction of contacts with individuals in the same sub-population, δ_{ij} is the Kronecker delta (i.e., 1 when $i = j$ and 0 otherwise), and

$$
f_j = (1 - \varepsilon_j)a_j N_j / \sum_k (1 - \varepsilon_k)a_k N_k, \quad j = 1, 2, \cdots, n.
$$

Clearly, unless all the sub-groups are isolated (i.e., no interactions between any groups), there must be some i with $\varepsilon_i < 1$. All parameters and their meanings are listed in Table 1. It is easy to verify that solutions of (3.1) remain nonnegative for all nonnegative initial conditions. Thus, the model is well posed.

For each sub-population i, if all contacts are with people within the same group (i.e., $c_{ii} = 1$ and $c_{ij} = 0$ for $i \neq j$), then the basic and control reproduction numbers for group i are, respectively,

$$
\mathcal{R}_{0i} = \frac{\beta a_i}{\mu + \gamma}, \quad \mathcal{R}_{vi} = \mathcal{R}_{0i}(1 - p_i), \quad i = 1, 2, \cdots, n.
\tag{3.4}
$$

When there are contacts between sub-populations, i.e., $c_{ii} < 1$ or $\varepsilon_i < 1$ for some i, we can derive the basic and control reproduction numbers for the metapopulation. These reproduction numbers will be functions of $\mathcal{R}_{0i}$ or $\mathcal{R}_{vi}$. Following the approach of [Diekmann *et al.* (1990)] we can obtain from model (3.1) the next generation matrix K_v (v for vaccination)

$$
K_v = \begin{pmatrix}
\mathcal{R}_{v1}c_{11} & \mathcal{R}_{v1}c_{12} & \cdots & \mathcal{R}_{v1}c_{1n} \\
\mathcal{R}_{v2}c_{21} & \mathcal{R}_{v2}c_{22} & \cdots & \mathcal{R}_{v2}c_{2n} \\
\vdots & \vdots & \ddots & \vdots \\
\mathcal{R}_{vn}c_{n1} & \mathcal{R}_{vn}c_{n2} & \cdots & \mathcal{R}_{vn}c_{nn}
\end{pmatrix}.
\tag{3.5}
$$

The control reproduction number $\mathcal{R}_v$ for the metapopulation is given by

$$\mathcal{R}_v = \rho(K_v) \tag{3.6}$$

where $\rho(K_v)$ denotes the dominant eigenvalue of K_v [Heesterbeek (2000)]. Note that $\mathcal{R}_v = \mathcal{R}_v(p_1, p_2, \cdots, p_n)$ is a function of vaccination fractions p_i. The basic reproduction number $\mathcal{R}_0$ for the metapopulation is given by $\mathcal{R}_v$ when $p_i = 0$ for all i, i.e., $\mathcal{R}_0 = \mathcal{R}_v(0, 0, \cdots, 0)$.

In the case of $n = 2$, an explicit formula for $\mathcal{R}_v$ can be obtained

$$\mathcal{R}_v = \frac{1}{2}\left[A + D + \sqrt{(A-D)^2 + 4BC}\,\right], \tag{3.7}$$

where $A = \mathcal{R}_{01}c_{11}(1-p_1)$, $B = \mathcal{R}_{01}c_{12}(1-p_1)$, $C = \mathcal{R}_{02}c_{21}(1-p_2)$, $D = \mathcal{R}_{02}c_{22}(1-p_2)$, and $\mathcal{R}_{0i}$ $(i = 1, 2)$ are given in (3.4). If $p_1 = p_2 = 0$, then $\mathcal{R}_v$ reduces to

$$\mathcal{R}_0 = \frac{1}{2}\left[\mathcal{R}_{01}c_{11} + \mathcal{R}_{02}c_{22} + \sqrt{(\mathcal{R}_{01}c_{11} - \mathcal{R}_{02}c_{22})^2 + 4\mathcal{R}_{01}c_{12}\mathcal{R}_{02}c_{21}}\,\right].$$

To study effects of vaccination strategies, assume that $\mathcal{R}_0 > 1$ in the absence of vaccination and

$$\mathcal{R}_{01} > 1, \qquad \mathcal{R}_{02} > 1. \tag{3.8}$$

Let

$$\Omega = \{(p_1, p_2)|\ 0 \le p_1 < 1,\ 0 \le p_2 < 1\}. \tag{3.9}$$

Then each point $(p_1, p_2) \in \Omega$ represents a vaccination strategy.

Because we are interested in the case when the two groups are not isolated, either $\varepsilon_1 < 1$ or $\varepsilon_2 < 1$. This will be assumed for the results below. It can be shown that, for each fixed $(p_1, p_2) \in \Omega$, $\mathcal{R}_v$ increases with both ε_1 and ε_2, i.e.,

$$\frac{\partial \mathcal{R}_v}{\partial \varepsilon_1} > 0, \quad \frac{\partial \mathcal{R}_v}{\partial \varepsilon_2} > 0 \qquad \text{for all } (\varepsilon_1, \varepsilon_2) \in \Omega. \tag{3.10}$$

The result in (3.10) is based on the inequality

$$\frac{\partial \mathcal{R}_v}{\partial \varepsilon_1} = \frac{1}{2}\left(\mathcal{R}_{01}(1-p_1)\frac{\left[(1-\varepsilon_2)a_2 N_2\right]^2}{\left[(1-\varepsilon_1)a_1 N_1 + (1-\varepsilon_2)a_2 N_2\right]^2}\right.$$

$$+ \mathcal{R}_{02}(1-p_2)\frac{(1-\varepsilon_2)^2 a_1 N_1 a_2 N_2}{\left[(1-\varepsilon_1)a_1 N_1 + (1-\varepsilon_2)a_2 N_2\right]^2}$$

$$\left. + \frac{\left[\mathcal{R}_{01}(1-p_1)a_1 N_1 - \mathcal{R}_{02}(1-p_2)a_2 N_2\right]^2 (1-\varepsilon_2)^2 a_2 N_2}{\left[(1-\varepsilon_1)a_1 N_1 + (1-\varepsilon_2)a_2 N_2\right]^3 \sqrt{(A-D)^2 + 4BC}}\right) > 0$$

and a similar one for $\frac{\partial \mathcal{R}_v}{\partial \varepsilon_2}$.

For each fixed $(\varepsilon_1, \varepsilon_2)$, there are different combinations of p_1 and p_2 that can reduce $\mathcal{R}_v$ to be below 1. For ease of presentation, consider the simpler case in which

$$\varepsilon_1 = \varepsilon_2 = \varepsilon,$$

and consider $\mathcal{R}_v = \mathcal{R}_v(\varepsilon)$ as a function of ε. Then, for each fixed $\varepsilon \in [0,1)$, the curve determined by $\mathcal{R}_v(\varepsilon) = 1$ divides the region Ω into two parts: one is the region

$$\Omega_\varepsilon = \{(p_1, p_2)|\ 0 \le \mathcal{R}_v(\varepsilon) < 1,\ (p_1, p_2) \in \Omega,\ 0 \le \varepsilon < 1\},$$

which includes all points above the curve (see Figure 3.5), and another is the region

$$D_\varepsilon = \{(p_1, p_2)|\ \mathcal{R}_v(\varepsilon) > 1,\ (p_1, p_2) \in \Omega,\ 0 \le \varepsilon < 1\},$$

which includes all points below the curve. It can be shown that the region Ω_ε decreases as ε increases and reduces to the region Ω^* as $\varepsilon \to 0$, while the region D_ε decreases as ε decreases and reduces to the region D^* as $\varepsilon \to 1$ (see Figure 3.5). All these curves intersect at a single point (p_{1c}, p_{2c}) with

$$p_{1c} = 1 - \frac{1}{\mathcal{R}_{01}}, \quad p_{2c} = 1 - \frac{1}{\mathcal{R}_{02}}. \tag{3.11}$$

We observe from Figure 3.5 that the region Ω^* (lighter-shaded) is determined by the two inequalities

$$p_{1c} < p_1 < 1, \quad p_{2c} < p_2 < 1, \tag{3.12}$$

where p_{1c} and p_{2c} are defined in (3.11). For region D^* (darker shaded), the upper bound is determined by the line

$$p_2 = -\mathcal{A}p_1 + \mathcal{B} \tag{3.13}$$

where

$$\mathcal{A} = \frac{\mathcal{R}_{01} a_1 N_1}{\mathcal{R}_{02} a_2 N_2}, \quad \mathcal{B} = \frac{(\mathcal{R}_{01} - 1)a_1 N_1 + (\mathcal{R}_{02} - 1)a_2 N_2}{\mathcal{R}_{02} a_2 N_2}. \tag{3.14}$$

The two regions intersect at the point (p_{1c}, p_{2c}).

This result suggests that there is a "lower bound" for vaccination efforts (p_1, p_2), above which the infection can be eradicated regardless of mixing patterns. Similarly, it provides an "upper bound" for vaccination efforts (p_1, p_2), below which the infection cannot be eradicated regardless of mixing patterns (see the definitions for p_1^* and p_2^* defined in (3.16) and see Figure 3.6 for an illustration of the lower and upper bound). For an "intermediate level" vaccination strategy (p_1, p_2), mixing parameters ε_1 and ε_2 can play an important role in influencing the effect of vaccination strategies on reducing $\mathcal{R}_v$. Thus, when designing vaccination strategies, one should take into consideration mixing patterns within and between sub-populations.

Notice that for given ε_1 and ε_2,

$$\frac{\partial \mathcal{R}_v}{\partial p_1} = -\frac{1}{2}\left[\mathcal{R}_{01} c_{11} + \mathcal{R}_{02} c_{22} + \frac{\mathcal{R}_{01} c_{11}(1 - p_1) + \mathcal{R}_{01}\mathcal{R}_{02}(1 - p_2)(1 + c_{12}c_{21})}{\sqrt{(A - D)^2 + 4BC}}\right] < 0,$$

$$\tag{3.15}$$

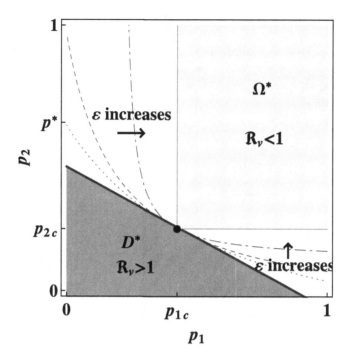

Fig. 3.5 Plot showing the regions Ω^* and D^*. Several curves of $\mathcal{R}_v(\varepsilon) = 1$ for different ε values are also shown, with the dashed curves corresponding to $0 < \varepsilon < 1$, the thin solid lines (boundary of Ω^*) corresponding to $\varepsilon = 1$, and the thick line corresponding to $\varepsilon = 0$ (the upper bound of the region D^*). The arrows indicate the direction of change of the curve $\mathcal{R}_v(\varepsilon) = 1$ as ε increases from 0 to 1. All of the $\mathcal{R}_v(\varepsilon) = 1$ curves intersect at the single point (p_{1c}, p_{2c}).

and similarly, $\frac{\partial \mathcal{R}_v}{\partial p_2} < 0$. When the curve $\mathcal{R}_v = 1$ lies between regions D^* and Ω^*, the curve intersects the p_1-axis and p_2-axis at $(p_1^*, 0)$ and $(0, p_2^*)$, respectively, where

$$p_1^* = 1 - \frac{1 - \mathcal{R}_{02}c_{22}}{\mathcal{R}_{01}c_{11}(1 - \mathcal{R}_{02}c_{22}) + \mathcal{R}_{01}\mathcal{R}_{02}c_{12}c_{21}},$$
$$p_2^* = 1 - \frac{1 - \mathcal{R}_{01}c_{11}}{\mathcal{R}_{02}c_{22}(1 - \mathcal{R}_{01}c_{11}) + \mathcal{R}_{22}\mathcal{R}_{01}c_{12}c_{21}}. \tag{3.16}$$

Because $\mathcal{R}_{0i} > 1$ for $i = 1, 2$, it is possible that $\mathcal{R}_{01}c_{11} > 1$ and/or $\mathcal{R}_{02}c_{22} > 1$. Thus, it is possible that $p_1^* > 1$ and/or $p_2^* > 1$. When $p_1^* > 1$, we know from (3.15) that $\mathcal{R}_v > 1$ for any vaccination strategy $(p_1, 0)$. Thus, it is impossible to eradicate the infection if only sub-population 1 is vaccinated.

The results described above are based on the control reproduction number. Figure 3.6 shows some simulation results illustrating the effect of vaccination on the prevalence of infection. Different preference levels are used: $\varepsilon_1 = 0.2$ and $\varepsilon_2 = 0.4$, i.e., group 2 has a higher preference contacting people in its own group. Other parameter values used are $\beta = 0.03$, $\gamma = 0.15$ (an infectious period of about 6 days), and $a_1 = 12$, $a_2 = 8$, $\mu = 0.00016$ (a duration of 17 years in school). These values correspond to $\mathcal{R}_{01} = 2.4$ and $\mathcal{R}_{02} = 1.6$. The initial conditions are $x_1(0) = S_1(0)/N_1(0) = 0.4$, $y_1(0) = I_1(0)/N_1(0) = 0.00002$, $x_2(0) = S_2(0)/N_2(0) = 0.6$, $y_2(0) = I_2(0)/N_2(0) = 0.00002$. For this set of parameters, $p_1^* = 0.77$ and $p_2^* \gg 1$. Figure 3.6(a) is for a vaccination strategy $(p_1, 0)$ with $p_1 = 0.2 < p_1^*$, for which the infection persists ($\mathcal{R}_v = 1.8 > 1$), while Figure 3.6(b) is for a vaccination strategy $(p_1, 0)$ with $p_1 = 0.8 > p_1^*$, in which case the infection dies out ($\mathcal{R}_v = 0.97 < 1$).

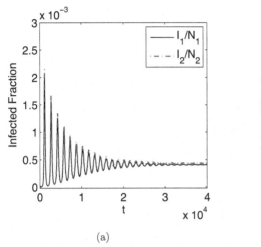
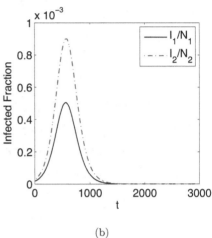

(a)   (b)

Fig. 3.6   When $p_1^* < 1$ and $p_2^* > 1$, the disease is eventually eradicated if the vaccination is applied to sub-population 1 alone at a level above p_1^*. (a) $(p_1, p_2) = (0.2, 0)$ and $p_1 < p_1^* = 0.77$, the disease persists ($\mathcal{R}_v = 1.8$); (b) $(p_1, p_2) = (0.8, 0)$ and $p_1 > p_1^* = 0.77$, the disease eventually disappears ($\mathcal{R}_v = 0.97$). Here, $x_1(0) = S_1(0)/N_1(0) = 0.4$, $y_1(0) = I_1(0)/N_1(0) = 0.00002$, $x_2(0) = S_2(0)/N_2(0) = 0.6$, $y_2(0) = I_2(0)/N_2(0) = 0.00002, a_1 = 12$, $a_2 = 8$, $\varepsilon_1 = 0.2$, $\varepsilon_2 = 0.4$, $p_1^* = 0.77$, and $p_2^* \gg 1$.

3.1.3   *Estimation of transmission parameters*·

The results presented in this section are from Glasser *et al.* [Glasser *et al.* (2012)]. In section 2.2.1, we presented a preferential mixing function (2.9). This function can be useful for parameter estimation in disease model. Here we use risks of infection, information typically available for transmission modeling, to derive probabilities of infection on contact, infection rates and reproduction numbers for two respiratory diseases. For purposes of illustration, we use the empirical contact matrices mentioned in section 2.2.1.

Patterns apparent in recently published studies of face-to-face conversations and periods in proximity with others motivated us to elaborate the preferential mixing model of Jacquez *et al.* [Jacquez *et al.* (1988)] to include contacts between parents and children and among co-workers as well as contemporaries. Unlike mixing among contemporaries, that between parents and children and among co-workers involves off-diagonal matrix elements. In cross-classified population models, the main diagonal is responsible for direct effects; other matrix elements are responsible for indirect ones. Thus, modelers using extant formulae, in which mixing between groups is proportional to their respective contacts, may have underestimated indirect effects.

Suppose that one wished to assess the impact of vaccinating young parents on pertussis among infants. Immunity-modified disease can be quite mild, but that among infants lacking maternal antibodies typically is severe and occasionally even fatal. Mothers with prolonged cough illnesses may not be very infectious, but mother-infant contacts are particularly intimate. Which mixing formula should one use? What about the impact of vaccinating older children against influenza on mortality among elderly adults, who occasionally die of pneumonia, which may complicate influenza? These groups are no more connected than any others in the existing formulation, but infants and young adults are directly connected in ours. And children and elderly adults are connected via their parents and children, respectively

Delta formulations are convenient mathematically, but do not allow the age range of one's contemporaries to vary as one ages (e.g., the range narrows perceptively among adolescents), much less differences between the age ranges of one's contemporaries and one's parents or children. By virtue of secular patterns in child-bearing, moreover, the age ranges of parents and children may change with age. But the Gaussian formulation allows such variation, reproducing the essential features of these observations (cf. Figure 3.7 (left panel)), and can inform-given constant or age-specific susceptibilities to infection on contact-infection rates for transmission modeling. Truncating the Gaussian, or using the lognormal or gamma distributions would ensure that ages were positive, but not that $a \leq L$. Infants' contemporaries have such a narrow age range and persons aged $a > L$ are so few that such complications would be an unwarranted distraction. Applications of continuous distributions to biological phenomena require common sense.

In the case of multiple age groups, the second equation for the i_{th} group in the classic SIR model, for example, is

$$I' = \lambda_i S_i - \gamma I_i \quad \text{where} \quad \lambda_i = a_i \beta_i \sum_{j=1}^{n} c_{ij} \frac{I_j}{N_j}, \tag{3.17}$$

and γ is the recovery rate, S_i, I_i, and R_i are the numbers susceptible, infected, and removed, respectively, and $N_i = S_i + I_i + R_i$. The contact functions c_{ij} and $c(\alpha, \alpha')$ are described in (2.9) and (2.11), respectively, These contact functions involve the preference parameters ε_{ij} and the parameters σ_i for the Gaussian kernels. THese parameters can be estimated from the Varicella-zoster virus data, which can then be used to compute the transmission parameter β_i.

Varicella-zoster virus is relatively stable, so increasing proportions by age are immune. We fit Farrington's [Farrington (1990)] model,

$$F(\alpha) = 1 - e^{-\int_0^\alpha \lambda(u)du} \quad \text{where} \quad \lambda(\alpha) = (a\alpha - c)e^{-b\alpha} + d$$

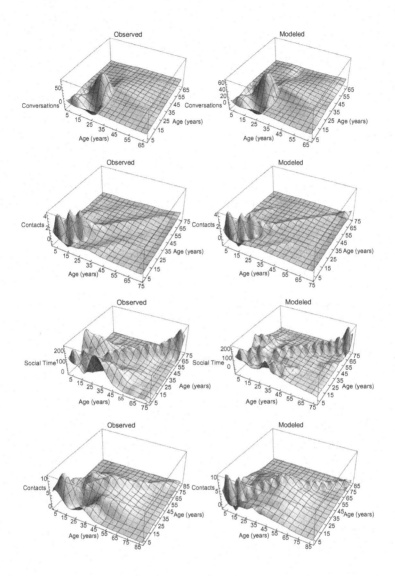

Fig. 3.7   Comparision of the mixing patterns generated by the four empirical studies [Wallinga *et al.* (2006); Mossong *et al.* (2008); Zagheni *et al.* (2008); Del Valle *et al.* (2007)] (left panel) and the fitted model (right panel).

with a, b, c being constants, to proportions of sera with protective antibody titers collected during the third National Health and Nutrition Survey 1988–1994. The best fit is obtained by using the FindFit function in the computing software MATHE-MATICA. FindFit uses singular value decomposition and the LevenbergMarquardt method for linear and nonlinear least-squares, respectively, and the FindMinimum methods described above otherwise. As the probability of remaining susceptible is

$$P_S(\alpha) = e^{-\int_0^\alpha \lambda(u)du},$$

its negative derivative,

$$-P_S'(\alpha) = \lambda(\alpha)e^{-\int_0^\alpha \lambda(u)du},$$

is the probability density function of first infection, which can be used to obtain $y_i = I_i/N_i$. Using these y_i and the equation (3.17) with λ_i being an average of the function $\lambda(\alpha)$ over age interval i, we calculate the β_i. Figure 3.7 shows the interpolating functions fitted to geometric means of corresponding row- and column-elements from the four empirical studies (left panel) and fits of our model (right panel) of Wallinga *et al.* [Wallinga *et al.* (2006)], Mossong *et al.* [Mossong *et al.* (2008)], Zagheni *et al.* [Zagheni *et al.* (2008)], and Del Valle *et al.* [Del Valle *et al.* (2007)], respectively.

The observed contact matrices are asymmetric, presumably because persons contacted need not also have participated in these studies (i.e., study populations were not closed). As this violates the third condition of Busenberg and Castillo-Chavez [Busenberg and Castillo-Chavez (1991)], we replaced reported elements by the geometric means of those from the corresponding rows and columns before estimating the ε's and σ's in the contact functions.

Figure 3.7 illustrates observations thus adjusted from these recent empirical studies (on the left) and our fitted models (on the right). The observed $C_{ij} = a_i c_{ij}$, together with influenza infection force (Figure 3.8), yield probabilities of infection on contact β_i (Figure 3.9). Using proportions protected and risks of infection derived from the cross-sectional survey of antibodies to varicella instead, the calculated β_i are similar.

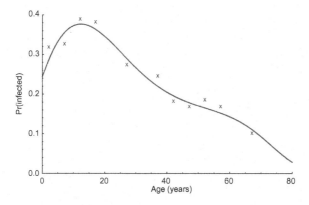

Fig. 3.8   Probability of infection from a prospective household study during the 1957 pandemic [Chin *et al.* (1960)] and the fitted regression equation (see [Glasser *et al.* (2012)]).

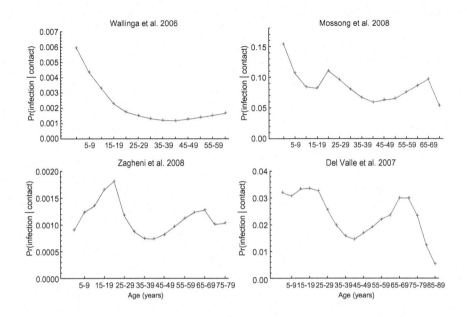

Fig. 3.9   Probabilities of infection with infuenza per conversation, period together, or effective contact, ordered left to right and top to bottom as in Figure 3.7. Patterns are similar despite quantitative differences reflecting average weekly [Wallinga *et al.* (2006)] or daily intimate and casual conversations [Mossong *et al.* (2008)], daily periods (minutes) with others [Zagheni *et al.* (2008)], and numbers of contacts probabilities of transmission per contact derived from mean contact durations [Del Valle *et al.* (2007)].

Given these β_i, we can write $a_i\beta_i c_{ij} = \beta_{ij}$, the rates of effective contact between members of groups i and j (alternatively, infection of susceptible members of group i by infectious members of group j) that are required for transmission. Noting that the β_{ij}/γ are elements of the next-generation matrix, whose largest eigenvalue is $\mathcal{R}_0$. Together with $1/\gamma$ of 3.8 days [Cauchemez *et al.* (2004)], these yield $\mathcal{R}_0 \approx 3.9$ for influenza. Carrat *et al.* [Carrat *et al.* (2008)] report viral shedding from healthy volunteers beginning the first day post-challenge and continuing for 4.8 days. People with varicella may be infectious from before lesions appear until scabs form, typically 6-7 days, for which $\mathcal{R}_0 \approx 12.6$.

The observed mixing matrices are qualitatively similar (Figure 3.7, left panel), but differ quantitatively. In particular, the sub- and super-diagonals from both time-use studies exceed those from both studies of face-to-face conversations. While people certainly can share spaces without conversing, careful analysis of similarities and differences between these and other encounters by which respiratory diseases might be transmitted would be illuminating. And, if protocols permitted comparison of contact patterns, as in the European countries studied by Mossong *et al.* [Mossong *et al.* (2008)], differences among societies could increase our understanding of the manner in which social structure mediates infectious disease transmission.

3.1.4 *Forecast for the fall wave of the 2009 H1N1 pendemic*

This example is taken from [Towers and Feng (2009)]. An extention of the classical SIR model is considered by using a seasonally forced transmission rate, which is consistent with the observed pattern for influenza. The model was intended to generate forecast for the fall wave of the 2009 H1N1, with available disease data for the summer.

With the recognition of a new, potentially pandemic strain of influenza A(H1N1) in April 2009, the laboratories at the US CDC and the World Health Organization (WHO) dramatically increased their testing activity from week 17 onwards (week ending 2 May 2009), as can be seen in Figure 3.10. In this analysis, we use the extrapolation of a model fitted to the confirmed influenza A(H1N1)v case counts during summer 2009 to predict the behavior of the pandemic during autumn 2009.

Below is the SIR epidemic model with periodic forcing for the transmission:

$$\frac{dS}{dt} = -\beta(t) \frac{SI}{N},$$
$$\frac{dI}{dt} = \beta(t) \frac{SI}{N} - \gamma I,$$
(3.18)

where $N = 305,000,000$ denote the total population in the United States (US). The R equation is omitted as it can be determined by $R = N - S - I$. The transmission rate is chosen to be

$$\beta(t) = \beta_0 + \beta_1 \cos(2\pi t/365)$$
(3.19)

where β_0 and β_1 are constants to be estimated from data. Assume that $I(t_0) = 1$ where t_0 is the initial time, which is another parameter to be estimated from data. The parameter γ is chosen so that the infectious period is $1/\gamma = 3$ days.

The three parameters (β_0, β_1, t_0) were estimated using the data shown in Figure 3.10 on influenza-positive tests reported to the US CDC by US WHO/NREVSS-collaborating laboratories, national summary, United States, 2009 until 26 September. To avoid bias due to increased testing for H1N1 around week 16, only data from week 21 to 33 (from 24 May to 22 August 2009) were used. From the past experience with influenza, the lower and upper bounds were chosen to be $\beta_0 \in (0.92\gamma, 2.52\gamma)$ and $\beta_1 \in (0.05\gamma, 0.8\gamma)$, and $t_0 \in (-8, 10)$ (weeks relative to the beginning of 2009). The best estimates were determined by fitting the model to data, and the parameter values that provided the best Pearson chi-square statistics are listed in Table 3.1.

Table 3.1   Estimates of parameters for the 2009 H1N1 model.

Parameter	Value	95% confidence interval
β_0	1.56	$(1.43, 1.77)$
β_1	0.54	$(0.39, 0.54)$
t_0	24 Feb	(8 Feb, 7 Mar)

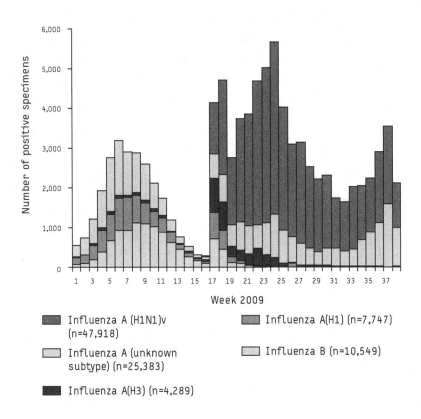

Fig. 3.10   Influenza-positve tests reported to the US CDC by US WHO/NERVSS-collaborating laboratories, national summary, United States, 2009 until 26 September.

Using the estimated parameter values, forecast of the time and size of the fall wave was produced by simulating the model under two scenarios: one is without vaccination and the other included the planned CDC vaccination program, which would begin with six to seven million doses being delievered by the end of the first full week in October (week 40), with 10 to 20 million doses being delievered weekly thereafter. It is assumed that for healthy adults, full immunity to H1N1 influenza is achieved about two weeks after vaccination with one dose of vaccine [Hannoun *et al.* (2004); Manuel *et al.* (2011)]. For simulations of the model when vaccinations are incorporated, the number of susceptible individulas in the simulations was decreased according to the appropriate proportion of immuned due to vaccination. The simulation outcomes are presented in Figure 3.11. The curves associated with to the darker and lighter areas correspond to the cases with and without vaccination. **The model predicts that in the absence of vaccine, the peak wave of infection will occur near the end of October in week 42** (95% CI: week 39,43). By the end of 2009, the model predicts that a total of 63% of the population will have been infected (95% CI: 57%, 70%). For the case when the planned vaccination program is considered, the model results suggest that a relative reduction of about 6% in the total number of people infected with influenza A(H1N1)v virus by the end of the year 2009 (95% CI: 1%, 17%).

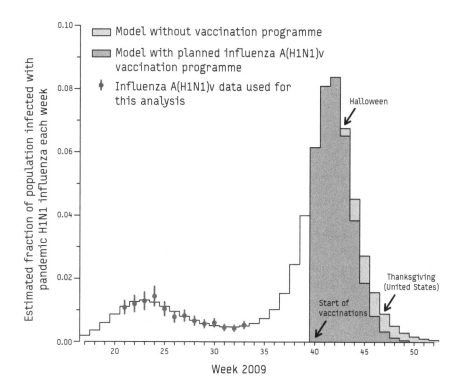

Fig. 3.11   Model of H1N1 influenza pandemic in the United States and prediction for autumn 2009 using the model (3.18).

The most striking feature of the model is that it accurately predicted the peak time of the pandemic. According to CDC 2009 H1N1 confirmed case count data (see [CDC (2011)]), the peak of the fall wave was reached at the end of October (which is between weeks 42 and 43, see the left hand plot in Figure 3.12), which is consistent with our model result. It is worth noting that the model used in the analysis is a simple SIR model with a seasonally forced infection rate. Although further examinations are certainly needed to study the applicability of the modeling approach to general scenarios, the model results in this analysis demonstrated clearly the advantage and capability of mathematical models in understanding disease dynamics.

The work received widespread national attention upon its release on October 15, 2009, with several national news agencies covering the story or referring to

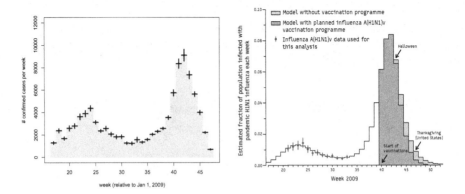

Fig. 3.12   The left figure illustrates the CDC 2009 confirmed H1N1 count data (the error bars represent variations calculated by the proposers that account for U.S. regional variablity in the timing of the pandemic). The right figure shows predictions by our model in [Towers and Feng (2009)].

the study including Washington times, Los Angeles Times and Chicago Tribune. Our published predictions were also mentioned during the October 21, 2009 hearing of the U.S. Senate Committee on Homeland Security and Government Affairs, entitled "H1N1 Flu: Monitoring the Nations Response". The Google flu trends (http://www.google.org/flutrends/us/#US) keeps track of flu activity for past and current years. Figure 3.13 is a snap shot of the data published at the website in February 2010 for the United States. It illustrates the time and size of the fall peak for years 2004 through 2009. We observe that for most of the years, the peak occurred around Feburary of the following year, but in 2009 the peak occurred near the end of October, which matches the prediction.

We point out again that most studies that compare model predictions with data regarding the timing and size of an epidemic or pandemic are published post the occurrence. In contrast, the model prediction made in [Towers and Feng (2009)] were published before the occurrence. It also demonstrated that for the particular questions addressed, the simple model (3.18) is capable of providing useful information.

3.1.5   *Effect of timing for implementing control programs*

To explore the influence of seasonally forced transmission in disease control and the final epidemic size of influenza, we consider an extension of the SIR epidemic model (1.1) by incorporating a periodic function $\beta(t)$ for the transmission rate (as in the model (3.18) as well as drug treatment and/or vaccination. This example is presented in [Feng *et al.* (2011)].

The model without treatment and vaccination is given by (3.18) with initial

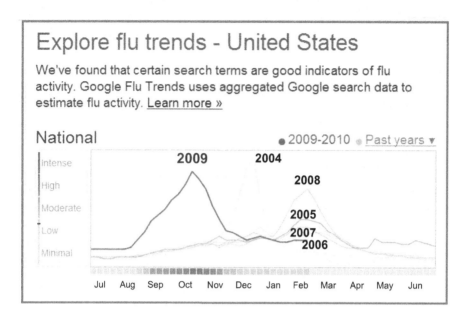

Fig. 3.13   A snap shot of data related to flu activities for the United States published by Google flu trends website. It demonstrates the time and size of the second peak of the flu epidemic/pandemic during each year from 2004 to 2009.

conditions

$$S(t_0) = N - I_0, \ I(t_0) = I_0, \ R(t_0) = 0, \tag{3.20}$$

where $I_0 > 0$ is a constant. In the initial conditions (3.20), $t_0 > 0$ denotes the time at which the infection is introduced into the population. Note that $t = 0$ corresponds to the first day of a calendar year and t_0 is the number of days from $t = 0$.

If vaccines are available before the epidemic starts, a certain level of population immunity can be achieved via vaccination. Let p_0 denote the level of population immunity at time t_0. Then the disease dynamics can be modeled by the equations in (3.18) with modified initial conditions:

$$S(t_0) = (1 - p_0)(N - I_0), \ I(t_0) = I_0, \ R(t_0) = 0. \tag{3.21}$$

We do not include vaccinated individuals in the R class because the value of $R(t)$ at the end of the disease outbreak will be used to measure the final epidemic size. The periodic transmission rate is the same as that in model (3.18) but written in a slightly different form:

$$\beta(t) = \beta_0 \big[1 + \epsilon \cos(2\pi t/365) \big], \tag{3.22}$$

where β_0 is a constant representing a background transmission rate, and ϵ is a constant related to the magnitude of the seasonal fluctuation. The function $\beta(t)$

given in (3.22) has its maximum at the beginning of a calendar year. When the transmission rate is expressed in this form, the time of introduction of the pathogen to the population t_0 is a crucial parameter of the seasonal model, because the final size and shape of the epidemic curve (including how many peaks the epidemic exhibits within one calendar year) can depend quite strongly on this parameter (whereas for models with constant β the final size and shape are independent of t_0). Further, the shape of the epidemic curve also strongly depends on the initial fraction of susceptibles in the population at t_0 when $\beta(t)$ is periodic.

The model can be further extended to include vaccination and drug treatment are considered. Let $f = f(t)$ denote the fraction of infected individuals who will receive treatment at time t; I_u and I_{tr} denote the numbers of untreated and treated infected individuals, respectively; and let the infectious period for an untreated individual is $1/\gamma_u$. Assume that treatment reduces infectiousness by a factor σ and reduces the infectious period to $1/\gamma_{tr} < 1/\gamma_u$. Following the approach of Lipsitch et al. [Lipsitch et al. (2007)] to model drug-treatment, we can extend the model (3.18) as

$$
\frac{dS}{dt} = -\beta(t)S\frac{I_u + \sigma I_{tr}}{N},
$$

$$
\frac{dI_u}{dt} = \beta(t)\big[1 - f(t)\big]S\frac{I_u + \sigma I_{tr}}{N} - \gamma_u I_u,
$$

$$
\frac{dI_{tr}}{dt} = \beta(t)f(t)S\frac{I_u + \sigma I_{tr}}{N} - \gamma_{tr} I_{tr}, \tag{3.23}
$$

$$
\frac{dR}{dt} = \gamma_u I_u + \gamma_{tr} I_{tr}
$$

with initial conditions

$$
S(t_0) = (1 - p_0)(N - I_{u0}), \ I_u(t_0) = I_{u0}, \ I_{tr}(t_0) = 0, \ R(t_0) = 0, \tag{3.24}
$$

where I_{u0} is a positive constant representing the initial number of infected people. A transition diagram between the epidemiological classes is shown in Figure 3.14.

Were there a delay in the supply of vaccines, we could still use equations in (3.23) with modifications. For example, let $t_v > t_0$ denote the time at which the vaccination program starts, and assume that a fraction p_0 of the remaining susceptibles will be vaccinated at the time t_v. We can first use equations in (3.23) and the initial conditions (3.24) with $p_0 = 0$ for simulations in the interval $t_0 \leq t < t_v$. For the time after t_v, we can continue to use the equations in (3.23), but with new initial conditions $(1 - p_0)S(t_v)$, $I(t_v)$ and $R(t_v)$.

We now consider the model (3.23). Denote the control reproduction number by $\mathcal{R}_c$. One objective of control programs is to reduce $\mathcal{R}_c$ to below 1, and the best strategy is usually the one that reduces $\mathcal{R}_c$ the most and the fastest for given resources. In the case when all rates are constant, i.e., $\beta(t) = \beta_0$ and $f(t) = f_0$ for all $t \geq t_0$, and when vaccines are available at t_0, the reproduction numbers for the

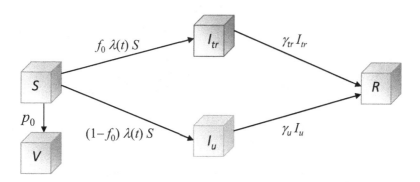

Fig. 3.14 **A transition diagram between the epidemiological classes for model (3.23).** The compartments are S (Susceptible), I_u (Infected untreated), I_{tr} (Infected treated), R (Recovered). The V class includes individuals who are immune to further infection due to vaccination. The λ function in the transmission term is defined by $\lambda(t) = \beta(t)(I_u + \sigma I_{tr})/N$, where $\beta(t)$ denotes the transmission rate and $\sigma \leq 1$ represents a reduction in infectiousness due to treatment. The constants p_0 and f_0 are the fractions of vaccinated susceptible and treated infectious individuals respectively. γ_u and γ_{tr} are the rates of recovery for the untreated and treated individuals respectively.

model (3.23) are given by

$$\mathcal{R}_0 = \frac{\beta_0}{\gamma_u} \quad \text{and} \quad \mathcal{R}_c = \frac{\beta_0(1-p_0)(1-f_0)}{\gamma_u} + \frac{\sigma\beta_0(1-p_0)f_0}{\gamma_{tr}}. \quad (3.25)$$

Note that $\mathcal{R}_c = \mathcal{R}_0$ in the absence of vaccination and treatment, i.e., when $p_0 = f_0 = 0$. For $p_0 > 0$ and/or $f_0 > 0$, because $\sigma \leq 1$ and $\gamma_{tr} \geq \gamma_u$,

$$\mathcal{R}_c \leq \mathcal{R}_0 \quad \text{for} \quad p_0 \geq 0, \ f_0 \geq 0.$$

In fact, $\mathcal{R}_c$ is a decreasing function of p_0 and f_0 as shown in Figure 3.15. An explicit formula for $\mathcal{R}_c$ such as the one given in (3.25) can be very helpful for designing control programs. For example, if the objective of a control measure is to reduce $\mathcal{R}_c$ to below a given level $\mathcal{R}_c^*$, then we can use this formula to determine the combined efforts of vaccination and treatment needed to achieve the goal $\mathcal{R}_c < \mathcal{R}_c^*$. The levels of vaccination p_0 and treatment f_0 that will lead to $\mathcal{R}_c < 1$ can be determined as shown in Figure 3.15. The curve on which $\mathcal{R}_c(p_0, f_0) = 1$ is shown, and we observe that all points (p_0, f_0) above the curve will lead to $\mathcal{R}_c < 1$, in which case the disease cannot take off. The right figure is a contour plot of $\mathcal{R}_c$ in the (p_0, f_0) plane. Several curves corresponding different $\mathcal{R}_c$ values are labeled with the thicker curve corresponding to $\mathcal{R}_c = 1$.

Note that the parameter p_0 in the function $\mathcal{R}_c(p_0, f_0)$ denotes the proportion of successfully vaccinated people. If the efficacy of vaccine is not 100%, then $p_0 = \eta\hat{p}$ where η and $\hat{p}$ denote respectively the efficacy of vaccine and the proportion of vaccinated population. In this case, the formula for $\mathcal{R}_c$ can also be considered as a function of η, $\hat{p}$ and f_0 given by

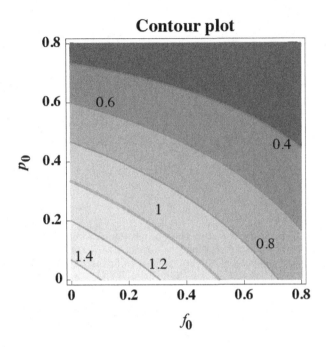

Fig. 3.15 **Contour plot showing the dependence of** $\mathcal{R}_c$ **on** p_0 **and** f_0. Several curves corresponding to different $\mathcal{R}_c$ values are labeled with the thicker curve corresponding to $\mathcal{R}_c = 1$. It shows also that $\mathcal{R}_c < 1$ for all points (p_0, f_0) above the curve in which case the disease cannot take off.

$$\mathcal{R}_c(\eta, \hat{p}, f_0) = \frac{\beta_0(1 - \eta\hat{p})(1 - f_0)}{\gamma_u} + \frac{\sigma\beta_0(1 - \eta\hat{p})f_0}{\gamma_{tr}}. \tag{3.26}$$

This formula allows us to evaluate the effect of improved vaccines. For example, in the absence of treatment (i.e., $f_0 = 0$) the control reproduction number $\mathcal{R}_c(\eta, \hat{p}, 0)$ is a function of η and $\hat{p}$ only. This function is plotted in Figure 3.16. The curve corresponding to the threshold condition, $\mathcal{R}_c(\eta, \hat{p}, 0) = 1$, is identified (the thicker curve). For a given proportion $\hat{p}$, we can use this curve to determine the minimum efficacy $\eta_{\min}$ such that $\mathcal{R}_c(\eta, \hat{p}, 0) < 1$ for all $\eta > \eta_{\min}$. The contour curves in the figure can also help us identify optimal combinations of η and $\hat{p}$ to achieve certain objectives. For example, suppose that the costs of vaccination and vaccine improvement are known and the total cost to successfully vaccinate a proportion $p_0 = \eta\hat{p}$ of the susceptibles can be determined, denoted by $C(\eta, \hat{p})$. Then for a pre-subscribed value $\mathcal{R}_c^*$, we can determine the optimal levels η_{opt} and $\hat{p}_{opt}$ that can minimize the total cost $C(\eta, \hat{p})$ under the constraint $\mathcal{R}_c(\eta_{opt}, \hat{p}_{opt}, 0) \leq \mathcal{R}_c^*$.

Besides the deterministic solutions, models such as (3.18) and (3.23), the models can also be used to simulate the situations in which the disease events (e.g., infection and recovery) may occur randomly (while being governed by the event rates

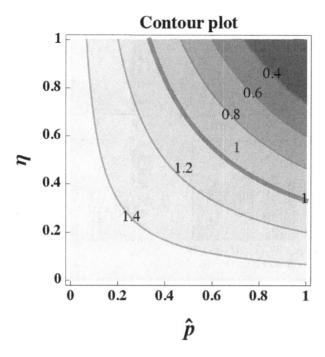

Fig. 3.16 **Contour plot showing the dependence of** $\mathcal{R}_c$ **on** η **and** $\hat{p}$ **with** $f_0 = 0$. This figure is similar to Figure 3.15 but with p_0 and f_0 replaced by η and $\hat{p}$. The thicker curve is the curves along which $\mathcal{R}_c = 1$. It shows that $\mathcal{R}_c < 1$ for all points $(\eta, \hat{p})$ above the curve, in which case the disease cannot take off.

determined by the equations. By incorporating the randomness of disease transmission process, the models will be capable of providing helpful information about how likely a certain disease outcome will be expected under a given condition. One of the approaches for stochastic simulations is the event-time approach [Renshaw (1993)]. Some simulation results are illustrated in Figure 3.17. Effects of model parameters (e.g., $\mathcal{R}_0$) on the disease outcomes can be examined by computing the statistics generated from the stochastic simulations. For example, among the 20 realizations (repeated runs with the same set of parameters), each of which showing different peak and final sizes and the time at the peak, the means are: epidemic peak=34% of the population, peak time=50 days, and the final size=40% of the population. The histograms show the variations of the 20 epidemic peak sizes and peak times around the respective means.

When model parameters vary with time, analytical results such as the threshold conditions described in (3.25) will be very difficult to obtain. Most results are based on numerical simulations. In this section, we present examples in which the transmission rate $\beta(t)$ has the form as in (3.62), with possible delayed vaccination and treatment uses, i.e., $t_v \geq t_0$ and $t_f \geq t_0$. These scenarios are based on the consideration that vaccines and antiviral drugs may not be available at the beginning

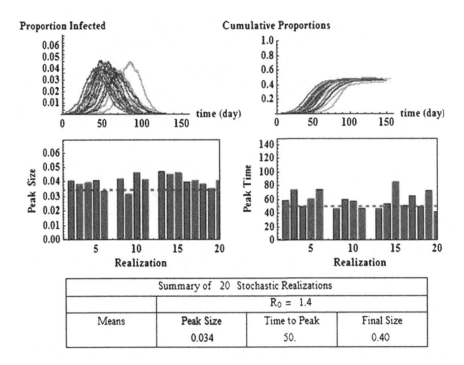

Fig. 3.17 **Stochastic simulations of model (3.18) with initial conditions (3.20).** It shows 20 repeated runs with the same parameter values. The top-left plot shows all 20 epidemic trajectories and the top-right plot shows all 20 curves of cumulative infections. The histograms illustrate the variations of peak sizes and the peak times relative to the respective means (the dashed line). A summary of the mean values of the 20 realizations is provided in the table on the bottom, showing that the mean value of the epidemic sizes is 34% of the population, the mean of peak times is 50 days, and the mean final sizes is 40% of the population.

of an epidemic. We will again focus on three important measures when assessing the effect of a control program: (a) peak size of the epidemic curve (the maximum number of infections during the course of a pandemic); (b) peak time (the time at which the peak occurs); and (c) final size (the total number of infections at the end of a pandemic). Main objectives of effective control strategies should include lowering the peak size to keep demand for facilities below available supply, lowering the final size to reduce morbidity, and delaying the peak to provide more time for response.

Unlike in the case of constant parameters, where vaccination and treatment will always help reduce morbidity (see Figure 3.15), the case a periodic transmission rate $\beta(t)$ may generate non-intuitive results. That is, the model can exhibit outcomes in which an increased use of vaccination or antiviral drugs will lead to a higher morbidity. To demonstrate this, we first consider the simpler case in which the model (3.18) with the initial conditions given in (3.21) is used. This represents the case when vaccination starts immediately when infection is introduced into the population. One such example is presented in Figure 3.18.

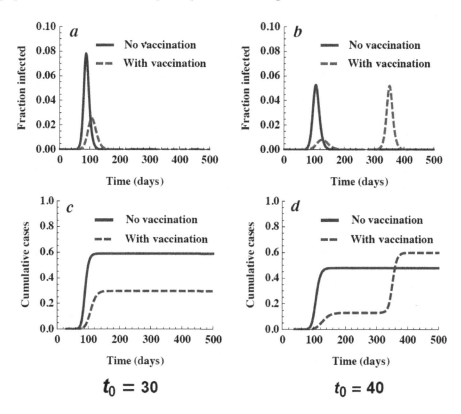

$$t_0 = 30 \qquad\qquad t_0 = 40$$

Fig. 3.18 **Plots of epidemic curves and cumulative infections for different t_0 values.**
These plots show the dramatic effect that the start time of an epidemic (t_0) can have on its
progression in a periodic environment given by the SIR model (3.18) with initial conditions (3.21).
From the plots in a and c we observe that when $t_0 = 30$, vaccination reduced both the peak size
and the peak time although there is not much delay in the peak time. However, plots in b and d
show that, although the size of the first epidemic peak is reduced, a second (and higher) peak is
also generated. More significantly, the final epidemic size is increased, showing the sensitivity of
the behavior of the epidemic to t_0.

In Figure 3.18, both the epidemic curve and curve representing the cumulative
infections are plotted. For the transmission function $\beta(t)$, the parameter values
used are those of an H1N1-like disease with $\beta_0 = 0.5$ and $\epsilon = 0.35$. We assume
$1/\gamma = 3$ days. These values are also used in other figures except when specified
differently. The figure shows two sets of simulations corresponding to two different
times t_0 of initial introduction of pathogens. One case is for $t_0 = 30$ and the other
is for $t_0 = 40$, and for demonstration purposes the vaccination level is chosen to be
$p_0 = 0.1$.

Some interesting observations can be made from Figure 3.18. We can first compare the outcomes between the model without vaccination (the solid curves) and

the model with vaccination (the dashed curves). It demonstrates the dependence of the model outcomes on the time of introduction (t_0) of initial infections. Particularly, we observe that although the peak and final sizes are both decreased for the case of $t_0 = 30$, the case of $t_0 = 40$ differs significantly. Although the first wave is reduced, the second wave is also generated; and moreover, the final epidemic size is even higher than that without vaccination, demonstrating a scenario in which the use of vaccine may have a detrimental consequence.

In the next example, we use the model (3.18) with initial conditions given in (3.21) to examine the interplay between t_0, t_v and p_0, and their influence on the effect of vaccination programs. In Figure 3.19, we consider two t_0 values: $t_0 = 30$ in panel a and $t_0 = 40$ in panel b. In each of these panels, model outcomes corresponding to different levels of vaccination (p_0) are compared. When $t_0 = 30$, panel a shows that the effect of vaccination use on the peak and final sizes are not monotone. That is, the peak and/or final sizes may either decrease or increase with p_0. When t_0 is increased from 30 to 40, panel b shows that, although the final sizes decrease monotonically with increasing p_0, the final epidemic size for the case of $p_0 = 0.1$ is even higher than that without vaccination use.

We now examine the simulation results from the model (3.23), which includes also drug treatment, with possible delays in vaccination and treatment use. Simulations illustrated in Figures 3.20 and 3.21 show how the peak and final epidemic sizes may change with the levels of vaccination and drug treatment (p_0, f_0) and the starting times of these control programs (t_v and t_f).

Figure 3.20 plots the epidemic curves and final epidemic sizes for the case of $t_0 = 55$ under various levels of vaccination (p_0) without treatment ($f_0 = 0$). Notice that for $t_0 = 55$ there is only a single peak in the epidemic curve in the absence of vaccination and treatment (see a). Simulation results for two different values t_v are demonstrated in the middle panel ($t_v = 60$) and the bottom panel ($t_v = 100$). In each panel, outcomes for three p_0 values are compared. A dramatic difference is that for the case of $t_0 = 60$ the peak and final epidemic sizes exceed that without vaccination use, whereas for the case of $t_0 = 100$ the peak and final size decrease monotonically with increasing p_0. We need to point out that, although this set of simulations suggest that an earlier start of vaccination use is not beneficial (an increased peak and final sizes), this may not be true for cases with other values of t_0 (simulations not shown here). That is, whether and how much vaccination can be helpful to reduce the severity of epidemics is dependent on not only the time when vaccination starts (t_v) but also the time of introduction of infections (t_0).

Figure 3.21 illustrates the effect of either drug treatment alone (i.e., $f_0 > 0$ and $p_0 = 0$) or combined with vaccination (i.e., $f_0 > 0$ and $p_0 > 0$). For demonstration purposes, it is chosen that $t_0 = 55$. Figure 3.21a is again for the case of no control ($f_0 = p_0 = 0$). By comparing the panels in the middle ($p_0 = 0$) and the bottom ($p_0 > 0$) we observe that, although antiviral use may increase the peak and final sizes when used alone, its benefit can be dramatically improved when combined

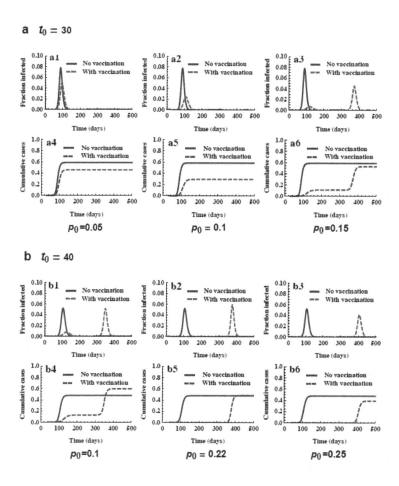

Fig. 3.19 **Comparison of epidemic severity for different values of t_0 and p_0 with $f_0 = 0$.**
It shows the joint influence of t_0 and p_0 on the effect of vaccination programs. Plots in panel a
show that the peak and final sizes are decreased when p_0 is increased from 0.05 (see a1 and a4) to
0.1 (see a2 and a5). However, when p_0 continues to increase from 0.1 to 0.15, one epidemic peak
becomes two with the second peak higher than the first (see a3 and a6), and the final epidemic
size is larger than the case of $p_0 = 0.1$. This differs from the case of $t_0 = 40$ shown in panel b. In
this case, for the lower p_0 value ($p_0 = 0.1$) two peaks occurred and the final size is higher than
that without vaccination use (compare b1 and b4). However, when p_0 exceeds 0.22 the peak and
final sizes will not be higher than those without vaccination use. Both peak and final sizes will
continue to decrease when p_0 increases (see b2-b3 and b5-b6).

with the use of vaccination.

As mentioned earlier, effective control strategies for pandemics should aim at
lowering the peak size to lest demand for facilities exceed supply, lowering the final
size to reduce morbidity, and delaying the peak to provide time for response. However, results presented here show that an increase in control efforts may not always
be beneficial for achieving these objectives. Particularly, the outcomes demon-

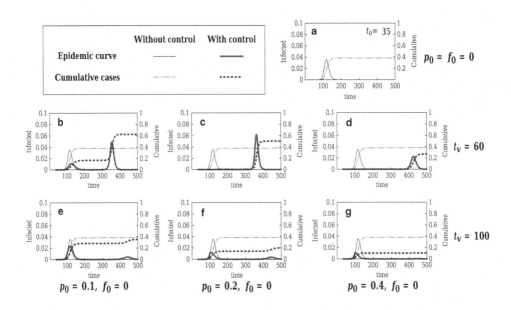

Fig. 3.20   **Comparison of epidemic severity for different values of t_v and p_0 with $f_0 = 0$.**
These plots of epidemic curves and final epidemic sizes are for the case of $t_0 = 35$ with two values
of t_v (the time when vaccination begins) under three different levels of vaccination (p_0) and no
treatment ($f_0 = 0$). Notice that there is only a single peak in the epidemic curve in the absence of
control (see a). Figure 3.20a is for the case of no vaccination ($p_0 = 0$). Two different starting times
of vaccination (t_v) are examined: Figures 3.20b-3.20d are for $t_v = 60$, whereas Figures 3.20e-3.20g
are for $t_v = 100$. For each t_v value, three levels of vaccination use are considered: $p_0 = 0.1$ (b
and e), $p_0 = 0.2$ (c and f), and $p_0 = 0.4$ (d and g). The peak of the epidemic curve and the final
epidemic size are displayed for each of the scenarios. The following observations can be made. **(i)**
For the earlier start of the vaccination program ($t_v = 60$), a lower level of vaccination ($p_0 = 0.1$)
may lead to an increased peak and final size (compare the sizes in Figure 3.20b with those in
3.20a), while a higher level of vaccination ($p_0 = 0.4$) can dramatically decrease the peak and the
final size (compare Figure 3.20d with 3.20a). **(ii)** For the later start of the vaccination program
($t_v = 100$), the use of vaccines did not increase the peak or final sizes (see Figures 3.20e-3.20g).
(iii) In all cases illustrated in Figures 3.20b and 3.20d, the time to epidemic peak is delayed, and
more delay in the peak time is expected for higher levels of vaccination use. Other parameter
values are the same as before.

strated in Figures 3.18–3.21 suggest that in some cases the uses of vaccine and
treatment may have a detrimental consequence (e.g., increasing peak and final epidemic sizes), and that whether and how much vaccination and treatment can be
helpful will depend critically on the time of introduction of infection (t_0), the starting time of vaccination use (t_v) and the levels of vaccination and treatment (p_0 and
f_0). Thus, it is very important that these control programs are designed appropri-
ately.

A possible explanation for the negative effects of these control measures is the
seasonal variation in disease transmission. Vaccination/treatment during the spring
wave of an epidemic essentially increases susceptibles at the end of the summer.

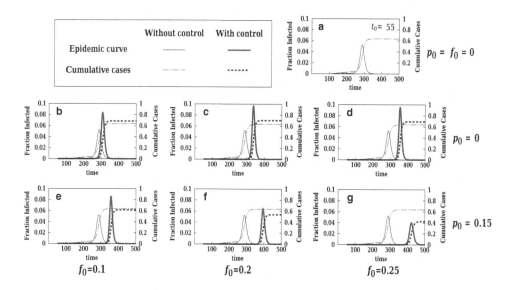

Fig. 3.21 **Comparison of epidemic severity for different combinations of p_0 and f_0.** This figure illustrates the effect of treatment when combined with vaccination. Figure 3.21a is again for the case of no control ($f_0 = p_0 = 0$) and for $t_0 = 55$. Figures 3.21b-3.21d are for the case of no vaccination ($p_0 = 0$), whereas Figures 3.21e-3.21g are for the case of with an intermediate level of vaccination ($p_0 = 0.15$) and a starting time $t_v = 150$. We observe that together with vaccination, antiviral drugs will be more beneficial.

This larger number of susceptibles can, in some cases, act as tinder for a much larger epidemic to occur in the more favorable influenza season of autumn than may have occurred if treatment had not been used in the summer months. Thus, to increase the likelihood of successful control programs, policy decisions should take into consideration the potential impact of these factors on disease transmission dynamics.

This modeling study suggests that the patterns of disease transmission (e.g., with or without seasonal variations) can have a significant influence on the effectiveness of control and intervention programs. When a disease exhibits periodic patterns in transmission, decisions of public health policies will be particularly important as to how control measures such as vaccination and drug treatment should be implemented. More importantly, mathematical models can be very helpful for understanding complex disease transmission dynamics and for identifying the optimal control strategies.

3.1.6 *Herald wave*

The example presented in this section is chosen from [Alfaro-Murillo *et al.* (2013)]. We describe a multiple strain Susceptible Infected Recovered deterministic model for the spread of an influenza subtype within a population. The model incorporates appearance of new strains due to antigenic drift, and partial immunity to reinfection with related circulating strains. It also includes optional seasonal forcing of the transmission rate of the virus, which allows for comparison between temperate zones and the tropics. Our model is capable of reproducing observed qualitative patterns such as the overall annual outbreaks in the temperate region, a reduced magnitude and an increased frequency of outbreaks in the tropics, and the herald wave phenomenon. Our approach to modeling antigenic drift is novel and further modifications of this model may help improve the understanding of complex influenza dynamics.

Influenza presents a significant morbidity and mortality burden in the world; a typical seasonal influenza epidemic kills up to 49,000 people per year in the United States, and from a quarter to half a million worldwide [WHO (2009)]. One of the key aspects of the influenza virus is its ability to reinfect hosts. Influenza is an RNA virus and lacks therefore a proofreading mechanism for its polymerase. As a consequence, it is prone to make mistakes every time a copy of its genome is made. Because these mutations usually do not change how the virus is identified by the immune system, infection with one influenza strain generally confers lasting immunity to an identical strain. However, if the virus has undergone sufficient significant changes in its surface proteins, then reinfection with influenza is possible. This process is termed antigenic drift. One of the main findings on this topic is that the immunity conferred after infection wanes with decreasing relatedness between the immunizing and challenging strains [Couch and Kasel (1983)]. Thus, once a few years have passed since the previous infection with a specific subtype of influenza, a human host may become almost completely susceptible to the currently circulating grand-daughter strain [Couch and Kasel (1983)].

The main challenge in influenza control is an early prediction of the strain that will be the dominant circulating one each season. This is needed so that a matching vaccine can be mass-produced on time. "Herald" waves of influenza (strains which begin to dominate very late in a previous season) have been successfully used many times in the past to develop the following year's vaccines [Glezen *et al.* (1982); Nakajima *et al.* (1992)]. A model of influenza dynamics that can provide useful information about the persistence of strains from season to season would be helpful to improve the success rates of vaccines. Such a model would ideally take into account several key biological features including the drift process (via the emergence of new strains), the co-circulation of related strains, and pre-existing immunity in the population. This work is an initial step towards this ideal model.

Influenza has been widely studied and characterized for temperate regions. The occurrence shows significant seasonal variations with marked peaks in the winter season each year [Viboud *et al.* (2006b)]. In contrast, influenza seasonality is less defined in the tropics, where a viral strain may persist in the population throughout the entire year at a constant intermediate level, with some peaks varying from country to country [Viboud *et al.* (2006a); Simmerman *et al.* (2004); Alonso *et al.*

(2007)], and no distinct annual patterns [Russell *et al.* (2008); Alonso *et al.* (2007); Lee *et al.* (2009)]. We applied our model to both regions and obtained similar results.

The model described here focuses on a certain subtype of influenza of type A (e.g., influenza A(H1N1), A(H3N2)), or influenza of type B. The main reason for this consideration is that we intend to study the dynamics of the drift process of influenza. The infection process is described by a Susceptible-Infected-Recovered model, in which individuals are divided into classes according to the history of infections by strains that had previously circulated or are currently circulating. The infection rate for individuals in each of the classes may change depending on how closely related the strains are, in an antigenic map. We assume that when there are enough individuals infected with a certain strain, a new mutation closely related to that circulating strain will predominate, at which time the model will be updated to incorporate the new strain.

Consider the case when n strains are currently co-circulating in a population. We divide the population into epidemiological classes according to the infection history of individuals. Let $\mathbf{v} = (v_1, v_2, \cdots, v_n)$ be a vector which keeps track of the infection history of an individual. The components of $\mathbf{v}$ are either one or zero; with $v_i = 1$ and $v_i = 0$ representing whether or not the individual has already been infected by the virus strain i, respectively. A vector of this type will be referred to as a history vector. Because each of the n positions can have two possible values (either 0 or 1), the total number of possible history vectors is 2^n. We will use $R_\mathbf{v}$ to denote the recovered class with history $\mathbf{v}$. Note that $R_\mathbf{0} = R_{(0,\ldots,0)}$ represents the completely susceptible class. To avoid any confusion between $R_\mathbf{0}$ and the usual notation for the basic reproduction ratio, we will use S to replace $R_{(0,\ldots,0)}$ in the model description.

For the infected classes, we need to keep track of the information on both the strain of current infection and the infection history. Let $I_\mathbf{v}^i$ denote the infected class which includes all individuals who are currently infected with strain i and have the history $\mathbf{v}$. For example, if $n = 3$, an individual in $I_{(1,0,1)}^1$ is currently infected with strain 1 and was previously infected with strain 3 but not strain 2. Note that if an individual is infected with strain i it necessarily has a one on position i of its history vector, thus, for example $I_{(0,1,1)}^1$ is not a possible class. So, in the case of n strains, there are $n2^{n-1}$ infected classes.

Figures 1-4 in [Alfaro-Murillo *et al.* (2013)] illustrate the model structure by exhibiting different scenarios for different numbers of co-circulating strains in a population. It is clear that the complexity of model structure increases significantly when the number of strains increases. However, the basic patterns are straightforward and the rules remain the same for arbitrary numbers of co-circulating strains. An individual in $I_\mathbf{v}^i$ (who is currently infected with strain i and has history $\mathbf{v}$) recovers and enters in the recovered class $R_\mathbf{v}$ with the same history $\mathbf{v}$. On the other hand, new infections entering in the class $I_\mathbf{v}^i$ have to come from the recovered class

$R_{\mathbf{v}'}$, where $\mathbf{v}'$ denotes the history vector that is the same as $\mathbf{v}$ except that $v_i = 0$. Here it is assumed that individuals that recover from an infection by strain i will have permanent immunity against the same strain.

Vital dynamics are also considered in our model. For simplicity, we assume a constant total population size. We denote by μ the natural death rate (equal to the birth rate), for all epidemiological classes in the model.

Using the notations defined above as well as in ealier sections, our system of differential equations for the case of n strains reads

$$\dot{S} = -\sum_{i=1}^{n} \sum_{\mathbf{w}:i\in H(\mathbf{w})} \left[\beta_i(t) \frac{I_{\mathbf{w}}^i}{N} S \right] + \mu(N - S),$$

$$\dot{I}_{\mathbf{v}}^i = \sum_{\mathbf{w}:i\in H(\mathbf{w})} \left[\rho_{\mathbf{v}'}^i \beta_i(t) \frac{I_{\mathbf{w}}^i}{N} R_{\mathbf{v}'} \right] - (\gamma_i + \mu) I_{\mathbf{v}}^i, \qquad (3.27)$$

$$\dot{R}_{\mathbf{v}} = \sum_{i\in H(\mathbf{v})} \gamma_i I_{\mathbf{v}}^i - \sum_{i\notin H(\mathbf{v})} \sum_{\mathbf{w}:i\in H(\mathbf{w})} \left[\rho_{\mathbf{v}}^i \beta_i(t) \frac{I_{\mathbf{w}}^i}{N} R_{\mathbf{v}} \right] - \mu R_{\mathbf{v}},$$

for $\mathbf{v} \neq \mathbf{0}$; where, "$\dot{x}$" denotes $\frac{dx}{dt}$ and N is the total population:

$$N = S + \sum_{i=1}^{n} \sum_{\mathbf{w}:i\in H(\mathbf{w})} I_{\mathbf{w}}^i + \sum_{\mathbf{v}\neq\mathbf{0}} R_{\mathbf{v}}, \qquad (3.28)$$

which remains constant for all time t. Because there are 2^n recovered classes (including S) and $2^{n-1}n$ infected classes, the total number of equations in this system is $2^{n-1}(n + 2)$ for n circulating strains.

Let $\lambda(i, j)$ denote the distance between strains i and j. Let ρ denote the factor of reduced susceptibility by strain i for an individual who had previously been infected with strain j, which is given by

$$\rho_j^i = 1 - e^{-\lambda(i,j)}. \qquad (3.29)$$

Thus, the transmission rate is 0 in the case of total immunity and it is $\beta_i(t)$ in the case of complete susceptibility.

To define the level of protection that a given class $\mathbf{v}$ has against strain i, we use the best protection in the history of previous infections of that class, that is:

$$\rho_{\mathbf{v}}^i = \begin{cases} \min_{k\in H(\mathbf{v})} \rho_k^i & \text{if } \mathbf{v} \neq \mathbf{0} \\ 1 & \text{if } \mathbf{v} = \mathbf{0}, \end{cases} \qquad (3.30)$$

where $H(\mathbf{v})$ is the set of all strains individuals with history $\mathbf{v}$ had previously been infected with, i.e.,

$$H(\mathbf{v}) = \{k : v_k = 1\}. \qquad (3.31)$$

The force of infection is assumed to be periodic with

$$\beta_i(t) = \hat{\beta}_i \left(1 + \varepsilon \cos \left(\frac{2\pi t}{365} \right) \right), \qquad (3.32)$$

where $\hat{\beta}_i$ is a constant background transmission rate of strain i, which also represents the constant transmission rate in the absence of seasonality (i.e., $\varepsilon = 0$). If $t = 0$ corresponds to January 1st of a calendar year, then the $\beta(t)$ function with $\varepsilon > 0$ corresponds to the northern hemisphere, implying that the transmission rate is higher in the winter, on dates that are close to January 1st. On the other hand, if $\varepsilon < 0$, then the transmission will be biggest on dates close to the middle of the year, which corresponds to the winter on the southern hemisphere. The case of $\varepsilon = 0$, or a value close to 0, corresponds to a model for the tropics.

For the initialization of model parameters, we simulate influenza transmission over many years in a population of ten million individuals in both the tropics ($\varepsilon = 0$ in Equation (3.32)) and northern temperate regions (positive values of ε). We start the simulation at time $t = 0$ with only one strain and a small number of infected individuals. The seasonal forcing of influenza is currently unknown and has not been extensively studied. However, a few data and modeling studies indicate that values of 0.20 to 0.35 for ε appear to be reasonable estimates [Ferguson *et al.* (2003); Towers and Feng (2009)]. Here we assume values of ε to vary between 0 and 0.20 and try to get an insight into the transition from a tropical region to a temperate one.

The basic reproduction ratio of influenza $\mathcal{R}_0$ has been measured in temperate climates to be between 1.1 to 1.7, with an average around 1.4 [Bonabeau *et al.* (1998); Flahault *et al.* (1988)]. We assume that these estimates occur in the winter when the transmission rate is maximal. We thus estimate the baseline transmission rate in Equation (3.32) to be $\hat{\beta} = \mathcal{R}_0\gamma/(1 + \varepsilon) \sim 0.4$, where γ is the recovery rate with $1/\gamma = 3$ (i.e., the infectious period is three days) [Colizza *et al.* (2007)].

For the criterion on the timing for the introduction of a new strain, based on the understanding that the probability of a significant mutation responsible for creating a new strain can be thought to be an independent process for each infection, it is reasonable to assume that the mean time for a new strain to appear depends mainly on the cumulative number of infections of a previously created strain i. This cumulative number, which we denote by $\mathcal{K}$, will be used to determine the time for a new strain to appear. Specifically, let $\mathcal{C}_i(a, b)$ denote the number of new infections of strain i between the time a and the time b, then

$$\mathcal{C}_i(t_1, t_2) = \int_{t_1}^{t_2} \left(\sum_{\mathbf{v}:i \in H(\mathbf{v})} \sum_{\mathbf{w}:i \in H(\mathbf{w})} \left[\rho_{\mathbf{v}'}^i \beta_i(t) \frac{I_{\mathbf{w}}^i}{N} R_{\mathbf{v}'} \right] \right) dt.$$

Let $t_1 > 0$ denote the time at which the last daughter strain of strain i was generated. Then a new strain will be created at time t_2 if $\mathcal{C}_i(t_1, t_2) = \mathcal{K}$; that is,

$$\mathcal{K} = \int_{t_1}^{t_2} \left(\sum_{\mathbf{v}:i \in H(\mathbf{v})} \sum_{\mathbf{w}:i \in H(\mathbf{w})} \left[\rho_{\mathbf{v}'}^i \beta_i(t) \frac{I_{\mathbf{w}}^i}{N} R_{\mathbf{v}'} \right] \right) dt. \tag{3.33}$$

The parameters for the threshold value $\mathcal{K}$ defined in Equation (3.33), and the antigenic distance λ are calibrated based on the following considerations. We select the parameter values for which sustained dynamic resonance is achieved in all populations, without an unnaturally high annual attack rate, or an over-abundance of co-circulating strains (which we found to be associated with a high attack rate. We take the unnaturally high attack rate to be 50% for the temperate regions, given that the maximal observed clinical attack rate is estimated to be 25% [Molinari *et al.* (2007)], but up to 60% of such infections are asymptomatic [King *et al.* (1988)]. We find that for $\mathcal{K}$ between 25% and 38% of the population, and λ between 1.5 and 1.8, the simulations yield sustained resonance in both the tropics and temperate regions.

In this study, we present a set of simulations from our model that are capable of capturing the observed patterns of influenza prevalence in the two regions, tropical and temperate. The parameter values used in the simulations are identical for all the regions except for the transmission rates, for which a constant (or close to constant) rate is used for the tropic region while a periodic function is chosen for the different temperate regions.. The simulations are for a period of 25 years.

To demonstrate the difference in influenza prevalence curves between tropical and temperate regions, we illustrate in Figure 3.22 the simulation results for various values of ε. Zero or smaller values of ε represent the tropical region whereas larger values represent temperate regions. The figure shows that the size of the epidemic peaks in the tropics is lower than in the temperate region, and that they occur more frequently than annually, matching the pattern that is seen in data (lower peaks more frequently [Russell *et al.* (2008); Alonso *et al.* (2007); Lee *et al.* (2009)]). The figure also presents diversity bottlenecks each summer, a pattern that is observed in data [Rambaut *et al.* (2008)]. The parameter values used are: $\gamma = 1/3$ days^{-1}, $\mathcal{K} = 0.27 \times 10^7$ (that is 27% of the total population), and $\lambda = 1.5$.

The herald wave phenomenon observed in data is reproduced in our model when applied to a temperate region (e.g., for the value of $\varepsilon = 0.20$), with the strain that dominates each season appearing late in the previous season (see Figure 3.23). The overall average annual attack rate in the temperate region with this set of parameters was around 45%, leading to an estimated clinical attack rate of around 18% [King *et al.* (1988)].

To examine how the seasonal forcing (ε) may affect the size of annual major peak of influenza prevalence and the timing of the peak, we plotted in Figures 3.24 and 3.25, for difference values of ε, the average peak and average date of the peaks, respectively, over a period of 20 years (starting four and half years after the initialization of the model). The standard deviation is also indicated for each fixed ε value. In these figures, the case of $\varepsilon = 0$ (or smaller ε values) corresponds to the tropics and the cases of larger positive ε correspond to the northern hemisphere. Figure 3.24 shows that the size of the major peak decreases first with ε and increases for larger ε. Figure 3.25 shows that as ε gets larger, the timing of the major peak per year shows less difference, naturally in the middle of the winter where the seasonal forcing is biggest. Due to the symmetric property of the model, the same results are

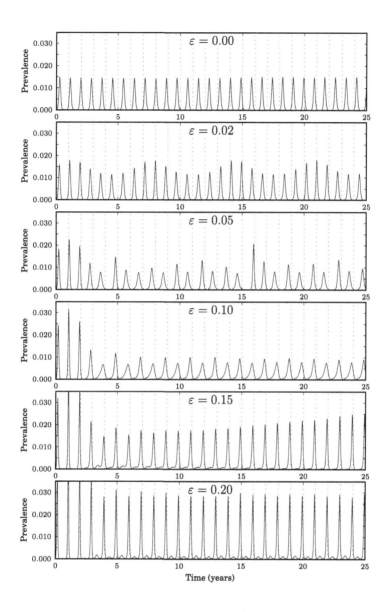

Fig. 3.22 Prevalence (measured in percentage of the population infected per day) for different values of non-negative ϵ, i.e., for the tropics and the northern hemisphere. The dashed vertical lines represent the middle of the winter of each year.

true, *mutatis mutandis*, for the southern hemisphere (negative values of ε). These results agree with the data presented in Alonso *et al.* [Alonso *et al.* (2007)], if we associate ε with latitude.

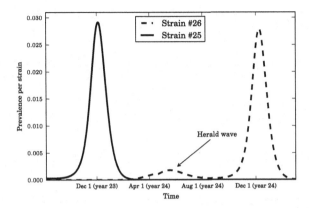

Fig. 3.23   The herald wave phenomenon. A summer wave of a new strain predicts the strain that will dominate over the following winter. The value of seasonal forcing used is $\epsilon = 0.20$. Prevalence for each strain is measured in percentage of the population infected per day.

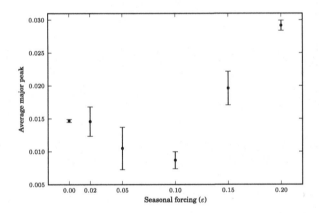

Fig. 3.24   Average size and standard deviation of the major annual peak for different non-negative values of ϵ, i.e., for the tropics and the northern hemisphere.

3.1.7   *Adverse effect of antiviral use due to drug-resistance*

The example considered in this section is adopted from Qiu and Feng [Qiu and Feng (2010)]. The model considered in [Qiu and Feng (2010)] is an extension of the model studied in [Lipsitch *et al.* (2007)] by including a vaccinated class and vital dynamics. As the main purpose is to look at the interaction between vaccination and drug use, prophylaxis is ignored. Let N denote the total number of the population which is divided into six sub-classes: susceptible (S), vaccinated (V), infected with the sensitive strain and untreated (I_{SU}) or treated (I_{ST}), infected with the resistant strain (I_R), and recovered (R). Assume that there is a constant recruitment rate Λ (into

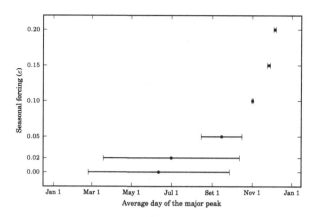

Fig. 3.25 Average day and standard deviation of the major annual peak for different non-negative values of ϵ, i.e., for the tropics and the northern hemisphere.

the susceptible class) and a per-capita natural death rate μ. A transition diagram between these epidemic classes is shown in Figure 3.26. Susceptible individuals are vaccinated at per-capita rate ν and the immunity wanes at per-capita rate σ. The functions $\lambda_S(t)$ and $\lambda_R(t)$ represent the rates at which a susceptible individual becomes infected with the sensitive and resistant strains, respectively. A fraction f of infected individuals with the sensitive strain receives treatment, and with a probability c an individual who received treatment will develop drug resistance. The transmission rate by an individual who received treatment will be reduced by a factor δ. An infected individual in the I_j $(j = ST, SU, R)$ class recovers at the rate k_j, and a recovered individual may lose immunity at the rate w ($w = 0$ in the case of permanent immunity). All parameters are positive except that $0 \leq f < 1$ and $0 \leq c < 1$.

Based on the transition diagram in Figure 3.26, the model is described by the following system of differential equations:

$$
\begin{cases}
\dfrac{dS}{dt} = \Lambda - (\mu + \nu)S - \lambda_S(t)S - \lambda_R(t)S + wR + \sigma V, \\[2mm]
\dfrac{dV}{dt} = \nu S - (\sigma + \mu)V, \\[2mm]
\dfrac{dI_{SU}}{dt} = (1 - f)\lambda_S(t)S - \mu I_{SU} - k_U I_{SU}, \\[2mm]
\dfrac{dI_{ST}}{dt} = f(1 - c)\lambda_S(t)S - \mu I_{ST} - k_T I_{ST}, \\[2mm]
\dfrac{dI_R}{dt} = \lambda_R(t)S + fc\lambda_S(t)S - \mu I_R - k_R I_R, \\[2mm]
\dfrac{dR}{dt} = k_T I_{ST} + k_U I_{SU} + k_R I_R - (\mu + w)R,
\end{cases}
\qquad (3.34)
$$

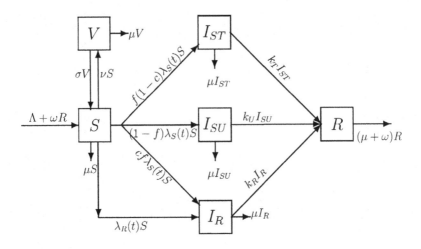

Fig. 3.26   Diagram of transitions between epidemiological classes.

where

$$\lambda_S(t) = \beta_S \frac{I_{SU} + \delta I_{ST}}{N}, \quad \lambda_R(t) = \beta_R \frac{I_R}{N},$$

$N = S + V + I_{SU} + I_{ST} + I_R + R$, and β_S and β_R denote the transmission coefficients for the sensitive and resistant strains respectively.

Simulations of the model show that an increase of treatment rate may not always help reduce the prevalence of infections. This is illustrated in Figure 3.27.

The two isolated control reproduction numbers for the sensitive and resistant straints are $\mathcal{R}_{SC}$ and $\mathcal{R}_{RC}$, respectively, given by

$$\mathcal{R}_{SC} = \frac{\sigma + \mu}{\sigma + \mu + \nu}\Big[(1 - f)\mathcal{R}_{SU} + f(1 - c)\mathcal{R}_{ST}\Big],$$

$$\mathcal{R}_{RC} = \frac{(\sigma + \mu)}{(\sigma + \mu + \nu)}\mathcal{R}_R, \tag{3.35}$$

where

$$\mathcal{R}_{SU} = \frac{\beta_S}{\mu + k_U}, \quad \mathcal{R}_{ST} = \frac{\beta_S \delta}{\mu + k_T}, \quad \mathcal{R}_R = \frac{\beta_R}{\mu + k_R}. \tag{3.36}$$

The biological interpretations of these quantities are as follows. $\mathcal{R}_{ST}$ and $\mathcal{R}_{SU}$ represent the numbers of secondary sensitive cases produced by a treated and untreated sensitive case, respectively, during the period of infection in a susceptible population. Notice that each sensitive case may either receive treatment with probability f or remain untreated with probability $1 - f$, and only a fraction $f(1 - c)$ of treated sensitive cases remains sensitive (the fraction fc becomes resistant). Notice also that $(\sigma + \mu)/(\sigma + \mu + \nu)$ is the fraction of the population that is susceptible. Thus, $\mathcal{R}_{SC}$ (S for sensitive and C for control) represents the number of secondary

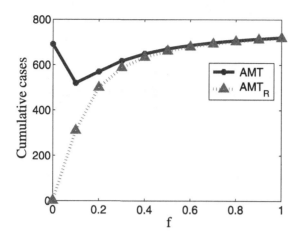

Fig. 3.27 Simulation results showing the dependence of cumulative cases (at a given time) on the level of treatment. The solid cuve represents the cumulative cases of both strains while the dashed curve represents the cumulative cases of the resistant strain only. It illustrates that it is possible that an increased treatment may lead to a high level of prevalence.

sensitive cases produced by a typical sensitive case during the period of infection in a population where control measures (vaccination and treatment) are implemented.

Similarly, $\mathcal{R}_R$ in (3.36) represents the number of secondary resistant cases produced by a resistant case during the period of infection in a completely susceptible population. Thus, $\mathcal{R}_{RC}$ (R for resistant and C for control) represents the number of secondary resistant cases produced by a typical resistant case, i.e., the control reproduction number for the resistant strain, during the period of infection in a population where the fraction of susceptibles is $(\sigma + \mu)/(\sigma + \mu + \nu)$. A detailed derivation of $\mathcal{R}_{SC}$ and $\mathcal{R}_{RC}$ are given in Appendix A in [Qiu and Feng (2010)].

As discussed in the last section that, in some cases, the level of infection may actually get higher when treatment rate is increased (see Figure 3.27). This possibility, however, cannot be reflected by the reproduction numbers $\mathcal{R}_{SC}$ and $\mathcal{R}_{RC}$ to completely determine the dynamics of the system. To see this, we consider the partial derivative of $\mathcal{R}_{SC}$ (see (3.35) and (3.36)):

$$\frac{\partial \mathcal{R}_{SC}}{\partial f} = (1 - c)\mathcal{R}_{ST} - \mathcal{R}_{SU},$$

which is a constant independent of f. From $0 < c < 1$, $\delta \leq 1$, and $k_T > k_U$, we know that $\mathcal{R}_{ST} < \mathcal{R}_{SU}$, and hence, $\frac{\partial \mathcal{R}_{SC}}{\partial f} < 0$. That is, $\mathcal{R}_{SC}$ is a decreasing function of f for a given value of ν. It is clear from (3.35) and (3.36) that $\mathcal{R}_{RC}$ does not depend on f.

The main reason for this discrepancy between the reproduction numbers ($\mathcal{R}_{SC}$ and $\mathcal{R}_{RC}$) and the epidemic size (AMT), concerning their dependence on drug treatment, can be explained as follows. When a person is infected with the drug-sensitive strain, the number of secondary cases consist of three components:

(i) $\dfrac{S^0}{N^0}(1-f)\mathcal{R}_{SU},$ (ii) $\dfrac{S^0}{N^0}f(1-c)\mathcal{R}_{ST},$ (iii) $\dfrac{S^0}{N^0}fc\mathcal{R}_R.$

The component (i) represents the number of new sensitive cases if the person is untreated. The component (ii) represents the number of new sensitive cases if the person is treated and did not develop resistance. The component (iii) represents the number of cases with acquired resistance due to antiviral use (*de novo* resistance). Clearly, the quantity $\mathcal{R}_{SC}$ is the sum of only the first two components. Therefore, $\mathcal{R}_{SC}$ underestimates the number of secondary infections by a sensitive case. For ease of reference, we denote the component (iii) by $\mathcal{R}_{AR}$ which will be referred to as *acquired* reproduction number (the subscript AR represent Acquired Resistance), i.e.,

$$\mathcal{R}_{AR} = \frac{S^0}{N^0}fc\mathcal{R}_R = \frac{\sigma+\mu}{\sigma+\mu+\nu}fc\mathcal{R}_R. \tag{3.37}$$

Let $\mathcal{R}_{TC}$ (T for total and C for control) denote the sum of all three components, i.e.,

$$\mathcal{R}_{TC} = \mathcal{R}_{SC} + \mathcal{R}_{AR}.$$

Clearly, $\mathcal{R}_{TC} = \mathcal{R}_{TC}(\nu, f)$ is also a function of ν and f. Notice that

$$\frac{\partial \mathcal{R}_{TC}}{\partial f} = \frac{\sigma+\mu}{\sigma+\mu+\nu}[-\mathcal{R}_{SU} + (1-c)\mathcal{R}_{ST} + c\mathcal{R}_R],$$

which is independent of f. Thus, $\mathcal{R}_{TC}$ is a linear function of f (see Figure 3.28). Furthermore, for a given value of ν, $\mathcal{R}_{TC}$ is a decreasing function of f if $-\mathcal{R}_{SU}+(1-c)\mathcal{R}_{ST}+c\mathcal{R}_R < 0$, and it is an increasing function of f if $-\mathcal{R}_{SU}+(1-c)\mathcal{R}_{ST}+c\mathcal{R}_R > 0$. Thus, $\mathcal{R}_{TC}$ is always a monotone function of f, and consequently, cannot capture the nonlinear relationship with f. This suggests that more generations of infections need to be considered.

For the purpose of presentation, denote $\mathcal{R}_{TC}$ by $\mathcal{R}_{TC}^{[2]}$, where the superscript [2] represents the second generation in which new infections are produced. Let $\mathcal{R}_{TC}^{[3]}$ denote the number of tertiary infected cases (including tertiary sensitive cases and tertiary resistant cases) produced by a sensitive case. (The square root of $\mathcal{R}_{TC}^{[3]}$ gives the average number of new infection in one generation. To simplify the notation, we will focus on the quantity without taking the squared root.) The biological meanings of these symbols are also listed in Table 3.2. If the total population size N^0 is sufficiently large so that the number of infected cases is relatively small, then $\frac{S(t)}{N(t)}$ can be closely approximated by $\frac{S^0}{N^0}$. Hence, $\mathcal{R}_{TC}^{[3]}$ can be expressed by

$$\mathcal{R}_{TC}^{[3]} = \mathcal{R}_{SC}\mathcal{R}_{SC} + (\mathcal{R}_{SC} + \mathcal{R}_{RC})\mathcal{R}_{AR}.$$

The first and second items in the above expression represent the numbers of tertiary sensitive and resistant cases, respectively, produced in the third generation by a typical sensitive case.

Table 3.2   Definition of various reproduction numbers.

Notation	Biological meaning
$\mathcal{R}_{SU}$	The number of secondary sensitive cases produced by an untreated sensitive case in the absence of vaccination and treatment, i.e., the basic reproduction number for the sensitive strains.
$\mathcal{R}_{ST}$	The number of secondary sensitive cases produced by a treated sensitive case in the absence of vaccination and treatment.
$\mathcal{R}_R$	The number of secondary resistant cases produced by a resistant case in the absence of vaccination and treatment, i.e., the basic reproduction number for the resistant strains.
$\mathcal{R}_{SC}$	The number of secondary sensitive cases produced by a sensitive case in the presence of vaccination and treatment.
$\mathcal{R}_{RC}$	The number of secondary sensitive cases produced by a sensitive case in the presence of vaccination and treatment, i.e., the control reproduction number for the resistant strains.
$\mathcal{R}_{TC}$ $(= \mathcal{R}_{TC}^{[2]})$	The number of total secondary cases (including the sensitive cases and the resistant cases) produced by a sensitive case in the presence of vaccination and treatment, i.e., the control reproduction number for the sensitive strains.
$\mathcal{R}_{TC}^{[n]}$ $(n \geq 2)$	The number of new infections in the nth generation (including the sensitive cases and the resistant cases) produced by a sensitive case in the presence of vaccination and treatment.

The derivative of $\mathcal{R}_{TC}^{[3]}$ with respect to f is

$$\frac{\partial \mathcal{R}_{TC}^{[3]}}{\partial f} = \left(\frac{\sigma+\mu}{\sigma+\mu+\nu}\right)^2 \Big\{ 2[-\mathcal{R}_{SU} + (1-c)\mathcal{R}_{ST}][-\mathcal{R}_{SU} + (1-c)\mathcal{R}_{ST} + c\mathcal{R}_R]f +$$
$$c\mathcal{R}_R^2 + \mathcal{R}_{SU}[-2\mathcal{R}_{SU} + 2(1-c)\mathcal{R}_{ST} + c\mathcal{R}_R] \Big\}.$$

Let

$$f^* = \frac{c\mathcal{R}_R^2 + \mathcal{R}_{SU}[-2\mathcal{R}_{SU} + 2(1-c)\mathcal{R}_{ST} + c\mathcal{R}_R]}{2[-\mathcal{R}_{SU} + (1-c)\mathcal{R}_{ST}][-\mathcal{R}_{SU} + (1-c)\mathcal{R}_{ST} + c\mathcal{R}_R]},$$

and let

$$f_c = \begin{cases} 0, & f^* \leq 0; \\ f^*, & 0 < f^* < 1; \\ 1, & f^* \geq 1. \end{cases} \tag{3.38}$$

We can show that if

$$[-\mathcal{R}_{SU} + (1-c)\mathcal{R}_{ST}][-\mathcal{R}_{SU} + (1-c)\mathcal{R}_{ST} + c\mathcal{R}_R] > 0,$$

then

$$\frac{\partial \mathcal{R}_{TC}^{[3]}}{\partial f} < 0 \text{ for } f < f_c,$$

$$\frac{\mathcal{R}_{TC}^{[3]}}{\partial f} > 0 \text{ for } f > f_c.$$

Similarly, if

$$[-\mathcal{R}_{SU} + (1-c)\mathcal{R}_{ST}][-\mathcal{R}_{SU} + (1-c)\mathcal{R}_{ST} + c\mathcal{R}_R] < 0,$$

then

$$\frac{\partial \mathcal{R}_{TC}^{[3]}}{\partial f} > 0 \ \text{ for } f < f_c,$$

$$\frac{\partial \mathcal{R}_{TC}^{[3]}}{\partial f} < 0 \ \text{ for } f > f_c.$$

Thus, the dependence of $\mathcal{R}_{TC}^{[3]}$ on f is nonlinear (see Figure 3.28(b)). Figure 3.28(b) shows that there may be a critical value f_c such that $\mathcal{R}_{TC}^{[3]}$ decreases for $0 < f < f_c$ and increases for $f_c < f < 1$.

The key difference between $\mathcal{R}_{TC}^{[3]}$ and $\mathcal{R}_{TC}^{[2]}$ in terms of their functional relationships with f suggest that $\mathcal{R}_{TC}^{[3]}$ can provide a more accurate description on how treatment might negatively impact the disease dynamics. Apparently, it would be desirable to be able to calculate the number of new infections for more generations, which is likely to improve the description more significantly. Let $\mathcal{R}_{TC}^{[n]}$ ($n \geq 2$) denote the reproduction numbers in n-th generation of infections. Then $\mathcal{R}_{TC}^{[n]}$, $n \geq 4$, can be approximately expressed by

$$\mathcal{R}_{TC}^{[n]} = (\mathcal{R}_{SC})^{n-1} + \Big(\sum_{i=1}^{n-2} (\mathcal{R}_{SC})^i (\mathcal{R}_{RC})^{n-2-i} \Big) \mathcal{R}_{AR}$$

$$= \Big(\frac{\sigma+\mu}{\sigma+\mu+\nu} \Big)^{n-1} \big((1-f)\mathcal{R}_{SU} + f(1-c)\mathcal{R}_{ST} \big)^{n-1} \tag{3.39}$$

$$+ \Big(\frac{\sigma+\mu}{\sigma+\mu+\nu} \Big)^{n-1} fc \sum_{i=1}^{n-2} \big((1-f)\mathcal{R}_{SU} + f(1-c)\mathcal{R}_{ST} \big)^i (\mathcal{R}_R)^{n-1-i}.$$

Although for the case of $n = 3$ we are able to identify analytically the critical value f_c at which $\mathcal{R}_{TC}^{[n]}$ switches its monotonicity (see (3.38)), it is not easy to do this for general n. Nonetheless, the analytical formula (3.39) may be helpful for further exploration of more suitable presentations of control reproduction numbers.

In this study, we studied an influenza model that includes both drug sensitive and drug resistant strains and vaccination. The main purposes of this study is to examine the joint impact of vaccination and treatment on the prevalence of the disease influenced by the development of resistance due to drug treatment. A detailed stability and persistence analysis is presented and extensive numerical simulations are conducted using parameter values relevant to influenza. The mathematical results are used to interpret the biological implications of various strategies for disease control and prevention.

The analysis shows that the qualitative behavior of the model are completely determined by three key quantities: $\mathcal{R}_{SC}$ (the sensitive reproduction number), $\mathcal{R}_{RC}$ (the resistant reproduction number), and $\mathcal{R}_C = \max\{\mathcal{R}_{SC}, \mathcal{R}_{RC}\}$. More specifically, the disease will die out if $\mathcal{R}_C < 1$, and it will spread if $\mathcal{R}_C > 1$. The competitive outcomes of the two strains are determined by the relative magnitudes of $\mathcal{R}_{SC}$ and $\mathcal{R}_{RC}$. These results are obtained by analyzing the stability of biologically feasible equilibria and the uniform persistence of the limiting system of (3.34) (see the system (D2) in [Qiu and Feng (2010)]).

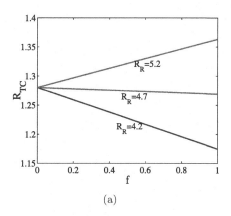

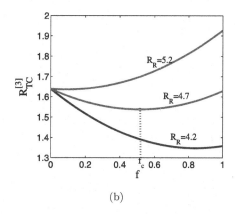

(a)   (b)

Fig. 3.28   Here are two figures side-by-side. (a) Plots of $\mathcal{R}_{TC}^{[2]} = \mathcal{R}_{TC}$ vs. treatment rate f for three different values of $\mathcal{R}_R$. It shows that the total control reproduction number in the second generation is a linear function of f. (b) Plots of the total reproduction number in the third generation, $\mathcal{R}_{TC}^{[3]}$, vs. f for three different values of $\mathcal{R}_R$. It shows that $\mathcal{R}_{TC}^{[3]}$ is a nonlinear function of f and can become increasing with f after reaching a minimum at some critical point f_c. Parameter values are $b = 0.003$, $\mu = 0.0005$, $\eta = 0.001$, $c = 0.23$, $\mathcal{R}_{SU} = 1.7$, and $\mathcal{R}_{ST} = 0.68$.

The stability results for E_0 (the disease-free equilibrium) and $\hat{E}$ (the boundary equilibrium at which only the resistant strain is present) are obtained analytically, and the stability of the coexistence equilibrium E^* is obtained via numerical simulations. These results provide important qualitative understanding of the effects of treatment and vaccination on the infection levels of both strains. For example, when $0 < f < 1$ the resistant strain will always persist in the population provided that $\mathcal{R}_{TC} > 1$ even if the reproduction number of the resistant strain $\mathcal{R}_R < 1$ is less than 1. This suggests that antiviral treatment tends to promote persistence of the resistant strain in the sense that the resistant strain can only invade the population in the presence of drug use (because the resistant infection cannot persist in the population alone when $\mathcal{R}_R < 1$).

One of the interesting findings in this study is that, despite the key role that the reproduction numbers $\mathcal{R}_{SC}$ and $\mathcal{R}_{RC}$ play in determining the qualitative behavior of the system as mentioned above, they do not provide appropriate measures for examining the effect of antiviral use (described by the treatment rate f) on the level of infection (cumulative cases of both strains). This is because the fact that $\mathcal{R}_{SC}$ is always a decreasing function of f (see (3.35)) and $\mathcal{R}_{RC}$ does not depend on f, whereas the size of total infection can increase with f in some cases (see Figure 3.28). The reason for this discrepancy between reproduction numbers and infection size is that $\mathcal{R}_{SC}$ represents only a fraction of secondary cases produced by one sensitive case. The other fraction of new cases that are resistant (represented by $\mathcal{R}_{AR}$, see (3.37)) is not accounted for.

To take account of $\mathcal{R}_{AR}$, we derived the new quantity $\mathcal{R}_{TC}^{[n]}$ (the subscript T for total and C for control) which represents the total number of new infections in the

n-th generation ($n > 2$) produced by one sensitive case. $(\mathcal{R}_{TC}^{[n]})^{1/n}$ gives the average reproduction number by one sensitive case per generation. We showed for $n = 3$ that, under certain conditions, there exists a critical value $f_c \in (0,1)$ such that $\mathcal{R}_{TC}^{[3]}$ decreases with f for $f < f_c$ but increases with f for $f > f_c$. This suggests that the number $\mathcal{R}_{TC}^{[n]}$ may provide a better quantity than $\mathcal{R}_{SC}$ or $\mathcal{R}_C$ for examining the effect of treatment on the level of infection.

Another contribution of this study is the derivation of the new epidemiological quantity, $\mathcal{R}_{TC}^{[n]}$. The concept of reproduction numbers have been very useful in the study of disease control and prevention. It has been shown for simple models that the basic and control reproduction numbers are directly linked to the final size of infections in epidemic models without demographics. However, it is not the case in our model which is an endemic model with demographics. Our results suggest that an inverse relationship between the usual reproduction number and the infection level may occur. The main reason for this is that a fraction of new cases produced by an initially sensitive case will be drug-resistant, which will then produce resistant cases at a different rate in the following generation. A consequence of this process is that the traditionally defined reproduction number may not provide an accurate description for changes in the final size of infection affected by antiviral use. This is indeed observed in the numerical simulations of our model, which demonstrated that although in most cases drug treatment may reduce the epidemic size and delay the time to the peak of an epidemic, it may in some cases lead to an increased cumulative incidence and a higher epidemic peak. Therefore, while the reproduction numbers calculated under the traditional definition can provide useful information for the qualitative dynamics, it may not be appropriate for assessing the effect of control measures on the disease prevalence.

3.2   SARS

The example included in this section is from [Feng *et al.* (2009)]. Early in 2003, a physician infected while treating patients with atypical pneumonia in Guangdong Province, China, infected other travelers in their Hong Kong hotel. On returning home to Singapore, Taiwan, Toronto, and Vietnam, they transmitted the pathogen causing the disease later named Severe Acute Respiratory Syndrome (SARS) to local residents. Global spread of this heretofore unknown pathogen led WHO to issue travel advisories and some national health authorities to quarantine travelers from affected areas. As subsequent infections were largely nosocomial, hospital infection-control procedures were increasingly enforced. Absent knowledge of the onset of infectiousness, local authorities also quarantined community contacts.

Faced with an infectious disease of unknown etiology, policymakers acted quickly [Gerberding (2003)]. Case-series were among the earliest sources of information (e.g., presenting symptoms, duration of distinguishable clinical stages, and outcomes). Upon isolation of the etiologic agent, experience with illnesses caused by related pathogens became germane. While these actions were thoughtful, in retrospect some were unnecessary. We have since developed a model with which policymakers could use similar information (which should be available in future outbreaks, especially of new diseases causing serious morbidity) to assess the likely impact of available interventions.

Because identifying infected people before they become ill is difficult, quarantine is inefficient (e.g., only 11 of 238 probable cases were identified in Singapore [Tan (2008)]; and 24 of 480 in Taiwan [Hsieh *et al.* (2005)]). Isolating people with symptoms that may herald disease, especially if exposed to someone since diagnosed, is more efficient. Impact is proportional to the product of efficiency and effectiveness, which must exceed $1 - 1/\Re_0$ for control, where $\Re_0$ is the average number of sufficiently intimate contacts for transmission while infectious. Consequently, infected people must be highly infectious or infectious long before becoming ill for quarantine to be a better strategy than encouraging those with early symptoms to seek care and ensuring their proper disposition.

To weigh possible interventions to control infectious disease outbreaks, models must faithfully represent transmission. Ours allows infected people to become infectious before or after becoming ill. It permits quarantine while asymptomatic or isolation on seeking medical care and being diagnosed, at rates and with efficiencies that depend on clinical stage. Because health communications could influence these social phenomena, our model also allows hospitalization rates during distinguishable clinical stages to evolve. How effectively cases are isolated, when authorities begin searching for their contacts, if ever, and the proportion found, also may vary temporally.

Policymaking is difficult enough with accurate and complete information, neither of which is available during outbreaks of new diseases. Using only the Hong Kong case series (see [Lee *et al.* (2003); Tsang *et al.* (2003)]) and admissions to Tan Tock Seng Hospital (TTSH) through 24 March, when quarantine began in Singapore, we estimated that infected people were not particularly infectious until acutely ill, a consensus reached months later [WHO (2003)]. As the distribution of infectiousness largely determines the intervention of choice [Fraser *et al.* (2004)], this information is critical.

To assist in controlling other newly emerging diseases, we modeled a generic respiratory illness caused by pathogens transmitted by close interpersonal contact. To evaluate actual interventions for SARS, we chose some parameters from the initial case series in Hong Kong, adjusted others to fit clinically-diagnosed probable cases in Singapore, and calculated outbreak sizes under alternative scenarios. Motivated by the enormous social cost of quarantine relative to its contribution to control in Singapore, we also derived analytical expressions policymakers could use with limited information to determine the most effective strategy for controlling future outbreaks of new human diseases.

The model considered in this section distinguishes the prodrome and acute respiratory phase and allow infected people to become infectious before or after becoming ill, but at rates (products of contact rates and probabilities of transmission on contact) that differ among stages. In addition, for the three infected stages, E (exposed), I_P (prodrome), and I_R (acute respiratory), the stage durations are assumed to have gamma distributions with parameters m, n, and l, respectively (see the previous chapter for the meaning of m, n, and l). The transition diagram is shown in Figure 3.29, and the model reads:

$$S' = \mu N - \lambda(t)S - \mu S,$$
$$E_1' = \lambda(t)S - (\chi + m\alpha + \mu)E_1,$$
$$E_j' = m\alpha E_{j-1} - (\chi + m\alpha + \mu)E_j, \quad j = 2, ..., m,$$
$$Q_1' = \chi E_1 - (m\alpha + \mu)Q_1,$$
$$Q_j' = \chi E_j + m\alpha Q_{j-1} - (m\alpha + \mu)Q_j, \quad j = 2, ..., m,$$
$$I_{P1}' = m\alpha E_m - (\phi_P + n\delta_P + \mu)I_{P1},$$
$$I_{Pj}' = m\delta_P I_{p(j-1)} - (\phi_P + n\delta_P + \mu)I_{Pj}, \quad j = 2, ..., n,$$
$$H_{P1}' = m\alpha Q_n + \phi_P I_{P1} - (n\delta_P + \mu)H_{P1},$$
$$H_{Pj}' = \phi_P I_{Pj} + n\delta_P H_{P(j-1)} - (n\delta_P + \mu)H_{Pj}, \quad j = 2, ..., n, \qquad (3.40)$$
$$I_{R1}' = n\delta_P I_{Pn} - (\phi_R + l\delta_R + \mu)I_{R1},$$
$$I_{Rj}' = l\delta_R I_{R(j-1)} - (\phi_R + l\delta_R + \mu)I_{Rj}, \quad j = 2, ..., l,$$
$$H_{R1}' = n\delta_P H_{Pn} + \phi_R I_{R1} - (l\delta_R + \mu)H_{R1},$$
$$H_{Rj}' = \phi_R I_{Rj} + l\delta_R H_{R(j-1)} - (l\delta_R + \mu)H_{Rj}, \quad j = 2, ..., l,$$
$$R' = l\delta_R I_{Rn} + l\delta_R H_{Rl} - \mu R,$$

$$\lambda(t) = \frac{1}{N}\Big\{ \beta_E[E + (1 - \rho_E)Q] + \beta_P[I_P + (1 - \rho_P)H_P]$$
$$+ \beta_R[I_R + (1 - \rho_R)H_R] \Big\}.$$

Here, $S = S(t)$, $E = \sum_{j=1}^m E_j$, $Q = \sum_{j=1}^m Q_j$, $I_p = \sum_{j=1}^n I_{Pj}$, $H_p = \sum_{j=1}^n H_{Pj}$, $I_R = \sum_{j=1}^l I_{Rj}$, and $H_R = \sum_{j=1}^l H_{Rj}$, and $R = R(t)$ denote numbers of susceptible, exposed (infected, but not yet symptomatic), quarantined, prodrome, acute respiratory, hospitalized (isolated) prodrome, hospitalized acute respiratory, and recovered (or immune) individuals at time t, respectively. Numbered subscripts denote sub-stages, of which there are m, n, and l for exposed, prodrome, and acute respiratory stages, respectively. The total population size is $N = S + E + Q + I_P + I_R + H_P + H_R + R$, whose size is constant by virtue of equal per capita rates of birth and death, μ. Other rates (reciprocals of mean sojourns) are: χ, quarantine; ϕ, hospitalization; α, progression during the incubation, and δ, successive stages of acute illness denoted by subscripts. In $\lambda(t)$, the force of infection, the $\beta_{E,P,R}$ are transmission coefficients and $\rho_{E,P,R}$ are fractional reductions due to quarantine at home or other suitable facilities and isolation in hospitals.

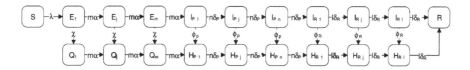

Fig. 3.29 Model diagram with sub-stage denoted by lettered and numbered subscripts, e.g., the incubation period is partitioned into $E_1, E_2, \cdots, E_m$; the prodrome into $I_{P1}, I_{P2}, \cdots, I_{P_n}$; and the acute respiratory phase into $I_{R1}, I_{R2}, \cdots, I_{R_n}$.

Our model's biological parameters are derived entirely from observations of early SARS patients in Hong Kong [Lee *et al.* (2003); Tsang *et al.* (2003)]. We estimated time-varying control parameters by minimizing sums of squared differences between cumulative model hospitalizations, $\chi E + \phi_p I_p + \phi_R I_R$ (where χE are those hospitalized from quarantine and ϕI from successive symptomatic stages, denoted by subscripts) and probable cases admitted to Tan Tock Seng Hospital (TTSH). We estimated these parameters twice, once from the first 30 days of hospital admissions (Table 3.3), before quarantine began, and once from admissions during the entire outbreak (Table 3.4).

As we were uncertain of many parameter values, we chose either gamma or triangular distributions for Latin hypercube sampling (Table 3.5). From the literature just mentioned, we obtained mean values for $1/\alpha$, $1/\delta_p$, and $1/\delta_R$ of 6, 4, and 8 days, respectively. Because the gamma's mean is the product of its parameters, here called A and B, we know that $B=\text{mean}/A$. Note that $A = 1$ corresponds to the exponential distribution, and that the gamma's variance decreases with increasing A. We chose $A = 3$ for all three distributions, and thus, $B = 6/3 = 2$, $4/3$, and $8/3$, respectively. ($A = 3$ also corresponds to 3 sub-stages in these respective states.) As for their upper and lower values, we chose ones for which the area under the probability density function (pdf) in between was $95 - 98\%$. Similarly, for the gamma distributions of β_E, β_P, and β_R, we used our best estimates as means (Table 3.4). We chose A values of 2, 2, and 3 (i.e., assumed the variance of β_E and β_P exceed that of β_R), so the B values are $0.024/2$, $0.152/2$, and $0.582/3$. Each range was again chosen so the area under the pdf in between was $95 - 98\%$. Results are similar if the A values are neither too large nor close to 1.

We compared final outbreak sizes with our best estimates of time-varying control parameters (i.e., quarantine and hospitalization rates and isolation efficiencies) in Singapore with hypothetical alternatives. Differences with and without interventions, or with advanced or delayed timing, are estimates of cases (and hence deaths) averted, conditional on others. Such assessments tacitly assume independence (e.g., quarantine of possible contacts did not affect the timeliness with which actual ones who developed compatible symptoms sought medical care). Social distancing may have increased with quarantine in Hong Kong [Lo *et al.* (2005)], but synergistic population responses in Singapore were more likely due to very effective health

Table 3.3   Parameter estimates from fitting predicted to observed cumulative admissions to TTSH during the first 30 days of the outbreak (see Figure 3.30). The ρ are factors by which isolation reduces contributions to the force of infection; the other parameters are per-capita rates.

Constant parameters	Estimates	
β_E	0.032	
β_P	0.259	
β_R	0.694	
Two epochs	$t \leq 3/14$	$t > 3/14$
ρ_E	0.049	0.12
ρ_P	0.133	0.65
ρ_R	0.15	0.745
ϕ_P	0.123	0.72
ϕ_R	0.25	0.759

Table 3.4   Parameter estimates from fitting predicted to observed cumulative admissions to TTSH during epochs bounded by τ_1, τ_2, and τ_3. Times correspond to 14 and 24 March, and 8 April, when control efforts changed (e.g., began or intensified). χ represents the per capita rate at which exposed individuals are quarantined.

Constant parameters	Estimates			
β_E	0.024			
β_P	0.152			
β_R	0.582			
Two epochs	$t \leq \tau_2$	$t > \tau_2$		
χ	0	0.0487		
Four epochs	$t < \tau_1$	$\tau_1 < t < \tau_2$	$\tau_2 < t < \tau_3$	$t > \tau_3$
ρ_E	0.049	0.195	0.432	0.575
ρ_P	0.157	0.361	0.627	0.757
ρ_R	0.162	0.5	0.957	1.0
ϕ_P	0.165	0.381	0.45	0.695
ϕ_R	0.245	0.465	0.579	0.877

communications.

Empirical results for SARS include parameter estimates from hospital admissions during the first 30 days and entire outbreak in Singapore by epoch; assessment of the reproduction number; and impact of control measures via sensitivity analysis and final outbreak sizes under hypothetical scenarios.

Trial periods in Table 3.3 correspond to before 14 March, when the second generation of cases began being reported, and between 14 and 24 March, when home quarantine began. In addition to these, Table 3.3's trial periods correspond to between 24 March and 8 April, when cumulative cases abruptly increased by about 10%, and after 8 April. Figure 3.30 illustrates model fits to hospital admissions with these two parameter sets.

The infection rates β, differ quantitatively (e.g., those during the acute phase

Table 3.5   Input parameters for uncertainty analysis. Assumed distributions before and after introduction of control measures on 3/24 (including not only quarantine and isolation, but exemplary health communications) are either gamma (G) or triangular (T).

Parameter	Before 3/24		After 3/24	
	Distribution	Range	Distribution	Range
$1/\alpha$	G(3,2)	(1, 15)	Same	
$1/\delta_P$	G(3, 4/3)	(0, 10)		
$1/\delta_R$	G(3, 8/3)	(0, 20)		
β_E	G(2, 0.024/2)	(0, 0.06)		
β_P	G(2, 0.152/2)	(0, 0.55)		
β_R	G(3, 0.582/3)	(0, 2)		
χ			T(0.063)	(0.057, 0.07)
ϕ_P	T(0.381)	(0.343, 0.419)	T(0.47)	(0.423, 0.517)
ϕ_R	T(0.465)	(0.419, 0.512)	T(0.579)	(0.521, 0.636)
ρ_E	T(0.195)	(0.176, 0.215)	T(432)	(0.388, 0.475)
ρ_P	T(0.361)	(0.325, 0.398)	T(0.627)	(0.564, 0.69)
ρ_R	T(0.5)	(0.45, 0.55)	T(0.957)	(0.928, 0.986)

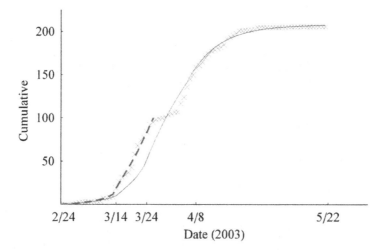

Fig. 3.30   Cumulative probable cases by date of symptom onset in Singapore (symbols) and fits to the first 30 days (dashed line) and entire outbreak (solid line). Parameter estimates differ quantitatively (i.e., they are higher initially), but are similarly qualitatively (Tables 3.3 and 3.5), particularly the ratio of infection rates during successive clinical phases.

are two to three times those during the prodrome and 21-25 times those during the incubation period, respectively subscripted R, P, and E) when estimated from admissions during the first 30 days or entire outbreak (Tables 3.4 and 3.5, respectively), but not qualitatively (i.e., infectiousness is negligible until patients are acutely ill). Isolation efficiency ρ among those hospitalized during the prodrome increased from

0.16 during the first epoch to 0.76 during the last. Among those hospitalized while acutely ill, efficiency increased from 0.16 to almost 1. Similarly, the hospitalization rate ϕ increased from 0.16 to 0.7 during the prodrome, while that during acute illness increased from 0.25 to 0.88. Both isolation efficiencies and hospitalization rates are respectively subscripted P and R. Estimates for intervening periods are intermediate (Table 3.4).

Given the transition probabilities (T_i and $\mathcal{T}_i$) and sojourns (D_i and $\mathcal{D}_i$) defined in Table 3.6, the control reproductive number $\mathcal{R}_C$ is given by

$$\mathcal{R}_C = \mathcal{R}_E + \mathcal{R}_Q + \mathcal{R}_{I_P} + \mathcal{R}_{I_R} + \mathcal{R}_{QH_P} + \mathcal{R}_{I_P H_P} + \mathcal{R}_{I_R H_R} + \mathcal{R}_{H_P H_R},$$

where

$$\mathcal{R}_E = \beta_E \mathcal{D}_E,$$
$$\mathcal{R}_Q = (1 - \rho_E)\beta_E(D_E - \mathcal{D}_E),$$
$$\mathcal{R}_{I_P} = \beta_P \mathcal{T}_E \mathcal{D}_{I_P},$$
$$\mathcal{R}_{I_P H_P} = (1 - \rho_P)\beta_P \mathcal{E}(D_{I_P} - \mathcal{D}_{I_P}),$$
$$\mathcal{R}_{QH_P} = (1 - \rho_P)\beta_P(T_E - \mathcal{T}_E)D_{I_P},$$
$$\mathcal{R}_{I_R} = \beta_R \mathcal{T}_E \mathcal{T}_{I_P} \mathcal{D}_{I_R},$$
$$\mathcal{R}_{I_R H_R} = (1 - \rho_R)\beta_R \mathcal{T}_E \mathcal{T}_{I_P}(D_{I_R} - \mathcal{D}_{I_R}),$$
$$\mathcal{R}_{H_P H_R} = (1 - rho_R)\beta_R(T_E T_{I_P} - \mathcal{E}\mathcal{T}_{I_P})D_{I_R}.$$

It is shown in [Feng *et al.* (2009)] that $\mathcal{R}_C$ descreases with all control parameters including χ, ϕ_i ($i = P, R$), and ρ_i ($i = E, P, R$). Absent control, the realized reproduction number reduces to the basic reproduction number, $\mathcal{R}_0 = \beta_E D_E + \beta_P T_E D_P + \beta_R T_E T_{I_P} D_{I_R}$. Their relationship is given by

$$\mathcal{R}_C = \mathcal{R}_0 - [\rho_E \beta_E(D_E - \mathcal{D}_E) + \rho_P \beta_P(T_E D_{I_P} - \mathcal{T}\mathcal{D}_{I_P})$$

$$+ \rho_R \beta_R(T_E T_{I_P} D_{I_R} - \mathcal{T}_E \mathcal{T}_{I_P}\mathcal{D}_{I_R})].$$

Clearly, $\mathcal{R}_C = \mathcal{R}_0$ when all control parameter values are zero. Also, $\mathcal{R}_C < \mathcal{R}_0$ because $D_i \geq \mathcal{D}_i$ and $T_i \geq \mathcal{T}_i$ which implies that the quantity inside the brackets is positive if some control parameters are not zero.

Our analytical results include an expression for the realized reproduction number $\mathcal{R}_C$, its partial derivatives with respect to control parameters and relationship to $\mathcal{R}_0$, the intrinsic reproduction number. A Mathematica^{TM} notebook that evaluates these expressions for user-supplied parameter values, comparing the impact of non-pharmaceutical interventions on any disease transmitted by close contact, is developed (see Chapter 4).

Our estimates of $\mathcal{R}_C$ and contributions of each infectious state before and after control measures were introduced, given the assumed distributions and ranges described in Table 3.5, are shown with their associated uncertainties in Table 3.7 and Figure 3.31. Our estimates of $\mathcal{R}_C$, 3.16 and 0.86 before and after 24 March are comparable to Wallinga's and Teunis' [Wallinga and Teunis (2004)] 3.1 and 0.7 before and after 12 March, respectively.

Table 3.6   Definition of transition probabilities and sojourns.

Variable	Definition
$T_E = (\frac{m\alpha}{m\alpha+\mu})^m$	Probability of surviving exposure to enter the prodrome
$\mathcal{T}_E = (\frac{m\alpha}{m\alpha+\chi+\mu})^m$	Quarantine adjusted probability of surviving exposure
$T_{I_p} = (\frac{n\delta_p}{n\delta_p+\mu})^n$	Probability of surviving the prodrome to become acutely ill
$\mathcal{T}_{I_p} = (\frac{n\delta_p}{n\delta_p+\phi_p+\mu})^n$	Quarantine adjusted probability of surviving the prodrome
$D_E = \frac{1}{m\alpha+\mu}\sum_{i=0}^{m-1}(\frac{m\alpha}{m\alpha+\mu})^i$	Mean death-adjusted sojourn in the exposed stage
$\mathcal{D}_E = \frac{1}{m\alpha+\chi+\mu}\sum_{i=0}^{m-1}(\frac{m\alpha}{m\alpha+\chi+\mu})^i$	Mean quarantine-and death-adjusted sojourn in the exposed stage
$D_{I_p} = \frac{1}{n\delta_p+\mu}\sum_{i=0}^{n-1}(\frac{n\delta_p}{n\delta_p+\mu})^i$	Mean death-adjusted sojourn in the prodrome
$\mathcal{D}_{I_p} = \frac{1}{n\delta_p+\phi_p+\mu}\sum_{i=0}^{n-1}(\frac{n\delta_p}{n\delta_p+\phi_p+\mu})^i$	Mean quarantine-and death-adjusted sojourn in the prodrome
$D_{I_R} = \frac{1}{l\delta_R+\mu}\sum_{i=0}^{l-1}(\frac{l\delta_R}{l\delta_R+\mu})^i$	Mean death-adjusted sojourn with acute respiratory symptoms
$\mathcal{D}_{I_R} = \frac{1}{l\delta_R+\delta_R+\mu}\sum_{i=0}^{l-1}(\frac{l\delta_R}{l\delta_R+\delta_R+\mu})^i$	Mean isolation-and death-adjusted sojourn with acute respiratory symptoms

Table 3.7   Results of uncertainty analysis. These $\mathcal{R}_i$ are contributions to $\mathcal{R}_C$, their sum, by infectious individuals in transit.

$\mathfrak{R}_i$	Before 3/24			After 3/24		
	Mean	Std. Dev.	$P(\mathfrak{R}_i > 1)$	Mean	Std. Dev.	$P(\mathfrak{R}_i > 1)$
$\mathfrak{R}_E$	0.147	0.147	0.002	0.114	0.103	0.0002
$\mathfrak{R}_Q$	0	0	0	0.017	0.024	0
$\mathfrak{R}_{I_P}$	0.259	0.2	0.007	0.175	0.137	0.0007
$\mathfrak{R}_{I_R}$	0.382	0.318	0.05	0.203	0.187	0.007
$\mathfrak{R}_{QH_P}$	0	0	0	0.052	0.061	0
$\mathfrak{R}_{I_PH_P}$	0.217	0.265	0.02	0.109	0.131	0.002
$\mathfrak{R}_{I_RH_R}$	0.644	0.806	0.19	0.036	0.049	0
$\mathfrak{R}_{H_PH_R}$	1.511	1.475	0.528	0.155	0.155	0.003
$\mathfrak{R}$	3.159	2.25	0.922	0.861	0.413	0.295

Sensitivity to control parameters is shown in Table 3.8.   Isolation efficiency during the incubation period hardly affects $\mathfrak{R}$, as infectiousness is negligible then. Because people did not seek care until acutely ill prior to 24 March, $\mathfrak{R}$ is also relatively insensitive to isolation efficiency during the prodrome or, by virtue of nosocomial transmission during that epoch [Leo *et al.* (2003)], hospitalization while acutely ill.

Final outbreak sizes (when birth rate and death rate are ignored, i.e., $\mu = 0$)

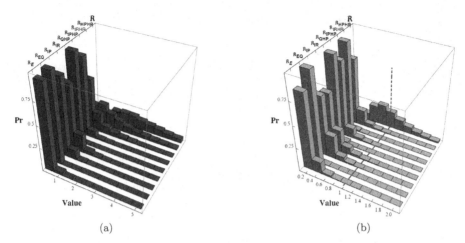

Fig. 3.31   Histograms of $\mathcal{R}_i$ before (in (a)) and after (in (b)) 24 March, when quarantine began and hospital infection control measures were more stringently enforced, from which the statistics in Table 3.7 were calculated.

Table 3.8   Sensitivity of $\mathcal{R}_C$ to uncertain parameters.

Parameter	Before 3/24		After 3/24	
	PRCC	p-Value	PRCC	p-Value
$1/\alpha$	0.164	< 0.0001	0.162	< 0.0001
$1/\delta_P$	0.309	< 0.0001	0.192	< 0.0001
$1/\delta_R$	0.906	< 0.0001	0.581	< 0.0001
β_E	0.19	< 0.0001	0.459	< 0.0001
β_P	0.552	< 0.0001	0.786	< 0.0001
β_R	0.919	< 0.0001	0.783	< 0.0001
χ			-0.029	0.004
ϕ_P	-0.032	0.002	-0.061	< 0.0001
ϕ_R	-0.013	0.185	-0.032	0.001
ρ_E	-0.004	0.726	-0.0004	0.97
ρ_P	-0.005	0.62	-0.051	< 0.0001
ρ_R	-0.124	< 0.0001	-0.255	< 0.0001

would have been larger had interventions been delayed (Figure 3.32 (a)); or had isolation efficiency declined, especially of acutely ill patients (Figure 3.32 (b)). Final size would have been smaller had greater proportions been quarantined (Figure 3.32 (c)) or hospitalized, particularly during the prodrome (Figure 3.32 (d)).

The model considered in this section is suitable for any directly transmitted pathogen (Figure 3.29). Its biological states are clinical, so sojourns are estimable from early case series. To model SARS, we set the incubation period at 6 days [Lee *et al.* (2003)] and prodrome and acute respiratory stages at 4 and 8 days [Tsang *et al.* (2003)]; these references were available at www.nejm.org on 4/7 and 3/31, respectively.

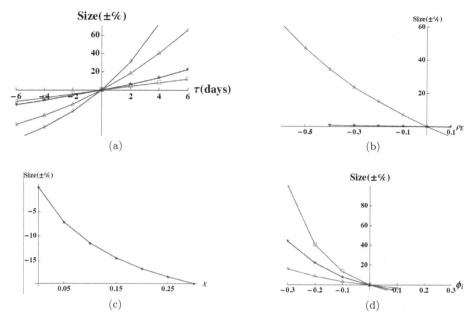

Fig. 3.32 Impact (percentage change) on (a) final outbreak size of intervention timing (stars are τ_1, triangles τ_2, squares τ_3, and diamonds τ), (b) efficiency of isolation during the prodrome and acute phases (stars and triangles, respectively), (c) proportion of contacts quarantined (stars), and (d) hospitalization rates during the prodrome and acute phases (stars and triangles, squares are both).

We assumed quarantined people spent half of their incubation periods at large, and varied only the proportion quarantined, but when infected people are quarantined must be a function of the stage at which those who infected them are diagnosed. Similarly, we combined the rate at which people sought medical care and their probability of diagnosis, but as these phenomena involve different social systems, they may evolve independently. Social responses are exceedingly important, but this model must be useful with insufficient information to fully characterize them.

Infection is a more familiar composite parameter than quarantine or hospitalization, being decomposed into constituent contact rates and probabilities of transmission on contact only as needed (e.g., in sexually-transmitted diseases, asymmetric transmission between genders or other partners may affect the impact of available interventions). Transmission of pathogens causing respiratory diseases may also be asymmetric (e.g., between children and adults), but the impact on interventions is less clear, so the composite is used.

Parameter estimates based on the first 30 days of hospital admissions indicate that (consistent with numbers of secondary infections among patients isolated on successive days after symptom onset [Lipsitch *et al.* (2003)] and observed viral loads [Peiris *et al.* (2003)]) infectiousness of patients with SARS was negligible until symptom onset and only one third during their prodrome as while acutely ill (Tables 3.4 and 3.5). Estimates based on the entire outbreak indicate temporal variation in the rates at which ill people sought care, probabilities of diagnosis and efficacy of isolation (Table 3.4).

Identifying infected people before they become ill is so difficult that quarantine is exceedingly inefficient. Consequently, unless infected people could infect many susceptible ones before becoming symptomatic (by virtue of being very infectious or infectious long before symptomatic) public health authorities should focus on identifying people with symptoms that may herald disease, encouraging them to seek care, especially if they might have been exposed to someone since diagnosed. Similarly, they should assist clinicians in diagnosing, and hospital infection control personnel in isolating effectively, those people before they become acutely ill.

To evaluate actual or hypothetical control measures, one must compare otherwise identical scenarios. Community intervention trials approach this ideal, but only modeling attains it. Not using all available measures that could be effective as expeditiously as possible during actual outbreaks of potentially lethal diseases would be unthinkable, but model outbreaks can be compared with and without interventions, alone or in various combinations, or with interventions at different times, to assess their impacts.

Our final size calculations with and without quarantine, but all else equal (i.e., experiments), indicate its impact would have been comparable to hospitalization during the prodrome (see Figures 3.32 (c) and (d)). But this assumes authorities could identify infected people during their incubation period. Reassuringly, our independent estimate of the rate at which exposed people were quarantined, Table 3.4 resembles the proportion of probable SARS patients actually quarantined, 11/238 [Tan (2008)]. In Taiwan, where this proportion was strikingly similar, 24/480 [Hsieh *et al.* (2005)], quarantined individuals were diagnosed as suspect earlier than others (within 1.2 versus 2.89 days, respectively), but not reclassified any sooner [Hsieh *et al.* (2005)]. Our final size calculations indicate the importance of timely interventions, but as only probable cases had priority for limited isolation facilities, evidently these shorter onset-diagnosis intervals contributed little.

In contrast, the rate at which probable cases were hospitalized during the prodrome, ϕ_P, increased from about 0.16 to 0.7 during the outbreak in Singapore (Table 3.4), contributing much more to control. The potential impact of quarantine has been so regularly confused with its actual contribution to SARS' control that one wonders if this longtime staple of public health [McNeill (1976); Rosner (1995)] has been rigorously evaluated before. Day *et al.* [Day *et al.* (2006)] refine Fraser *et al.*'s [Fraser *et al.* (2004)] conditions for quarantine to be effective: the number of asymptomatic infections and fraction preventable (essentially the number of pre-symptomatic infections) must be large and the probability of infected individuals being quarantined high. Short of everyone staying home after a household member is diagnosed, the last condition will be difficult to meet for respiratory diseases.

As the temporal distribution of infectiousness largely determines the optimal public health response [Fraser *et al.* (2004); Day *et al.* (2006)], this is among the most important epidemiological unknowns early in outbreaks of new human diseases. If patients could be infectious before developing symptoms, quarantine must be considered. But identifying people whose contacts were sufficiently intimate for infection is extremely difficult, especially given uncertainty about the mode of transmission. The gain in efficiency (by ensuring instead that people with early signs and symptoms seek medical care, clinicians diagnose and infection-control personnel isolate them effectively) may more than compensate for any loss of efficacy due to infections during the prodrome. Modeling can help to determine this distribution and, conditional on that essential information, to evaluate the relative impact of various possible public health interventions.

3.3 Tuberculosis

In this section, two examples are presented based on the results published in [Castillo-Chavez and Feng (1997); Jung *et al.* (2002); Roeger *et al.* (2009)] are presented. One is the application of the two-strain TB model considered in section 2.3.1 to determining the best combination of treatment for people with latent and active TB using the approach of optimal control. The other example focuses on the influence of HIV on the prevalence of TB.

3.3.1 *Optimal treatment strategies for TB*

This example is based on the two-strain TB model considered in section 2.3.1. The results are from [Castillo-Chavez and Feng (1997)] and [Jung *et al.* (2002)].

Consider first the two-strain TB model presented in section 2.3.1. For the purpose of examining the policy on the uses of prophylaxis and drug treatment, we introduce two parameters, p_1 and p_2, which represent the fractions of individuals in the L_1 and I_1 classes who will receive the treatment. We can examine how the control efforts described by p_1 and p_2 may affect the condition under which E_2 is stable, which is determined by the function g (see (2.22)). Replace the parameters r_i by $p_i r_i$ ($i = 1, 2$). Thus, $g = g(p_1, p_2)$ is a function of p_1 and p_2. Then from the analysis presented in section 2.3.1, E_2 is stable if and only if

$$\mathcal{R}_2 > g(p_1, p_2). \tag{3.41}$$

Figure 3.33 illustrates how p_1 and p_2 may affect the condition (3.41). In the left figure, it shows a 3D plot (left) of the function $g(p_1, p_2)$ and the constant plane $\mathcal{R}_2$. The intersection curve of the two surfaces identifies the p_1 and p_2 values for which $g(p_1, p_2) = \mathcal{R}_2$. This curve is also shown in the contour plot on the right (the dashed curve). The values of other parameters are similar to those used in section 2.3.1. In the region above (below) the dashed curve, $g(p_1, p_2) < \mathcal{R}_2$ ($g(p_1, p_2) > \mathcal{R}_2$), and thus resistant TB can (cannot) replace sensitive TB.

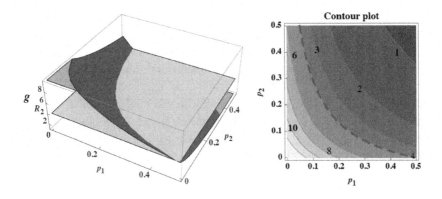

Fig. 3.33   Plot of the function $g(p_1, p_2)$ (left) and the contour plot of the curve determined by $g(p_1, p_2) = \mathcal{R}_2$ (right).

Let $u_1(t)$ and $u_2(t)$ denote the time-dependent control, which represent the fractions of individuals in L_1 and I_1 classes receiving prophylaxis and drug treatment, respectively, at time t. The state system considered in [Jung *et al.* (2002)] with controls $u_1(t)$ and $u_2(t)$ reads:

$$\dot{S} = \Lambda - \beta_1 S \frac{I_1}{N} - \beta^* S \frac{I_2}{N} - \mu S,$$

$$\dot{L}_1 = \beta_1 S \frac{I_1}{N} - (\mu + k_1)L_1 - u_1(t)r_1 L_1 + (1 - u_2(t))pr_2 I_1 + \beta_2 T \frac{I_1}{N} - \beta^* L_1 \frac{I_2}{N},$$

$$\dot{I}_1 = k_1 L_1 - (\mu + d_1)I_1 - r_2 I_1,$$

$$\dot{L}_2 = (1 - u_2(t))qr_2 I_1 - (\mu + k_2)L_2 + \beta^*(S + L_1 + T)\frac{I_2}{N},$$

$$\dot{I}_2 = k_2 L_2 - (\mu + d_2)I_2,$$

$$\dot{T} = u_1(t)r_1 L_1 + (1 - (1 - u_2(t))(p + q))r_2 I_1 - \beta_2 T \frac{I_1}{N} - \beta^* T \frac{I_2}{N} - \mu T$$

$$(3.42)$$

with $S(0), L_1(0), I_1(0), L_2(0), I_2(0), T(0)$ denoting the initial conditions.

The control functions, $u_1(t)$ and $u_2(t)$, are bounded, *Lebesgue* integrable functions. The "case finding" control, $u_1(t)$, represents the fraction of typical TB latent individuals that is identified and will be put under treatment (to reduce the number of individuals that may be infectious). The coefficient, $1 - u_2(t)$, represents the effort that prevents the failure of the treatment in the typical TB infectious individuals (to reduce the number of individuals developing resistant TB). When the "case holding" control $u_2(t)$ is near 1, there is low treatment failure and high implementation costs.

Our objective functional to be minimized is

$$J(u_1, u_2) = \int\limits_0^{t_f} [L_2(t) + I_2(t) + \frac{B_1}{2} u_1^2(t) + \frac{B_2}{2} u_2^2(t)] dt \qquad (3.43)$$

where we want to minimize the latent and infectious groups with resistant-strain TB while also keeping the cost of the treatments low. We assume that the costs of the treatments are nonlinear and take quadratic form here. The coefficients, B_1 and B_2, are balancing cost factors due to size and importance of the three parts of the objective functional. We seek to find an optimal control pair, u_1^* and u_2^*, such that

$$J(u_1^*, u_2^*) = \min_{\Omega} J(u_1, u_2) \qquad (3.44)$$

where $\Omega = \{(u_1, u_2) \in L^1(0, t_f) \mid a_i \leq u_i \leq b_i, i = 1, 2\}$ and $a_i, b_i, i = 1, 2$, are fixed positive constants.

In this analysis, we assume $\Lambda = \mu N, d_1 = d_2 = 0$. Thus the total population N is constant. We can also treat the nonconstant population case by these techniques, but we choose to present the constant population case here.

The necessary conditions that an optimal pair must satisfy come from Pontryagin's Maximum Principle [Pontrëïlagin (1962)]. This principle converts (3.42)-(3.44) into a problem of minimizing pointwise a Hamiltonian, H, with respect to u_1 and u_2:

$$H = L_2 + I_2 + \frac{B_1}{2} u_1^2 + \frac{B_2}{2} u_2^2 + \sum_{i=1}^{6} \lambda_i g_i \qquad (3.45)$$

where g_i is the right-hand side of the differential equation of the ith state variable. By applying Pontryagin's Maximum Principle [Pontrëïlagin (1962)] and the existence result for the optimal control pairs from [Fleming and Rishel (1975)], we know that there exists an optimal control pair u_1^*, u_2^* and corresponding solution, S^*, L_1^*, I_1^*, L_2^*, I_2^*, and T^*, that minimizes $J(u_1, u_2)$ over Ω. Furthermore, there exists adjoint functions, $\lambda_1(t), \ldots, \lambda_6(t)$, such that

$$\dot{\lambda}_1 = \lambda_1 (\beta_1 \frac{I_1^*}{N} + \beta^* \frac{I_2^*}{N} + \mu) + \lambda_2 (-\beta_1 \frac{I_1^*}{N}) + \lambda_4 (-\beta^* \frac{I_2^*}{N}),$$

$$\dot{\lambda}_2 = \lambda_2 (\mu + k_1 + u_1(t) r_1 + \beta^* \frac{I_2^*}{N}) + \lambda_3 (-k_1) + \lambda_4 (-\beta^* \frac{I_2^*}{N}) + \lambda_6 (-u_1^*(t) r_1),$$

$$\dot{\lambda}_3 = \lambda_1 (\beta_1 \frac{S^*}{N}) + \lambda_2 (-\beta_1 \frac{S^*}{N} - (1 - u_2^*(t)) p r_2 - \beta_2 \frac{T^*}{N}) + \lambda_3 (\mu + d_1 + r_2)$$

$$+ \lambda_4 (-(1 - u_2^*(t)) q r_2) + \lambda_6 (-(1 - (1 - u_2^*(t))(p + q)) r_2 + \beta_2 \frac{T^*}{N}),$$

$$\dot{\lambda}_4 = -1 + \lambda_4 (\mu + k_2) + \lambda_5 (-k_2),$$

$$\dot{\lambda}_5 = -1 + \lambda_1 (\beta^* \frac{S^*}{N}) + \lambda_2 (\beta^* \frac{L_1^*}{N}) - \lambda_4 (\beta^* \frac{S^* + L_1^* + T^*}{N}) + \lambda_5 (\mu + d_2) + \lambda_6 (\beta^* \frac{T^*}{N}),$$

$$\dot{\lambda}_6 = \lambda_2 (-\beta_2 \frac{I_1^*}{N}) + \lambda_4 (-\beta^* \frac{I_2^*}{N}) + \lambda_6 (\beta_2 \frac{I_1^*}{N} + \beta^* \frac{I_2^*}{N} + \mu), \qquad (3.46)$$

with transversality conditions

$$\lambda_i(t_f) = 0, \quad i = 1,\ldots,6 \tag{3.47}$$

and $N = S^* + L_1^* + I_1^* + L_2^* + I_2^* + T^*$. Moreover, the characterization holds:

$$u_1^*(t) = min(max(a_1, \tfrac{1}{B_1}(\lambda_2 - \lambda_6)r_1 L_1^*), b_1)$$

$$u_2^*(t) = min(max(a_2, \tfrac{1}{B_2}(\lambda_2 p + \lambda_4 q - \lambda_6(p+q)r_2 I_1^*)), b_2). \tag{3.48}$$

Due to the a *priori* boundedness of the state and adjoint functions and the resulting *Lipschitz* structure of the ODEs, we obtain the uniqueness of the optimal control for small t_f. The uniqueness of the optimal control follows from the uniqueness of the optimality system, which consists of (3.42) and (3.46)–(3.48). There is a restriction on the time interval in order to guarantee the uniqueness of the optimality system. This smallness restriction on the length on the time interval is due to the opposite time orientations of (3.42), (3.46), and (3.47); the state problem has initial values and the adjoint problem has final values. This restriction is very common in control problems (see [Fister *et al.* (1998)] and [Kirschner *et al.* (1997)]).

The optimal treatment is obtained by solving the optimality system, consisting of the state and adjoint equations. An iterative method is used for solving the optimality system. We start to solve the state equations with a guess for the controls over the simulated time using a forward fourth order Runge-Kutta scheme. Because of the transversality conditions (3.47), the adjoint equations are solved by a backward fourth order Runge-Kutta scheme using the current iteration solution of the state equations. Then, the controls are updated by using a convex combination of the previous controls and the value from the characterizations (3.48). This process is repeated and iteration is stopped if the values of unknowns at the previous iteration are very close to the ones at the present iteration.

For the figures presented here, we assume that the weight factor B_2 associated with control u_2 is greater than or equal to B_1 which is associated with control u_1. This assumption is based on the following facts: The cost associated with u_1 will include the cost of screening and treatment programs, and the cost associated with u_2 will include the cost of holding the patients in the hospital or sending people to watch the patients to finish their treatment. Treating an infectious TB individual takes longer (by several months) than treating a latent TB individual. In these three figures, the set of the weight factors, $B_1 = 50$ and $B_2 = 500$, is chosen to illustrate the optimal treatment strategy. Other epidemiological and numerical parameters are presented in [Jung *et al.* (2002)].

In the top frame of Figure 3.34, the controls, u_1 (solid curve) and u_2 (dashdot curve), are plotted as a function of time. In the bottom frame, the fractions of individuals infected with resistant TB, $(L_2 + I_2)/N$, with control (solid curve) and without control (dashed curve) are plotted. Parameters $N = 30000$ and $\beta^* = 0.029$ are chosen. Results for other parameters are presented in [Jung *et al.* (2002)]. To minimize the total number of the latent and infectious individuals with resistant

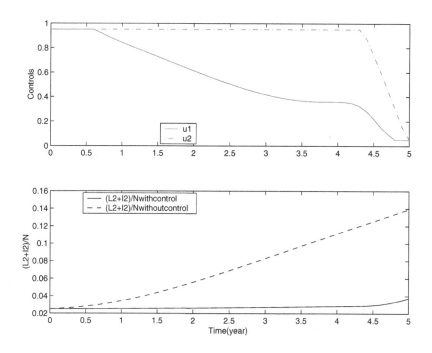

Fig. 3.34　The optimal control strategy for the case of $B_1 = 50$ and $B_2 = 500$.

TB, $L_2 + I_2$, the optimal control u_2 is at the upper bound during almost 4.3 years and then u_2 is decreasing to the lower bound, while the steadily decreasing value for u_1 is applied over the most of the simulated time, 5 years. The total number of individuals $L_2 + I_2$ infected with resistant TB at the final time $t_f = 5$ (years) is 1123 in the case with control and 4176 without control, and the total cases of resistant TB prevented at the end of the control program is 3053 ($= 4176 - 1123$).

Figure 3.35 illustrates how the optimal control strategies depend on the parameter β^*, which denotes the transmission rate of primary infections of resistant TB. The value of β^* is usually given by the product of the number of contacts (with an infectious individuals with resistant TB) per person per unit of time and the probability of being infected with resistant TB per contact. This value varies from place to place depending on many factors including living conditions. In Figure 3.35, the controls, u_1 (dark color curves) and u_2 (light color curves), are plotted as a function of time for the 4 different values of β^*, 0.0131, 0.0217, 0.0290, and 0.0436. It shows that u_1 plays an increasing role while u_2 remains almost the same as β^* decreases (that is why only one u_2 graph is shown). This is an expected result because when β^* is smaller, the new cases of resistant TB arise more from infections acquired from L_1 and I_1 due to treatment failure than from primary infections. In this case, identifying and curing latently infected individuals with sensitive TB becomes more important in the reduction of new cases of resistant TB.

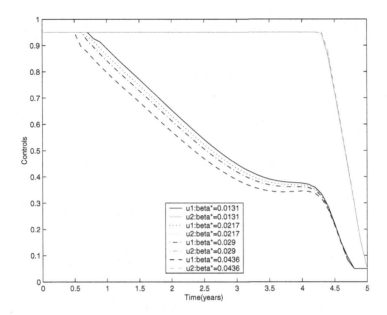

Fig. 3.35   The controls μ_1 are plotted as a function of time for the 4 dierent values of β^*, 0.0131, 0.0217, 0.0290, and 0.0436 and the only one control μ_2 (top curve) is plotted because μ_2 remains almost the same as β^* increases.

In Figure 3.36, the controls, u_1 and u_2, are plotted as a function of time for $N = 6000$, 12000, and 30000 in the top and bottom frame, respectively. Other parameters except the total number of individuals and $\beta^* = 0.029$ are fixed for these three cases. These results show that more effort should be devoted to "case finding" control u_1 if the population size is small, but "case holding" control u_2 will play a more significant role if the population size is big. Note that, in general, with B_1 fixed, as B_2 increases, the amount of u_2 decreases. A similar result holds if B_2 is fixed and B_1 increases.

In conclusion, our optimal control results show how a cost-effective combination of treatment efforts (case holding and case finding) may depend on the population size, cost of implementing treatments controls and the parameters of the model. We have identified optimal control strategies for several scenarios. Control programs that follow these strategies can effectively reduce the number of latent and infectious resistant-strain TB cases.

3.3.2   *Influence of HIV on TB prevalence*

This example is from [Roeger *et al.* (2009)]. In this study, a system of differential equations is introduced to model the joint dynamics of TB and HIV. The total population is divided into the following epidemiological subgroups: S, susceptible;

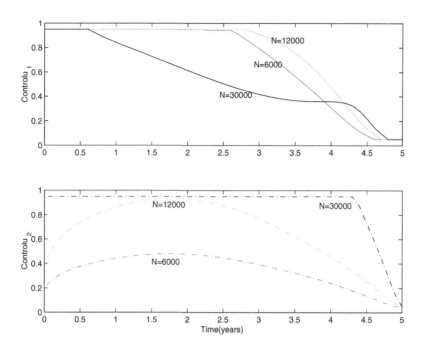

Fig. 3.36 The controls, μ_1 and μ_2, are plotted as a function of time for $N = 6000, 12000$, and 30000 in the top and bottom frame, respectively.

L, latent with TB; I, infectious with TB; T, successfully treated with TB; J_1, HIV infectious; J_2, HIV infectious and TB latent; J_3, infectious with both TB and HIV; and A, "full-blown" AIDS. The compartmental diagram in Figure 3.37 illustrates the flow of individuals as they face the possibility of acquiring specific-disease infections or even co-infections.

The TB/HIV model is given by the following systems of eight ordinary differential equations:

$$TB \begin{cases} \frac{dS}{dt} = \Lambda - \lambda_T(t)S - \lambda_H(t)S - \mu S, \\ \frac{dL}{dt} = \lambda_T(t)(S+T) - \lambda_H(t)L - (\mu + k + r_1)L, \\ \frac{dI}{dt} = kL - (\mu + d + r_2)I, \\ \frac{dT}{dt} = r_1 L + r_2 I - \lambda_T(t)T - \lambda T \frac{J^*}{R} - \mu T, \end{cases}$$

$$HIV \begin{cases} \frac{dJ_1}{dt} = \lambda_H(t)(S+T) - \lambda_T(t)J_1 - (\alpha_1 + \mu)J_1 + r^* J_2, \\ \frac{dJ_2}{dt} = \lambda_H(t)L + \lambda_T(t)J_1 - (\alpha_2 + \mu + k^* + r^*)J_2, \\ \frac{dJ_3}{dt} = k^* J_2 - (\alpha_3 + \mu + d^*)J_3, \\ \frac{dA}{dt} = \alpha_1 J_1 + \alpha_2 J_2 + \alpha_3 J_3 - (\mu + \delta)A, \end{cases}$$

(3.49)

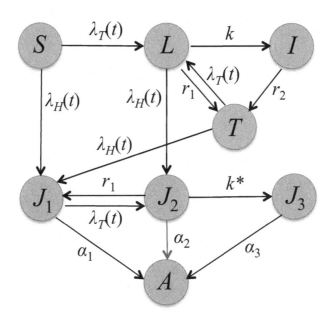

Fig. 3.37   Transition diagram for the interaction between TB and HIV.

where the forces of infection for TB and HIV are

$$\lambda_T(t) = \beta\frac{(I+J_3)}{N}, \qquad \lambda_H(t) = \lambda\frac{J^*}{R}, \tag{3.50}$$

respectively, and

$$N = S + L + I + T + J_1 + J_2 + J_3 + A,$$
$$R = N - I - J_3 - A = S + L + T + J_1 + J_2, \tag{3.51}$$
$$J^* = J_1 + J_2 + J_3.$$

The variable R denotes the "active" population that is the subgroup of individuals who do not have active TB or AIDS.

One of the key parameters in the model to consider is k^*, which is the rate of TB progression in individuals who are co-infected with both HIV and latent TB. It has been reported that TB carriers who are infected with HIV are 30 to 50 times more likely to develop active TB than those without HIV. This suggests that $k^* \geq k$, and in some cases, $k^* \gg k$. Our numerical studies indicate that only in certain cases, this factor may play an important role for explaining the effect of HIV epidemics on the increased prevalence level of TB. This can be done either by examining the reproduction numbers or compare the levels of TB prevalence. The reproduction numbers for TB and HIV (in the absence of the other disease) are

$$\mathcal{R}_1 = \frac{\beta k}{(\mu + k + r_1)(\mu + d + r_2)} \tag{3.52}$$

and

$$\mathcal{R}_2 = \frac{\lambda}{\alpha_1 + \mu}, \tag{3.53}$$

respectively. These reproduction numbers $\mathcal{R}_1$ and $\mathcal{R}_2$ are directly related to the infection levels of the respective diseases (in the absence of other diseases). Thus, we consider the impact of HIV on TB by first examining the effect of $\mathcal{R}_2$ on the prevalence of TB. Notice that both $\mathcal{R}_1$ and $\mathcal{R}_2$ are independent of the parameters k^*, α_3, or δ. Thus, we can fix these parameters and consider various values of $\mathcal{R}_1$, and look at changes in the levels of TB infections as $\mathcal{R}_2$ increases.

The system has several possible equilibrium points including the infection-free (E_0), TB only (E_T), HIV only (E_H), and interior equilibrium points. For this example, we focus on the existence and stability of interior equilibrium points (i.e., an equilibrium at which both diseases are present). It was shown in [Roeger *et al.* (2009)] that when both reproduction numbers are greater than 1, i.e., $\mathcal{R}_1 > 1$ and $\mathcal{R}_2 > 1$, E_T and E_H both exist and E_0 is unstable. In this case, the numerical simulations of the model show that it is possible that all three boundary equilibria are unstable and solutions converge to an interior equilibrium point. Although explicit expressions for an interior equilibrium are very difficult to compute analytically, we have managed to obtain some relationships that can be used to determine the existence of an interior equilibrium.

Let $\hat{E} = (\hat{S}, \hat{L}, \hat{I}, \hat{J}_1, \hat{J}_2, \hat{J}_3, \hat{A})$ denote an interior equilibrium with all components positive, and let x and y denote the fractions in the incidence terms:

$$x = \frac{\hat{I} + \hat{J}_3}{\hat{N}} > 0 \quad \text{and} \quad y = \frac{\hat{J}^*}{\hat{R}} > 0. \tag{3.54}$$

Note that x and y correspond to the levels of disease prevalence for TB and HIV, respectively. It is shown that all components of $\hat{E}$ can be expressed in terms of x and y. Although the explicit expressions for x and y are difficult to obtain, it can be shown that x and y are the intersections of the two curves determined by the equations $F(x, y) = 1$ and $G(x, y) = 1$, where the two functions F and G are given in [Roeger *et al.* (2009)]. We will consider $\hat{x}$ as a measure for the TB prevalence. The intersection property of the two curves given by $F(x, y) = 1$ and $G(x, y) = 1$ are illustrated in Figure 3.38.

Figure 3.38 plots the intersection point $(\hat{x}, \hat{y})$ of the contour plots of $F(x, y) = 1$ (dashed curve) and $G(x, y) = 1$ (solid curve) for several values of $\mathcal{R}_2$ with $\mathcal{R}_1$ being fixed ($\mathcal{R}_1 = 1.5$ corresponding to $\beta = 12$). Again, an interior equilibrium $\hat{E}$ can be determined by $\hat{x}$ and $\hat{y}$ if $0 < \hat{x} < 1$ and $\hat{y} > 0$. This figure illustrates how $\hat{x}$ changes with increasing $\mathcal{R}_2$. We have chosen $k^* = 5k$ (i.e., the progression rate to active TB in individuals with both latent TB and HIV is five times higher than that in individuals with latent TB only), $\alpha_3 = 5\alpha_1$ (i.e., the progression to AIDS in individuals with active TB is five times higher than that in individuals without TB). For this set of parameter values, the value of $\mathcal{R}_2$ in Figures 3.38(A)-(C) are 3.6, 4.6, and 7, respectively. It shows that when $\mathcal{R}_2$ increases from 3.8 to 4.6, the

$F(x, y) = 1$ curve does not change much while the right-end of the $G(x, y) = 1$ curve moves to the right of the $F = 1$ curve. This leads to an intersection point of the two curves (see (A) and (B)), which corresponds to an interior equilibrium $\hat{E}$. When $\mathcal{R}_2$ is further increased to 7, the $G(x, y) = 1$ curve changes from decreasing to increasing (see (C)). Although there is still a unique intersection point, the $y = \hat{J}^*/\hat{R}$ component may become greater than 1. This is still biologically feasible as J/R can exceed 1 (see (C)). The intersection points in (A)-(C) are $(\hat{x}, \hat{y}) = (\frac{\hat{I}+\hat{J}_3}{\hat{N}}, \frac{\hat{J}^*}{\hat{R}}) = (0.15, 0.07), (0.25, 0.4), (0.33, 1.25)$, respectively. We observe that $\hat{x}$ increases with $\mathcal{R}_2$ from 0.15 to 0.33. This implies that the prevalence of HIV may have significant impact on the infection level of TB.

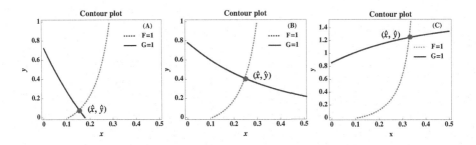

Fig. 3.38   Contour plots showing the intersection points of the curves $F(x, y) = 1$ (dashed curve) and $G(x, y) = 1$ (solid curve) for various values of $\mathcal{R}_2$ with $\mathcal{R}_1$ being fixed at 1.5 ($\beta c \doteq 12$). The value of $\mathcal{R}_2$ in (A)-(C) are 3.6, 4.6, and 7, respectively (corresponding to $\lambda \sigma = 0.41, 0.52$, and 0.8). The axes are $x = (I + J_3)/N$ and $y = J^*/R$, representing the factors in the incidence functions for TB and HIV, respectively. The intersection $(\hat{x}, \hat{y}) = (\frac{\hat{I}+\hat{J}_3}{\hat{N}}, \frac{\hat{J}^*}{\hat{R}})$ determines components of the interior equilibrium $\hat{E}$ if $0 < x < 1$ and $\hat{y} > 0$. It shows that as $\mathcal{R}_2$ increases, the $F(x, y) = 1$ curve changes little, while the right-end of the $G(x, y) = 1$ curve change dramatically. This leads to the change of the intersection point of the two curves (see (A)-(C)), which represents an interior equilibrium $\hat{E}$. The right-end of the $G = 1$ curve moves further up as $\mathcal{R}_2$ is increased from 3.6 to 4.6, there is still a unique interior equilibrium with a larger x component (see (B)). Finally, when $\mathcal{R}_2$ is very large, the right-end of the $G(x, y) = 1$ curve continues to rise and it changes from decreasing to increasing. Although the y component of the unique intersection point is greater than one, it is still biologically feasible as $\hat{y} = \hat{J}/\hat{R}$ can exceed 1 (see (C)).

Figure 3.39 examines changes in infection levels over time. It plots the time series of $[I(t) + J_3(t)]/N(t)$ (fraction of active TB) and $J^*(t)/R(t)$ (activity-adjusted fraction of HIV infectious) for fixed $\mathcal{R}_1$ and various $\mathcal{R}_2$. The top two figures are for the case when the reproduction number for TB is less than 1 ($\mathcal{R}_1 = 0.96 < 1$ or $\beta c = 7.5$), and the reproduction number for HIV is $\mathcal{R}_2 = 0.9 < 1$ (or $\lambda c = 0.105$) in (a) and $\mathcal{R}_2 = 1.3 > 1$ (or $\lambda c = 0.15$) in (a). It illustrates in Figure 3.39(a) that TB cannot persist if $\mathcal{R}_2 < 1$. However, if $\mathcal{R}_2 > 1$ then it is possible that TB can become prevalent even if $\mathcal{R}_1 < 1$ (see Figure 3.39(b)). The bottom two figures are for the case when the reproduction number of TB is greater than 1 ($\mathcal{R}_1 = 1.2$, or $\beta c = 9.1$), and $\mathcal{R}_2 = 2$ (or $\lambda c = 0.23$) in (c) and $\mathcal{R}_2 = 3$ (or $\lambda c = 0.34$) in (d).

It demonstrates that an increase in $\mathcal{R}_2$ will lead to an increase in the level of TB prevalence as well. All other parameters are the same as in Figure 3.38 except that $k^* = 3k$.

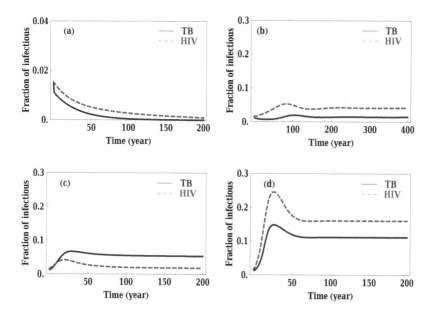

Fig. 3.39   Time plots of prevalence of TB and HIV. The TB curves (solid) represents the fraction of active TB $((I + J_3)/N)$, and the HIV curve (dashed) represents the activity-adjusted fraction of HIV (J^*/R). In the top two figures, the reproduction number for TB is fixed and less than 1 ($\mathcal{R}_1 = 0.96$ or equivalently $\beta c = 7.5$), and the reproduction number for HIV is either less than 1 (see (a), $\mathcal{R}_2 = 0.9$ and equivalently $\lambda c = 0.105$) or greater than 1 (see (b), $\mathcal{R}_2 = 1.3$ and $\lambda c = 0.15$). Figure (a) illustrates that TB cannot persist if $\mathcal{R}_2 < 1$. Figure (b) shows that if $\mathcal{R}_2 > 1$ then it is possible that TB can become prevalent even though it cannot persist in the absence of HIV (as $\mathcal{R}_1 < 1$). The bottom two figures are for the case when the reproduction number of TB is greater than 1 ($\mathcal{R}_1 = 1.2$, or $\beta c = 9.1$), whereas the reproduction number for HIV is greater than 1 but either small (see Figure (c), $\mathcal{R}_2 = 2$ or $\lambda c = 0.23$) or large (see Figure (d), $\mathcal{R}_2 = 3$ or $\lambda c = 0.34$). It illustrates that an increase in $\mathcal{R}_2$ can lead to an increase in the prevalence level of TB.

Another way to look at the role of HIV on TB dynamics is to compare the outcomes between the cases where HIV is absent or present (instead of varying the value of $\mathcal{R}_2$). This result is presented in Figure 3.40. The reproduction numbers are identical in Figures 3.40(A)(B): $\mathcal{R}_1 = 0.98 < 1$ ($\beta = 7.7$) and $\mathcal{R}_2 = 1.2 > 1$ ($\lambda = 0.137$). Other parameter values are the same as in Figure 3.39 except that $k^* = k$. The variables plotted are $(I + J_2)/N$ and J^*/N. Figure 3.40(A) is for the case when HIV is absent by letting $J^*(0) = 0$. It shows that TB cannot persist. In Figure 3.40(B), the initial value of HIV is positive (i.e., $J^*(0) > 0$). It shows that both TB and HIV coexist.

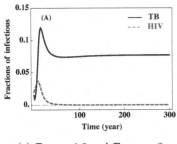
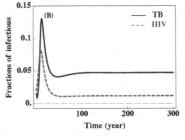

(a) $\mathcal{R}_{TB} = 1.3$ and $\mathcal{R}_{HIV} = 2$ (b) $\mathcal{R}_{TB} = 1.3$ and $\mathcal{R}_{HIV} = 3$

Fig. 3.40   Demonstration of similar properties as shown in Figure 3.39 using a different approach. Instead of changing $\mathcal{R}_2$ as in Figure 3.39, different initial values of HIV are used in these figures. The reproduction numbers remain the same for Figure 3.40: $\mathcal{R}_1 = 0.98 < 1$ and $\mathcal{R}_2 = 1.2 > 1$. Other parameter values are the same as in Figure 3.39 except that $k^* = k$. The variables plotted are the fractions of active TB $(I + J_2)/N$) and HIV $((J_1 + J_2 + J_3)/N)$. In (A), HIV is absent by letting $J^*(0) = 0$. It shows that TB cannot persist. In (B), the initial value of HIV is positive (i.e., $J^*(0) > 0$). It shows that both TB and HIV will coexist.

Examples of other mathematical models on dynamics of TB/HIV co-infections include [Kirschner (1999); Naresh and Tripathi (2005); Porco *et al.* (2001); Raimundo *et al.* (2003); Schulzer *et al.* (1994)].

3.4   Schistosomiasis

This section includes applications of the model results presented in section 2.3.2. The examples are chosen from [Zhang *et al.* (2007a); Yang *et al.* (2012); Xu *et al.* (2012)].

3.4.1   *Age-targeted drug treatment in humans*

Due to the fact that the highest infection rate occurs in the age-group 11–20 [Sturrock (2001); Massara *et al.* (2004)], the model used here is from [Zhang *et al.* (2007a)], which is also a model with an age-structure in human hosts (see section 2.3.2 for the definition of variable and parameters used in the model). The structure in the snail host is relatively simpler by considering only one type of snail host and one parasite strain. However, the infection-age dependent cercaria releasing rate in the snails is included.

To examine the effect of various strategies of age-targeted drug-treatment in humans on the level of parasite infection, we compare the parasite loads in humans under different treatment programs. Although the highest infection rate occurs in the age-group 11-20 [Sturrock (2001); Massara *et al.* (2004)], targeting at this age group for treatment may or may not be the best strategy (i.e., leading to the lowest mean parasite load). The parameter function represeting treatment effort is $\sigma(a)$, which is actually a product of several factors including the fraction of treated humans in a given age group and how soon an infected person gets treated. Here we assume that an increase in σ is due to an increased fraction of treated humans in the corresponding age-group. The model reads:

$$
\begin{cases}
\dfrac{\partial}{\partial t}n(t,a) + \dfrac{\partial}{\partial a}n(t,a) = -\mu_h(a)n(t,a), \\[2mm]
\dfrac{\partial}{\partial t}p(t,a) + \dfrac{\partial}{\partial a}p(t,a) = \beta(a)Cn(t,a) - [\mu_h + \mu_P + \sigma(a)](a)p(t,a), \\[2mm]
\dfrac{d}{dt}S(t) = b(S,I) - \mu_s S(t) - \rho\gamma S(t)\displaystyle\int_0^\infty p(t,a)da, \\[2mm]
\dfrac{\partial}{\partial t}x(t,\tau) + \dfrac{\partial}{\partial \tau}x(t,\tau) = -(\mu_s + \delta_s)x(t,\tau), \\[2mm]
P(t) = \int_0^\infty p(t,a)da, \;\; C(t) = \displaystyle\int_0^\infty r(\tau)x(t,\tau)d\tau, \;\; I(t) = \int_0^\infty x(t,\tau)d\tau, \\[2mm]
n(t,0) = \Lambda_h, \;\; p(t,0) = 0, \;\; x(t,0) = \rho\gamma S(t)\displaystyle\int_0^\infty p(t,a)da, \\[2mm]
n(0,a) = n_0(a), \;\; p(0,a) = p_0(a), \;\; S(0) = S_0, \;\; I(0) = I_0, \;\; x(0,\tau) = x_0(\tau),
\end{cases}
\tag{3.55}
$$

where τ denotes the infection-age of snails and $x(t,\tau)$ denotes the infection-age density of snails at time t. The total number of infected snails I is therefore $I(t) = \int_0^\infty x(t,\tau)d\tau$.

For the age-dependent parameters (β and σ), we choose piecewise constant functions. For ease of demonstration, we assume that the population of all ages can be divided into the following age groups: $a_{i-1} < a \le a_i$ (group i) where $a_i = 10i$ (e.g., $a_0 = 0, a_1 = 10$, etc.), and that $\beta(a)$ is a step function with $\beta(a) = \beta_i$ (where β_i is a constant) for $a_{i-1} < a \le a_i$. Similarly, $\sigma(a)$ is also assumed to be a step function with $\sigma(a) = \sigma_i$ (where σ_i is a constant) for $a_{i-1} < a \le a_i$. That is,

$$
\beta(a) = \begin{cases} \beta_i & \text{if } a_{i-1} < a \le a_i, \quad i = 1, 2, \cdots, 7, \\ 0 & \text{if } a > 70, \end{cases}
$$

$$
\sigma(a) = \begin{cases} \sigma_i & \text{if } a_{i-1} < a \le a_i, \quad i = 1, 2, \cdots, 7, \\ 0 & \text{if } a > 70, \end{cases}
\tag{3.56}
$$

where β_i and σ_i are constants.

The measure we use here to assess the effectiveness of age-dependent treatment strategies is the mean parasite load within humans. The overall mean parasite load is defined as the ratio of the total number of parasites to the total number of humans, which we denote by M_k for a strategy indexed by k. We also look at the mean parasite load within each age group which is denoted by m_{ki}, with the index i denoting the ith age group.

Of course, an ideal scenario would be to treat all infected humans. However, in many cases this is impossible to do due to limited access and/or resources. The problem we study here concerns a different scenario. Suppose that there is a current treatment program, and that health officials want to choose one or more age groups to increase their treatment rate. Which age group(s) should be targeted? To answer this question, we compare the outcomes (in terms of the mean parasite load–both the overall and the age-specific) under different age-targeted treatment strategies. Here, the overall effort for a treatment program is simply defined as the sum of the treatment rates $\sum_{i=1}^{3} \sigma_i$ (which is independent of the age-distribution of humans), and we compare the mean parasite loads under different age-targeted treatment programs that have the same level of overall effort.

Some questions associated with cost-effectiveness will also be considered by comparing treatment programs that have different levels of overall effort. This can be done by looking at the percentage of reduction in the mean parasite load in relation to the percentage of increase in the treatment effort of certain age groups. We remark that our goal here is not to identify the optimal treatment strategy given a constraint on the effort (though this would be a useful optimization problem) in which case the effort for an age-dependent control program should involve the age-distribution of humans.

Assume that a baseline treatment program is described by $\sigma_i = 0.2$ for $i = 1, 2, 3$. First, we consider the case in which only one age group will be devoted a higher treatment effort. For example, for one group σ is increased to 0.4 while for other age groups σ remains at 0.2. Table 3.9 lists three such treatment strategies that have the same overall effort. The function $\beta(a)$ is chosen such that the age group 11–20 has a higher value than other age groups.

The overall mean parasite load is obtained by numerically integrate the system (3.55) for a given set of parameter values. After the system has stabilized we calculate the total number of parasites and the total number of humans under a given age-targeted treatment program, which will give us the values of M_k and m_{ki}. Figure 3.41 shows the outcomes of these three strategies listed in Table 3.10. Figure 3.41(a) shows that the strategy that targets at the age group 11–20 is the most effective one (in the sense that it results in a lowest overall mean parasite load), while the strategy that targets at the age group 0–10 is the least effective one. More detailed age-dependent distributions of the parasite load corresponding to these treatment programs are given in Figure 3.41(b), which shows that the strategy II results in a more uniform parasite distribution among all age groups.

Next we consider treatment programs that target at several age groups (i.e., these targeted age groups will have a higher treatment rate while other groups have the same baseline treatment rate $\sigma = 0.2$, see Table 3.10). In this case, as the overall treatment efforts are different among these programs, instead of looking at which program will produce a lower mean parasite load, we consider questions associated with cost-effectiveness such as what percentage reduction of the mean parasite load

Table 3.9   Control strategies targeted at only one age group.

	Treatment value $\sigma(\cdot)$		
	Age 0-10	Age 11-20	Age 21-30
Strategy 1	0.4	0.2	0.2
Strategy 2	0.2	0.4	0.2
Strategy 3	0.2	0.2	0.4

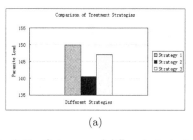   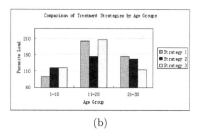

(a)   (b)

Fig. 3.41   Outcomes of different strategies when infection rate is higher between age 11 and 20. In (a) it shows the mean parasite load among all individuals between age 0 and 30. In (b) it shows the mean parasite load in each age group under different strategies.

will result from a given percentage increase in the treatment effort of certain age groups.

Table 3.10   Control strategies targeted at only one age group.

	Treatment value $\sigma(\cdot)$		
	Age 0-10	Age 11-20	Age 21-30
Strategy I	0.2	0.2	0.2
Strategy II	0.2	0.4	0.2
Strategy III	0.2	0.4	0.4
Strategy IV	0.4	0.4	0.2
Strategy V	0.4	0.4	0.4

Table 3.10 lists five treatment programs with the first one being the baseline treatment. We assess the effect of each of the four programs ($k = 2, 3, 4, 5$) by looking at the reduction in the parasite load relative to the baseline treatment program (Strategy I). That is, we look at the ratio of the overall mean parasite loads M_k/M_1 and the ratio of age-specific mean parasite loads m_{ki}/m_{1i} ($i = 1, 2, 3$). These ratios are plotted in Figure 3.42. Figure 3.42(a) shows that Strategy II, which targets at the single age group 11–20, can further reduce the mean parasite load by about 13% ($M_2/M_1 \approx 0.87$). When the treatment coverage is doubled (i.e., two of the age groups will receive an additional treatment effort–Strategies III and IV), the resulted additional reduction in the mean parasite load is around 20% ($M_k/M_1 \approx 0.80$ for $k = 3, 4$). When the treatment coverage is tripled (i.e., three

age groups will receive an additional treatment effort–Strategies V), the additional reduction in the mean load is around 28% ($M_5/M_1 \approx 0.72$). Figure 3.42(b) presents a more detailed distribution of the mean parasite load within each age group under the five treatment programs. It seems that the reduction in the parasite load within these age groups will follow the same pattern as the distribution of treatment efforts.

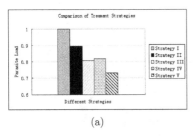
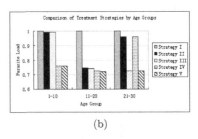

(a)   (b)

Fig. 3.42   Outcomes of the treatment strategies I through V listed in Table 3.10.

3.4.2   *Evolution dynamics of hosts and parasites*

In section 2.3.2 we presented some of the model properties for the human-schistosome-snail system, which are reported in [Yang *et al.* (2012)] and [Xu *et al.* (2012)]. These model results can also be helpful for understanding the effect of various control strategies on other aspects such as evolution of pathogens and/or hosts.

In [Yang *et al.* (2012)], an example is given to demonstrate how age-dependent drug treatment may affect the evolution of host's resistance to infection. It focuses on the intermediate snail host and assume that there are two competing snail host types, one is sensitive and the other resistant to parasite infection. It illustrates that outcomes may depend on the intensity of drug-treatment in human hosts, age-dependent infection rates, and the level of drug-resistance of the parasites.

Consider the type 1 snail host as the resident type which is more susceptible to parasite infection (e.g., with a higher ρ_1 value), and consider the type 2 snail host to be more resistant to parasite infection (i.e., $\rho_2 < \rho_1$). The trade-off for type 2 hosts is that a higher resistance may result in a reduced birth rate (which is represented by the parameter b_{12}, i.e., $b_{12} < b_{11}$). We will investigate how the changes in ρ_k and b_{1k} ($k = 1, 2$) may affect population outcomes. More specifically, we will examine how the threshold quantities determined by the invasion reproduction numbers, $\mathcal{R}_{21} = 1$ and $\mathcal{R}_{12} = 1$, may depend on changes in the two quantities:

$$\frac{\rho_1 - \rho_2}{\rho_1} \quad \text{and} \quad \frac{b_{11} - b_{12}}{b_{11}}. \tag{3.57}$$

To do this, we need to specify the forms of age-dependent functions in the human hosts such as $\beta(a)$ and $\sigma(a)$, as well as the forms of parasite production as a function of drug-resistance, $\gamma(\theta)$. Here, the age-dependent transmission rate $\beta(a)$ and drug-treatment rate $\sigma(a)$ are chosen as step functions similar to that in (3.56). The function $f(\sigma(\theta))$ denotes the death rate of parasites due to drug treatment with treatment rate $\sigma(\theta)$ for parasites with a drug-resistant level $\theta \geq 1$ ($\theta = 1$ for drug-sensitive parasites). In our simulations, this death rate is assumed to be

$$f(\sigma(a)) = \frac{\sigma(a)}{\theta}, \quad \theta \geq 1. \tag{3.58}$$

Based on experimental studies [Gower and Webster (2005)], we assume that parasites with drug-resistance level θ have a reduced reproduction rate $\gamma(\theta)$ which reflects the cost of drug-resistance. Two particular forms of $\gamma(\theta)$ considered here are given by

$$\gamma(\theta) = \gamma_0 \left(\frac{9 - \theta}{8} \right), \tag{3.59}$$

$$\gamma(\theta) = \gamma_0 \left(\frac{16 - \theta}{15} \right)^{1/5}, \tag{3.60}$$

where $\gamma_0 > 0$ is a constant representing the background replication rate of drug-sensitive parasites.

We can compare two treatment strategies that target at different age-groups. Figure 3.44(a) is for the case when the age-dependent functions $\beta(a)$ and $\sigma(a)$ following the distributions shown in Figure 3.43, and Figure 3.44(b) is for the treatment strategy that targets only at the age-group 10-30, i.e., $\sigma(a) \neq 0$ only for $10 < a \leq 30$. For the two $\sigma(a)$ functions described above, the areas under the $\sigma(a)$ curve are kept the same.

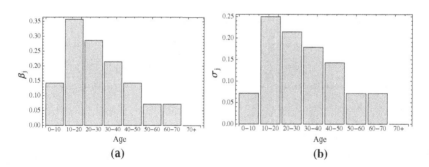

(a) (b)

Fig. 3.43   Distributions for $\beta(a)$ and $\sigma(a)$ used to generate the bifurcation diagram in 3.44(a).

Given these functional forms and assumptions specified above, and for fixed values of θ, β_0 and σ_0, we can generate a bifurcation diagram in the parameter plane $\left(\frac{\rho_1 - \rho_2}{\rho_1}, \frac{b_{11} - b_{12}}{b_{11}} \right)$ as shown in Figure 3.44. In this figure, the infection rate and growth rate for the resident snail host (b_{11} and ρ_1) are fixed, and the corresponding rates for the mutant type snail host (b_{12} and ρ_2) can vary. The fraction $(b_{11} - b_{12})/b_{11}$ provides a measure for the cost of parasite resistance in the snail host, and the fraction $(\rho_1 - \rho_2)/\rho_1$ describes the level of parasite resistance in the snail host. We plotted three curves determined by the equations $\mathcal{R}_{12} = 1$, $\mathcal{R}_{21} = 1$ and $\mathcal{R}_2 = 1$, which divide the plane in three regions. In the region below (above) the $\mathcal{R}_{12} = 1$ curve, the mutant type host will out-compete (be excluded by) the resident type host. In the region between the three curves (i.e., above the $\mathcal{R}_{12} = 1$ curve and below the two curves determined by $\mathcal{R}_{21} = 1$ and $\mathcal{R}_2 = 1$), coexistence will be expected. This can be seen in Figure 2.13(a) in section 2.3.2, which is for the case when $\mathcal{R}_{12} > 1$, $\mathcal{R}_{21} > 1$ and $\mathcal{R}_2 > 1$, that is, the parameters are in the coexistence region.

We observe in Figure 3.44 that for smaller values of $(b_{11} - b_{12})/b_{11}$ (i.e., when cost of resistance is low), mutant snails with a wider range of resistance (represented by the values of $(\rho_1 - \rho_2)/\rho_1$) can out-compete the resident snails. On the other hand, when cost of resistance is high (larger values of $(b_{11} - b_{12})/b_{11}$), it is more likely that the resident type will exclude the resistance type. Coexistence of both resident and mutant snails is favored only when the cost is at an intermediate level.

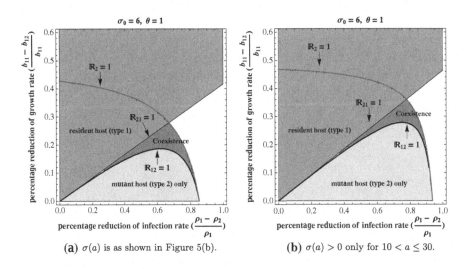

(a) $\sigma(a)$ is as shown in Figure 5(b).   **(b)** $\sigma(a) > 0$ only for $10 < a \le 30$.

Fig. 3.44   Competitive outcomes between the parasite-susceptible (resident) strain and the parasite-resistant (mutant) strain of snail hosts in the trade-off space characterized by $\left(\frac{\rho_1 - \rho_2}{\rho_1}, \frac{b_{11} - b_{12}}{b_{11}}\right)$. Effects of two age-dependent treatment strategies, determined by the function $\sigma(a)$, are compared. The plot in (a) is for the case when $\sigma(a)$ has the form described in Figure 3.43, whereas the plot in (b) is for the case in which only the age-group $10 < a \le 30$ is targeted (i.e., $\sigma(a) = 0$ only for $10 < a \le 30$). It demonstrates that the targeted treatment strategy makes it more likely for the mutant host type (parasite-resistant) to out-compete the resident type (parasite-susceptible).

We need to point out that the coexistence region described above may not allow for an evolutionary branching point in adaptive dynamics terms. Such an example is presented in [Bowers *et al.* (2005)]. The approach they presented requires that one can solve the invasion equations (both mutant and resident), which are needed in order to consider their derivatives and the derivatives of trade-offs. A branching point can then be determined by observing in which region the trade-off curve enters the singular TIP (trade-off and invasion plots). For our model, it is not easy to determine the branching point as the invasion equations are difficult to solve analytically. It is possible that the coexistence state in the coexistence region is temporary and unstable, depending on the properties of the evolutionarily singular strategy that are determined by the trade-off assumptions on hosts' benefits and costs (see [Yang *et al.* (2012)]).

The next example is from [Xu *et al.* (2012)] concerning the model (2.23) presented in section 2.3.2. The invasion criterion based on the invasion reproduction numbers $\mathcal{R}_{ij}$ (see (2.26)) will be used to investigate the evolutionary outcomes influenced by the interactions among virulence, coinfection, and drug resistance. Several trade-off functions are considered including the one about the relationship between the reproduction rate γ and the drug resistance, which assumes that high drug resistance is associated with low reproduction rates γ_i in the definitive host [Webster *et al.* (2008)] and describe the rate γ_i by a decreasing function

$$\gamma_i = \gamma(\theta_i) = \gamma_0 \bar{\gamma}(\theta_i), \quad i = 1, 2,$$

where $\gamma_0 > 0$ is a constant. We will consider three types of functions for $\bar{\gamma}(\theta)$: linear, convex-concave and concave. See Figure 3.45 for the shape of the function.

Consider first the case in which the coinfection functions are differentiable. In the case when the coinfection rates ρ_{ij} (given by the equation (15) in [Xu *et al.* (2012)] for $n = 2$) are differentiable, together with the trade-off relationships are described in section 4 of [Xu *et al.* (2012)], numerical simulations and the pairwise-invasibility plots (PIPs) show that there exists a convergent evolutionary stable strategy (ESS) θ^* (see Figure 3.46). That is, in evolutionary time the drug resistance level θ^* tends to be approached.

To introduce PIPs, let us consider the case of the linear trade-off curve $\gamma(\theta)$ (see Figure 3.45 for trade-off curves). When there is no coinfection (i.e., $\eta = 0$) (Figure 3.46A), any resident strain of parasites with drug resistance θ less than the ESS θ^* can be invaded by some more drug-resistant strains, not by any less resistant strain. Any resident strain with drug resistance level $\theta > \theta^*$ can be invaded by less resistant strains. Only θ^* is not invadable and thus it is evolutionarily convergent and stable. In this case, θ^* is the evolutionary endpoint (in the sense of evolutionary convergence and stability) and dimorphism or polymorphism is not possible. Figure 3.46B is for the case of facilitation ($\xi \geq 1$) and when coinfections are allowed (i.e., $\eta > 0$). It illustrates that the ESS θ^* is invadable and hence evolutionarily unstable. In this case, an initially monomorphic population will become dimorphic or even polymorphic, and the ESS θ^* is called a branching point. Figure 3.46C is for the case of competitive suppression ($\xi < 1$), and it shows that the ESS θ^* is evolutionarily stable and hence is an evolutionary endpoint.

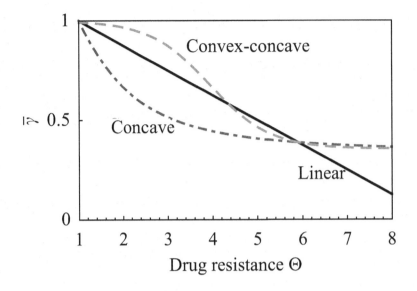

Fig. 3.45 Plot of the function $\bar{\gamma}(\theta)$ versus the drug resistance level θ for the reproduction rate of adult parasites in human hosts $\gamma_i = \gamma_0\bar{\gamma}$. In numerical simulations, $\bar{\gamma}(\theta) = (9 - \theta)/8$ for the linear trade-off, $\bar{\gamma}(\theta) = (0.55 + 1/(8e^{1.5\theta-8} + 1))/1.55$ for the convex-concave trade-off, and $\bar{\gamma}(\theta) = 4/3 - \theta^2/(2 + \theta^2)$ for the concave trade-off and $\gamma_0 = 9000$ per year.

When the trade-off curve γ is linear, the PISs reveals the following observations (Figures 3.46A–3.46D). The first observation is that variations in the coinfection efficiency η or in ξ do not affect the drug resistance level θ^* at ESS (see Figures 3.46A–3.46C). However, high coinfection efficiencies (η) and reproduction ($\xi \geq 1$) can alter the evolutionary stability of θ^* and hence make it more likely to have dimorphism or polymorphism (Figure 3.46B). Our numerical simulations also show that, in the case of competitive suppression ($\xi < 1$), an increase in the coinfection efficiency η or a decrease in ξ tends to stabilize the ESS point θ^* and inhibit dimorphism as an evolutionary endpoint (e.g. Figure 3.46C). The second observation is that if two strains exhibit facilitation within intermediate hosts ($\xi \geq 1$), the drug treatment rate σ tends to stabilize the ESS point θ^* and promote monomorphism as the final evolutionary result (Figures 3.46B and 3.46D). In contrast, our simulations show that drug treatment does not change the stability of the ESS point in the case of competitive suppression $\xi < 1$.

Similar observations can be made for the case of convex-concave trade-off (Figures 3.46E–3.46H). A noticeable difference between the linear and convex-concave trade-offs for low treatment rates (i.e., Figures 3.46A–3.46C vs. Figures 3.46E–3.46G, or see Figure 3.47) is that the resistance level θ^* at the ESS is higher in Figures 3.46E–3.46G. Recall that the convex-concave trade-off corresponds to a relatively lower cost for drug resistance (see Figure 3.45). This suggests that even

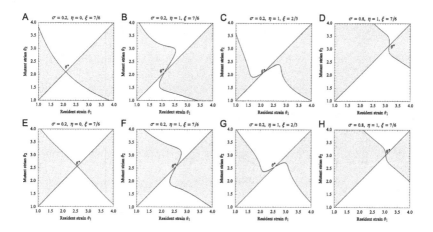

Fig. 3.46 Pairwise invadability plots (PIP) when the coinfection function ρ_{ij} is differentiable ($n = 2$) with respect to θ_2 at $\theta_2 = \theta_1$. The invasion fitness $\mathcal{R}(m, r)$ is greater than 1 in the shaded regions and less than 1 in the unshaded regions. AD are for the linear trade-off function $\gamma(\theta)$ while EH are for the convex-concave trade-off function $\gamma(\theta)$. Except parameter values listed in the graphs, all others are $\Lambda_h = 8, \mu_h = 0.014, \Lambda_s = 25, \rho_0 = 2x10^9, \mu_s = 0.5, c_0 = 28000, \beta_0 = 0.000027, \delta_0 = 3.5$.

under moderate treatment rates, higher levels of drug resistance will still be likely to develop if the cost for drug resistance is low.

When the coinfection functions ρ_{ij} are not differentiable, we use the linear trade-off γ to illustrate the evolution of drug resistance. Unlike in the previous cases in which there is only a single ESS point, there are now two ESS points, which we denote by θ_* and θ^* as shown in Figure 3.48. We observe from the plots in Figure 3.48 that if a resident parasite strain has a drug resistance level below θ_*, then it can be invaded by any parasite strain that has a higher level of drug resistance. If a resident strain has a resistance level above θ^*, then it can be invaded by any parasite strain that has a lower resistant level. Therefore, θ_* and θ^* are evolutionarily convergent and attainable either from left or from right.

Figure 3.48A is for the case of $\xi > 1$ (competitive facilitation). It illustrates that the strategies θ_* and θ^* can be invaded by mutant strains with resistant levels between θ_* and θ^*. Hence, the final evolutionary outcome can be dimorphism or even polymorphism with drug resistance levels being in the range $[\theta_*, \theta^*]$. Different outcomes can be expected for the case of $\xi < 1$ (competitive suppression) as demonstrated in Figures 3.48B–3.48D. Figures 3.48B and 3.48C are for different values of η (coinfection efficiency). It shows in Figure 3.48B that a resident strain with resistance level $\theta \in [\theta_*, \theta^*]$ is capable of preventing an invasion by mutant strains, and hence the evolutionary endpoint can be a monomorphism with a resistance level in $[\theta_*, \theta^*]$. As η increases (Figure 3.48C), θ_* and θ^* can become invadable. Figures 3.48C and 3.48D compare the outcomes for different values of σ (drug treatment),

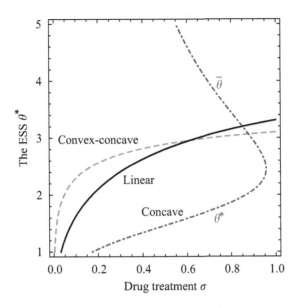

Fig. 3.47   Plot of the drug resistance level at the evolutionarily stable strategy (ESS) Θ^* against the drug treatment rates when the coinfection function ρ_{ij} given in (15) is differentiable ($n = 2$). The ESS $\bar{\Theta}$ in the case of the concave trade-off curve $\gamma(\theta)$ is evolutionarily unstable, functioning as an evolutionary repeller. The parameter values are the same as in Figure 3.46.

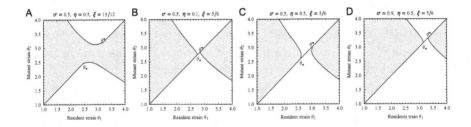

Fig. 3.48   Similar to the PIP plots in Figure 3.46 but for the case of a linear trade-off function $\gamma(\theta)$ and a non-differentiable coinfection function ρ_{ij}. In this case, there are two ESS points, θ_* and θ^*. Both θ_* and θ^* are evolutionarily attainable either from left or from right. But they may or may not prevent an invasion by mutant strains (see explanations in the text). The plot in A is for the case of $\zeta \geq 1$, and plots in BD are for the case of $\zeta < 1$ with different values of η or σ. Parameters are the same as in Figure 3.46 except those specied on the graphs.

and they show that an increase in σ tends to stabilize θ_* and θ^*.

In Figure 3.49, we examine how the length of the interval $[\theta_*, \theta^*]$ may depend on other factors such as η and σ. The curves in Figure 3.49 show the changes of the ESS points θ_* and θ^* with η (A and C) or σ (B and D). For both the case $\xi > 1$ (A and B) and the case $\xi < 1$ (C and D), it shows that the interval $[\theta_*, \theta^*]$ increases with η but decreases with σ (except for very small σ). All parameter values are the same as in Figure 3.46 except those listed on the graphs.

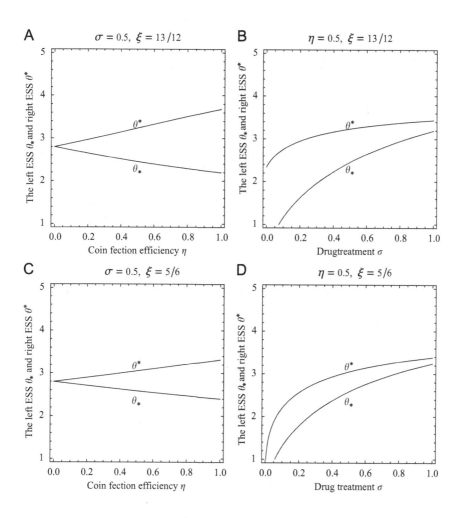

Fig. 3.49   Plots of the ESS points (θ_* and θ^*) versus Z (A and C) or σ (B and D) in the case of a linear trade-off function $\gamma(\theta)$ and a non-differentiable coinfection function ρ_{ij}. A and B illustrate the case of facilitation ($\zeta > 1$), whereas C and D for the cases of competitive suppression ($\zeta < 1$). We observe that the length of the interval $[\theta_*, \theta^*]$ increases with η but decreases with σ (except for very small σ). All parameter values are the same as in Figure 3.46 except those listed on the graphs.

3.5   Synergy between HSV-2 and HIV

In section 2.2.2 we presented a model that couples HIV and HSV-2 transmission dynamics, and derived threshold conditions that determine the outcomes of disease prevalence and control. These results are helpful for the investigation of the synergy between HIV and HSV-2, as well as the effect of HSV-2 therapy on the prevalence and control of HIV.

Influence of HSV-2 therapies on HIV prevalence and control

To illustrate the synergy between HIV and HSV-2 at the population level, particularly the effect of HSV-2 therapy on the epidemiological synergy of HIV, we use two measures. One is the population attributable fraction (PAF), defined by

$$PAF(t) = \frac{\text{Incidence of HIV directly due to HSV-2 at time t}}{\text{Total incidence of HIV at time t}}, \tag{3.61}$$

which measures the direct effect of HSV-2 on HIV incidence at each time, i.e., the fraction of new HIV cases that are caused by HSV-2 directly [Abu-Raddad *et al.* (2008)]. This differs from the indirect effect of HSV-2 on HIV incidence, secondary HIV infections (i.e., transmitted by HIV cases that were directly caused by HSV-2) [Abu-Raddad *et al.* (2008)]. For the indirect synergy of HSV-2 and HIV, we use the excess prevalence of HIV [Abu-Raddad *et al.* (2008)], defined as the difference between HIV prevalence when the two diseases interact and the counterfactual scenario in which they do not.

We also examine how HSV-2 may affect the prevalence of HIV (PoV), defined as

$$PoV(t) = \frac{\sum\limits_{i=m,f_1,f_2} (H_i(t) + P_i(t) + Q_i(t))}{\sum\limits_{i=m,f_1,f_2} N_i(t)}.$$

To compare the impact of different HSV-2 treatment strategies on HIV control, we also use the reduction in cumulative HIV cases per HSV-2 case treated. Let $RCT(t)$ denote this quantity at time t:

$$RCP(t) = \frac{\begin{array}{c}\text{The difference in HIV cumulative cases at time t between the scenarios}\\ \text{with and without HSV-2 treatment}\end{array}}{\text{Total numbers of HSV-2 treated cases at time t}}$$

This quantity expresses the impact of the current HSV-2 treatment strategy on HIV control at each time t, the reduction in cumulative HIV cases per HSV-2 treated case.

In the simulations presented here, the parameter values associated with HIV and HSV-2 transmissions are all fixed with

$$\beta^A_{mf_2} = 0.40, \quad \beta^A_{f_2m} = 0.20, \quad \beta^A_{mf_1} = 1.00, \quad \beta^A_{f_1m} = 1.00;$$

$$\beta^H_{f_1m} = 0.1, \quad \beta^H_{mf_1} = 0.2, \quad \beta^H_{f_2m} = 0.02, \quad \beta^H_{mf_2} = 0.04. \tag{3.62}$$

At the end of this section, we provide a sensitivity and uncertainty analysis to examine the sensitivity of outcomes to the uncertainty of the parameter values (including the transmission parameter and some other parameters) using a method based on Latin Hypercube Sampling.

In Figure 3.50, we present some simulation results illustrating the influence of HSV-2 on HIV prevalence using the quantities mentioned above. This figure also demonstrates how the sexual structure (preference levels of males for females of different risk groups measured by c) may affect the synergy between HIV and HSV-2. Figure 3.50(a) is for the case when $c = 0.9$, i.e., the preference for females of groups 1 and 2 are 0.9 and 0.1, respectively. Figure 3.50(b) is for $c = 0.5$, in which case the male preferences for the two female groups are identical. We observe in both (a) and (b) that, for the set of parameter values used, between $10\% - 15\%$ of new HIV infections (at the endemic steady-state) may be attributed directly to HSV-2 (see the PAV curve). The HIV excess prevalence (thicker dashed curve) indicates that HSV-2 has played an important role in fueling HIV indirectly, with the excess prevalence stabilizing at 12% and 8% in the cases of $c = 0.9$ and $c = 0.5$, respectively. From the PoV curves with and without interactions between HIV and HSV-2 (broken and dotted curves respectively), we observe that HSV-2 may increase HIV prevalence while reducing the time-to-peak. If we compare Figures 3.50(a) and (b), the magnitude of PoV (with or without interaction) for $c = 0.5$ is dramatically higher than that for $c = 0.9$. This suggests that, when male preference for high-risk females increases, the synergy between HIV and HSV-2 may diminish while HIV prevalence increases. It also indicates that the preference parameter (c) may significantly influence HIV prevalence and synergy between the two pathogens. This is in fact confirmed by the sensitivity and uncertainty analysis (see section 3.5).

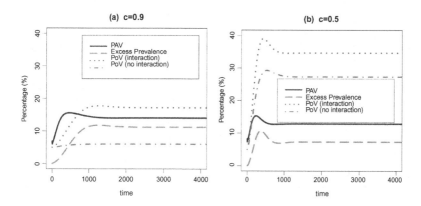

Fig. 3.50 Demonstrations of the synergy between HIV and HSV-2 for different levels of male preference for females in the two risk groups. The plot in (a) is for the case of a higher preference to the low-risk group ($c = 0.9$) whereas the plot in (b) is for the case of equal preference of both female groups ($c = 0.5$).

To examine the impact of treating HSV-2 on transmission of HIV, we consider different treatment strategies. Figures 3.51 and 3.53 illustrate how various HSV-2 therapy programs may affect directly and indirectly the epidemiological synergy and prevalence of HIV. In these simulations, we assume that the effective treatment rates of P_i and A_i are equal, i.e., $\theta_i^A = \theta_i^P = \theta_i$ for all $i = m, f_1, f_1$, and choose

$$\gamma_i^L(\theta_i) = \gamma_i^L(0)\frac{\alpha_i^L}{\alpha_i^L + \theta_i},$$

$$\gamma_i^Q(\theta_i) = \gamma_i^Q(0)\frac{\alpha_i^Q}{\alpha_i^Q + \theta_i},$$

in which the factor $\frac{\alpha_i^L}{\alpha_i^L+\theta_i}$ or $\frac{\alpha_i^Q}{\alpha_i^Q+\theta_i}$ represents reduced reactivation to class A_i or P_i due to anti-viral treatment. In the following, we set $\alpha_i^L = \alpha_i^Q = 2$ for all $i = m, f_1, f_2$, and consider these three strategies:

1) Fix $\theta_{f_1} = \theta_{f_2} = 0.0$ and increase θ_m;
2) Fix $\theta_m = \theta_{f_2} = 0.0$ and increase θ_{f_1};
3) Fix $\theta_m = \theta_{f_1} = 0.0$ and increase θ_{f_2}.

Simulations corresponding to strategies 1) – 3) are illustrated in Figures 3.51 and 3.53. Figure 3.51 is for the strategy 1), i.e., treatment is applied only to the male population (θ_m is varied). In this figure, time plots shown are (a) the fractions of new HIV infections that are directly caused by HSV-2, (b) the excess prevalence of HIV, (c) HIV prevalence, and (d) the reduction in cumulative HIV cases per HSV-2 infection treated. We observe that the fractions of new HIV infections that are caused by HSV-2, directly or indirectly, decrease as treatment rates θ_i increase. For example, the PAF and HIV excess prevalence for $\theta_m = 0$ are roughly 14% and 11%, respectively, compared to 6% and 4% for $\theta_m = 2.0$ (see (a) and (b)). We observe also that HSV-2 therapy can delay the time-to-peak and decrease both the peak size of HIV epidemics and the endemic equilibrium level (see Figure 3.51 (c)), and that the RCP decreases monotonically with the level of treatment (see Figure 3.51 (d)).

The time plots of simulations for strategies 2) and 3) (treating females of groups 1 and 2) show similar characteristics, which we omit here. A comparison of the three strategies is illustrated in Figure 3.53. We observe that for the first three measures (see (a)-(c)), the effect of treating the male and female group 1 are similar. They are linearly decreasing functions of θ, while there is almost no effect when treatment of female group 2 is increased. Thus, in terms of the first three measures, increasing treatment in the male or female group 1 is more effective than in the female group 2. However, the situation is different for the RCP measure (see (d)), in which RCP is increased from 0.28 to 0.41 when θ_{f2} increases from 0.4 to 2, but the change in RCP is much smaller for all levels of treatment θ_i for $i = m$ or f_1. This suggests that, in terms of RCP, focusing HSV-2 therapy on high-risk females is more effective than focusing on low-risk females or males.

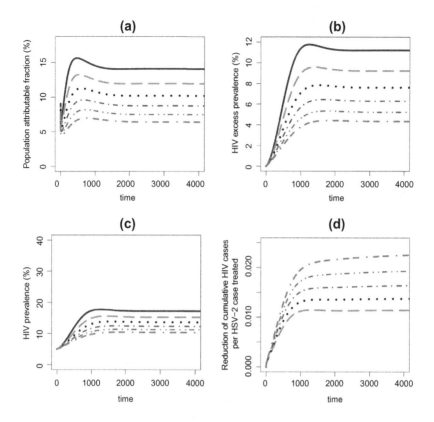

Fig. 3.51 Numerical simulations of the system (2.1) showing the dependence of HIV prevalence on HSV-2 therapy when only males are receiving treatment, i.e., $\theta_{f_1} = 0.0, \theta_{f_2} = 0.0$ and θ_m varies. The plot in (a) shows the fraction of incident HIV infections that are directly caused by HSV-2; (b) shows HIV excess prevalence; (c) shows HIV prevalence; and (d) displays the reduction of cumulative HIV cases per HSV-2 infection treated. These simulations are for the case when the preference parameter is $c = 0.9$. The curves corresponding to different values of θ_m, 0 (solid), 0.4 (dash), 0.8 (dot), 1.2 (dot-dash), 1.6 (dot-dot-dash), and 2 (dash-dash-dot).

Simulations presented in Figure 3.52 are similar to those in Figures 3.51 except that $c = 0.5$ (equal preference to the two female groups). We observe that, for the first two measures (PAF and HIV excess prevalence), the magnitude of the curves is higher when $c = 0.9$ than $c = 0.5$ (see Figures 3.52(a) and (b)). However, for the HIV prevalence, (see Figures 3.52(c)) the magnitude of the curves is much lower when $c = 0.9$ (between 10 and 17) than $c = 0.5$ (between 30 and 34), but the variation in prevalence between different levels of treatment is greater when $c = 0.9$ than $c = 0.5$. Similarly, for the RCP measure (see Figure 3.52(d)), the magnitude of the curves is much lower when $c = 0.9$ (between 0.01 and 0.021) than $c = 0.5$ (between 0.005 and 0.01).

The time plots of simulations for strategies 2) and 3) (treating females of groups

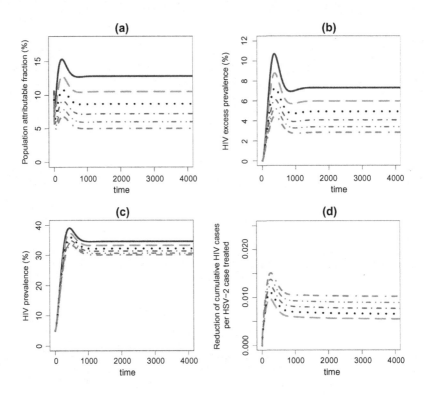

Fig. 3.52   Similar to 3.51 except that c = 0.5 (equal preference for the female groups). This figure illustrates the dependence of HIV prevalence on HSV-2 therapy when only males are receiving treatment, i.e., $\theta_{f_1} = \theta_{f_2} = 0$ and θ_m varies. It shows that in comparison with the case of c = 0.9 (see FIgure 3.51), the effects of treating HSV-2 on HIV prevalence are reduced for the measures presented in (a), (c), and (d), but increased for the measure presented in (c). Moreover, the variations between different levels of treatment (θ) are smaller in the case of c = 0.5 than c = 0.9. The curves corresponding to different values of h m: 0 (solid), 0.4 (dash), 0.8 (dot), 1.2 (dot-dash), 1.6 (dot-dot-dash), and 2 (dash-dash-dot).

1 and 2) show similar characteristics, which we omit here. A comparison of the three strategies is illustrated in Figure 3.53. For the case of c = 0.9, we observe that for the first three measures (see Figures 3.53 (a)-(c)), the effect of treating the male and female group 1 are similar. They are linearly decreasing functions of θ, while there is almost no effect when treatment of female group 2 is increased. Thus, in terms of the first three measures, increasing treatment in the male or female group 1 is more effective than in the female group 2. However, the situation is different for the RCP measure (see Figure 3.53(d)), in which RCP is increased from 0.28 to 0.41 when θ_{f2} increases from 0.4 to 2, but the change in RCP is much smaller for all levels of treatment θ_i for $i = m$ or f_1. This suggests that, in terms of RCP, focusing HSV-2 therapy on high-risk females is more effective than focusing on low-risk females or males.

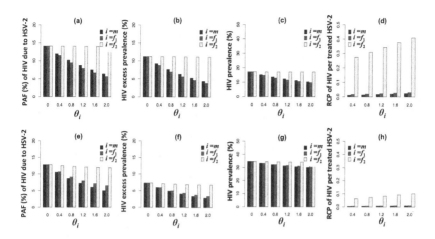

Fig. 3.53 Comparison of effects of HSV-2 on HIV prevalence when different population groups receive HSV-2 treatment (θ represents the level of treatment) for the case of $c = 0.9$ (a-d) and $c = 0.5$ (e-h). Four measures described in section 3.5 are considered. In each of the plots, the height of a bar represents the value of the corresponding quantity at the endemic equilibrium. It illustrates the degree at which HIV prevalence are reduced by treating HSV-2 for the measures presented.

For the case of $c = 0.5$ we observe Figure 3.53 that, in terms of the first two measures (see Figures 3.53(e) and (f)), focusing HSV-2 therapy on low-risk females or males would be more effective than focusing on high-risk females, which is similar to the case of $c = 0.9$. However, between the low-risk female group and the male group, the case of $c = 0.5$ shows that it is more effective to treat males, while the case of $c = 0.9$ shows the opposite. Another difference between these two cases is HIV prevalence. In the case of equal preference ($c = 0.5$), HIV prevalence does not change much when θ_i varies for $i = m, f_1, f_2$, whereas in the case of higher preference for low-risk females ($c = 0.9$), HIV prevalence decreases with increasing θ_i for $i = m$ and f_1.

Sensitivity and uncertainty analysis

Figures 3.54 and 3.55 present our sensitivity and uncertainty analysis by computing the PRCC of the reproduction number using the Latin Hypercube Sampling method [Blower and Dowlatabadi (1994); Helton *et al.* (2006); Marino *et al.* (2008)]. It shows how uncertainty in model parameters may influence the threshold quantities $\mathcal{R}_0^A$, $\mathcal{R}_0^H$, $\mathcal{R}_A^H$ and $\mathcal{R}_H^A$. We observe in Figure 3.54 that $\mathcal{R}_0^A$ and $\mathcal{R}_0^H$ are very sensitive to the preference parameter c and the contact rate b_m (note that b_{f_1} and b_{f_2} depend on b_m and N_i), but not sensitive to the population size N_m (note that N_{f_1} and N_{f_2} vary with N_m). Among the various transmission probabilities (β_i), that from female group 2 to male (β_{f_2m}) is most influential for both $\mathcal{R}_0^A$ and $\mathcal{R}_0^H$. For the treatment parameter (θ_i), treating males is more effective in reducing $\mathcal{R}_0^A$.

Figure 3.55 illustrates that the preferential parameter c and contact rate b_m are

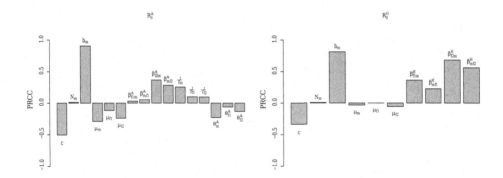

Fig. 3.54   Sensitivity analysis of the reproduction numbers $\mathcal{R}_A^H$ and $\mathcal{R}_H^A$. Based on the PRCC values, both $\mathcal{R}_A^H$ and $\mathcal{R}_H^A$ are very sensitive to the preference parameter c and contact rate b_m (note that b_{f_1} and b_{f_2} depend on b_m and N_i) but insensitive to the population size N_m (note that N_{f_1} and N_{f_2} vary with N_m). Among the various transmission probabilities (β_i), that from female group 2 to males (β_{f_2m}) is most influential for both $\mathcal{R}_0^A$ and $\mathcal{R}_0^H$. For the treatment parameter (θ_i), treating males is more effective in reducing $\mathcal{R}_0^A$.

most influential for the invasion reproduction numbers $\mathcal{R}_A^H$ and $\mathcal{R}_H^A$. Moreover, the HIV invasion reproduction number $\mathcal{R}_A^H$ is more sensitive to the transmission rates between males and low-risk females $(\beta_{f_2m}$ and $\beta_{mf_2})$ than between males and high-risk females $(\beta_{f_1m}$ and $\beta_{mf_1})$, and much less sensitive to treatment θ_i. However, the HSV-2 invasion reproduction number $\mathcal{R}_H^A$ is more sensitive to treatment θ_i.

Our results in this study confirmed the hypothesis that HSV-2 has almost certainly facilitated HIV epidemics indirectly. The main contributions of this study include the results that help understand how non-random mixing between males and females due to different risk levels in the female population, which have not been considered in other studies. This includes both the derivation of the reproduction numbers $(\mathcal{R}_0^A, \mathcal{R}_0^H, \mathcal{R}_H^A, \mathcal{R}_A^H)$ and the numerical results based on various measures for the synergy between HIV and HSV-2 (PAF, PoV, RCP). For example, the effect of HSV-2 on HIV prevalence decreases as the treatment rate of the male and low-risk female groups increases, and HSV-2 therapy of high-risk females is more effective than that of either low-risk females or males in terms of reducing HIV prevalence (RCP). Most importantly, from the sensitivity/uncertainly analysis of the reproduction numbers as well as numerical simulations, the assumption that males prefer low-risk females (represented by the preference parameter c) can dramatically alter the model outcomes. These insights can be very helpful for understanding the synergy between HIV and HSV-2, and cannot be obtained from models without multiple risk groups.

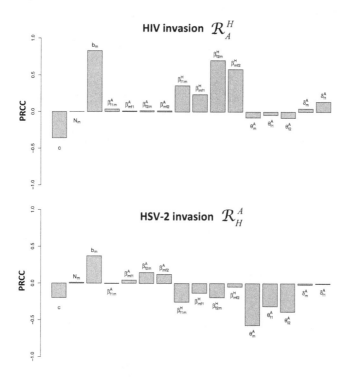

Fig. 3.55 Sensitivity analysis of the invasion reproduction numbers $\mathcal{R}_A^H$ and $\mathcal{R}_H^A$. We observe that both are very sensitive to c and b_m, as in the case for $\mathcal{R}_0^H$ and $\mathcal{R}_0^A$. For $\mathcal{R}_A^H$, the transmission rate $\beta_{f_2m}^H$ is most influential among all β_i^H, while for $\mathcal{R}_H^A$, treating males (θ_m) is more effective than treating other groups in reducing $\mathcal{R}_H^A$.

Chapter 4

Development of interactive tools to assist public health policymaking

4.1 An interactive tool for policymaking in disease control

The modeling results presented in chapters 1-3 suggest that whether or not an infectious disease can be prevented or controlled may depend on factors that play an critical role in the disease dynamics. These factors are represented by parameters involved in the models. It can be very helpful for decision making if the dependence of disease outcomes on these parameters are made easier to visualize by policymakers. One of the powerful computational tools is the Wolfram MATHEMATICA. It allows modelers to create interactive tools for non-modelers to use and to experiment scenarios. The user can also visualize the possible consequences under various actions. We demonstrate this by using several models considered in the previous sections. Sample MATHEMATICA codes that are used to create some of the interface outputs are also included.

4.1.1 *Simple examples*

In this section, we present a user-friendly MATHEMATICA notebook that includes a collection of interactive model outputs. Executions of the notebook allow users to change inputs and observe the changes simultaneously.

This notebook has been used to generate some of the figures in section 3.1.5. For example, Figure 3.18 shows solution behavior for the following SI model in the absence of vaccination:

$$\frac{dS}{dt} = -\beta(t)\,\frac{SI}{N},$$

$$\frac{dI}{dt} = \beta(t)\,\frac{SI}{N} - \gamma I, \qquad (4.1)$$

$$\frac{dR}{dt} = \gamma I$$

with initial conditions

$$S(t_0) = N - I_0 - R_0,\ I(t_0) = I_0,\ R(t_0) = R_0. \qquad (4.2)$$

The parameter t_0 is the time of introduction of the infection, $1/\gamma$ is the infectious period, $\beta(t)$ is the periodic function given in (3.22). If a vaccination program starts at time $t_v > t_0$ with a fraction p of the susceptibles being vaccinated, we will first run the system (4.1) with initial conditions (4.2) until time t_v. Denote the solution values at t_v for susceptible, infectious, and recovered individuals by S_v, t_v, and R_v, respectively. Then from the time t_v, we will run the system (4.1) with the new initial conditions

$$S(t_v) = (1-p)S_v, \ I(t_v) = I_v, \ R(t_v) = R_v. \tag{4.3}$$

In the notebook, we fix the function $\beta(t)$ (i.e., fix the parameters involved in the function) and allow t_0, $1/\gamma$, R_0, I_0, p, and t_v to take on different values.

Figures 4.1–4.3 illustrate the snapshots of the simulation outcomes. Figures 4.1 and 4.2 compare the effect of different t_0 with other parameter values fixed.

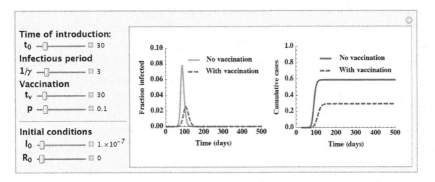

Fig. 4.1   A snapshot of the output from the MATHEMATICA notebook for the SIR model (4.1) with vaccination starting at time $t_v = 30$. The parameter values that generate this figure are shown in the panel on the left. Note that the time of introduction of infection is $t_0 = 30$.

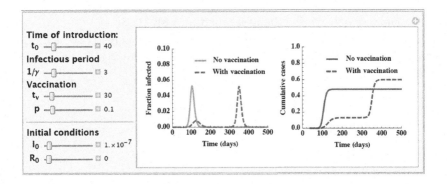

Fig. 4.2   Similar to Figure 4.1 except that it is for the case of $t_0 = 40$. Other parameter values are the same as in Figure 4.1. It shows different consequence of vaccination in both the epidemic peak and the final epidemic size in comparison to that for the case of $t_0 = 30$.

Figure 4.3 compares the effect of t_v when all other parameter values are fixed. The three t_v values are 50, 75, and 100. It illustrates that the timing of vaccination can have a significant impact on the epidemic peak(s) and the final size. Particularly, it may not be better to start vaccination early. These conclusions are dependent on the choice of other parameter values. For example, if the fraction of vaccination p has a different value, then it is possible that early vaccination (smaller t_v) might be helpful to delay and lower the epidemic peak, while reducing the final size. The codes used to generate these snapshots are included in Appendix B.

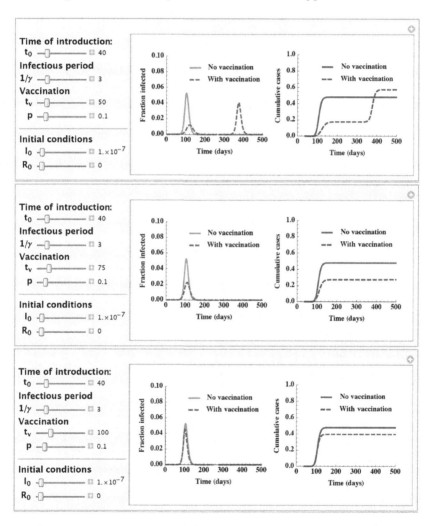

Fig. 4.3   Similar to Figures 4.1 and 4.2 except that this set of plots is to compare the effect of t_v on the final and cumulative epidemic sizes. The three t_v values are 50, 75, and 100.

As pointed in section 3.1.5, besides the deterministic solutions for systems such as (3.18) and (3.23), the models can also be used to simulate the situations in which the disease events (e.g., infection and recovery) may occur randomly (while being governed by the event rates determined by the equations). One of the approaches for stochastic simulations is the event-time approach [Renshaw (1993)]. Consider the model (4.1) with constant transmission rate β. There are only two possible events at time $t_i (i = 0, 1, 2, \cdots)$; a new infection which occurs at the rate $r_i(t_i) = \beta S(t_i)I(t_i)/N$ and a recovery which occurs at the rate $r_2(t_i) = \gamma I(t_i)$. Let $r_T(t_i) + r_2(t_i)$ denote the total event rate at time t_i. Then the "time to next event", T_i, can be determined by

$$T_i = -\frac{\ln(Y_1)}{r_T(t_i)},$$

where Y_1 is a random number generated from the exponential distribution with parameter $r_T(t_i)$. To determine which of the two events occurs at time $t_{i+1} = t_i + T_i$, we use another random number, Y_2, generated from a uniform distribution on $[0, 1]$. For example, if

$$0 \leq Y_2 \leq \frac{r_1(t_i)}{r_T(t_i)}$$

then the first event (infection) occurs, in which case

$$S(t_{i+1}) = S(t_i) - 1, \quad I(t_{i+1}) = I(t_i) + 1, \quad R(t_{i+1}) = R(t_i).$$

Some of these results are illustrated in Figure 3.17. Here, we present a similar figure but with a control panel included (see Figure 4.4). The users can vary the values of the parameters shown in the left panel, which include the transmission rate (β), infectious period ($1/\gamma$), length of the time steps for the simulation (T), the number of realizations (m), the total population size (N), and the initial number of infectious individuals (I_0). We observe that, although the epidemic did not take off for some realizations, the mean values of the 20 realizations provide useful information about the severity of the epidemic, including the size of epidemic peak, time at which the peak is reached, and the final size of the epidemic. The Mathematica code for this output is provided in the section 4.1.2.

4.1.2 *Notebook for the SARS model in section 3.2*

In section 3.2, we presented several figures to demonstrate modeling outcomes concerning the spread and control of SARS in 2003. In this section, we present a collection of MATHEMATICA notebooks used to create the figures including both those presented in 3.2 and some additional figures discussed in [Feng *et al.* (2009)].

The cover page provides the contents showing the type of coutputs included in the notebooks. The pages that follow demonstrate model outcomes such as the dependence of the control reproduction number on various parameters, parameter estimation using real data, and the effect of alternative control measures on the final disease outcomes. Each page has also a brief explanation about the outputs,

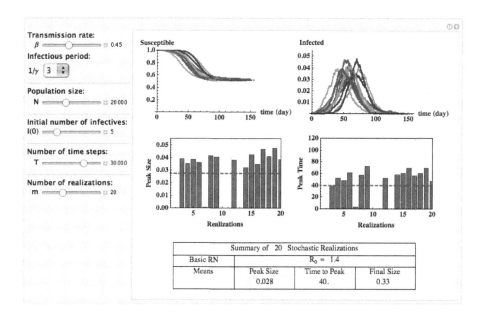

Fig. 4.4 Similar to Figures 3.17 except that a panel of parameters is included. The plot is generated using the parameter values indicated in the panel. As the values of parameters are varied, the plots on the right, as well as the numbers in the table on the bottom, will change simultaneously.

and at the bottom there is a hyperlink which allows the access to the code used to generate the outcome. In each of the output pages the buttons in the left panel allow users to vary the parameter values, and the figure on the right shows the outcomes resulted from the parameter changes. For example, for the plot titled "Calculation of the reproduction number, $\mathcal{R}_c$", the parameters shown in the panel on the left can be varied. The ranges of these parameters are determined according to their biological constraints. Integer values $1, 2, 3$, etc. are possible for m, n, l, which are the number of sub-stages in the gamma distribution. The parameters representing quarantine (χ) and isolation $(\phi_P$ and $\phi_R)$ can take on values in $(0, 1)$. The effect of these control measures on the reduction in $\mathcal{R}_c$ can be observed as the bars corresponding to the parameters are moved within the specified ranges. The MATHEMATICA codes are similar to to those included in Appendix B.

Table of Contents

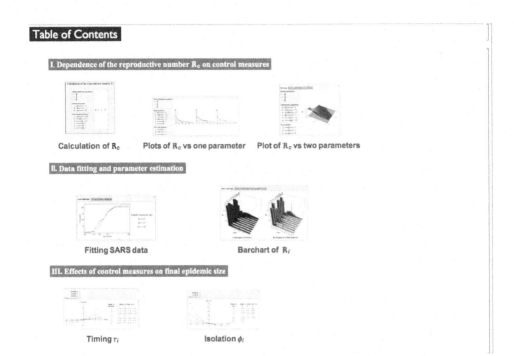

I. Dependence of the reproductive number R_c on control measures

Calculation of R_c Plots of R_c vs one parameter   Plot of R_c vs two parameters

II. Data fitting and parameter estimation

Fitting SARS data   Barchart of R_i

III. Effects of control measures on final epidemic size

Timing τ_i Isolation ϕ_i

Change of the value of R_c with respect to parameters

Three groups of parameters can be varied as shown in the left panel below. The group on the top includes the numbers of sub-stages in the gamma distribution. The values for *m*, *n*, and *I* can be changes by clicking on the arrows. The middle and bottom groups are for other parameters whose values can be varied by moving the bars to the left (decrease) or right (increase) with the corresponding value displayed at the end of each bar. The value for R_c is shown on the right, which will change when any of the parameter values is changed. We observe that for the deflut setting of the parameter values, when *m* changes from 1 to 3 the value of R_c is decreased, whereas when *I* is changed from 1 to 3 the value of R_c is increased.

Calculation of the reproductive number, R_c

Gamma distribution parameters:
m = 1
n = 1
I = 3

Control parameters:
χ — 0.1
ϕ_P — 0.1
ϕ_R — 0.1

$$R_c = 1.86$$

Other model parameters:
λ_E — 0.05
λ_P — 0.3
λ_R — 0.5
ρ_E — 1
ρ_P — 1
ρ_R — 1

(see code)   Back to top

Plots of R_c vs. various control parameters (the case of a single control)

R_c is plotted against three control parameters, χ, ϕ_P and ϕ_R separately. Several other parameters are allowed to change. The bottons on top provide seletion of plots for different control measures. Each plot can be viewed by cliking the corresponding botton. The dotted line marks the value of control parameter for which $R_c=1$. We observe that for larger $1/\delta_P$ medication of people in the prodrome stage (ϕ_P) is more effective (e.g., $1/\alpha=4$, $1/\delta_P=6$), whereas for smaller $1/\delta_P$ isolation (ϕ_R) is more effective (e.g., $1/\alpha=3$, $1/\delta_P=3$), and for larger $1/\alpha$ (e.g., $1/\alpha=8$, $1/\delta_P=4$) quarantine (χ) is more effective. Here, more effective means that a smaller level of control is required to reduce R_c to below 1.

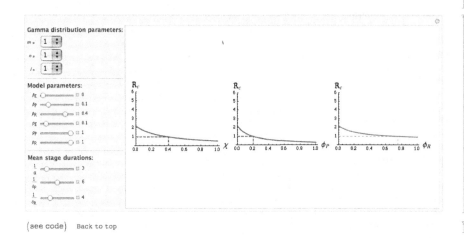

(see code)   Back to top

Plot of R_c vs two cntrol parameters

R_c is plotted jointly against two control parameters repressenting quarantine and isolation. A type of plot can be selected by cliking on the botton on the top. The first plot is for the case of $\phi_P=\phi_R=\phi$ and R_c is plotted against χ and ϕ. The second plot is R_c vs χ and ϕ_R with fixed ϕ_P=0.1. Several other parameters are allowed to change. The plane with light blue color correspons to R_c=1, and the surface shows R_c as a function of two control variables. The intersection curve of the surface and the plane marks the combination of χ and ϕ (or combination of χ and ϕ_R) values for which R_c=1. From the plot of R_c vs χ and ϕ we observe that the effect of χ and ϕ change dramatically when ρ_R is reduced from 0.8 to 0.75 or increased to 0.85 (see bookmarks). Similar changes can also be observed by varying ρ_P. Furthermore, if we decrease durations (to 2,4,6 respectively) then we observe that isolation is much more efficient than quarantine (see bookmark). For example, for ϕ=0.3, it is very difficult to bring R_c below one with quarantine, whereas for χ=0.3, it is much easier to bring R_c below one with isolation. For the plot R_c vs χ and ϕ_R, observe the changes when ρ_P varies.

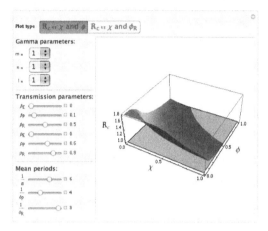

(see code) Back to top

Data fitting and estimation of model parameters (*Mathematica*)

When a data set is chosen (e.g., the first 30 days or the entire 3 months), we can fit the model to the data and estimate some of the model parameters such as the transmission rates β_P and β_R, and the durations $1/\alpha$, $1/\delta_P$ and $1/\delta_R$. The fitted curves are shown on the left
and the estimated parameter values are listed on the right.

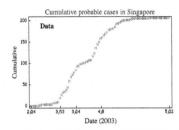

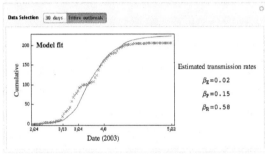

Estimated transmission rates

$\beta_E = 0.02$

$\beta_P = 0.15$

$\beta_R = 0.58$

(see code) Back to top

Sensitivity of R_c to variations of parameters

Histograms of R_i ($i=1, .., 8$) and before and after 24 March, when quarantine began and hospital infection control measures were more stringently enforced, from which the statistics in the Table 4 were calculated.

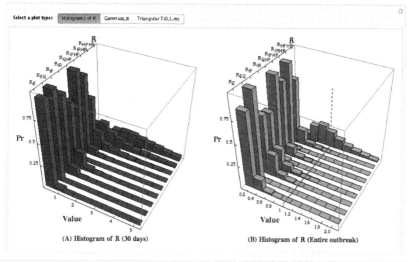

Select a plot type: Histograms of R   Gamma(α, β)   Triangular T(0,1,∞)

(A) Histogram of R (30 days)   (B) Histogram of R (Entire outbreak)

(see code) Back to top

Effect of timing of the controls on the final epidemic size

We observe that during the 3 months period of the outbrake there were 3 time points τ_1, τ_2, and τ_3, at which control measures are taken. What wold the final size change if one or more of these time points are taken a few days earlier or later? The plots on the left shows the percentages of change in the final size with respect to the change in days of actions. The Table on the right list the numerical values of the changes.

change τ_1 only ☑
change τ_2 only ☑
change τ_3 only ☑
$\tau_1 = \tau_2 = \tau_3 = \tau$ ☑

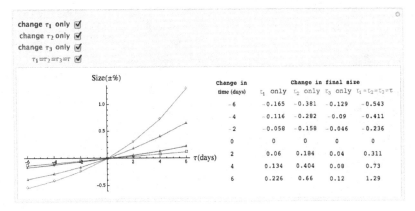

Change in time (days)	Change in final size			
	τ_1 only	τ_2 only	τ_3 only	$\tau_1 = \tau_2 = \tau_3 = \tau$
-6	-0.165	-0.381	-0.129	-0.543
-4	-0.116	-0.282	-0.09	-0.411
-2	-0.058	-0.158	-0.046	-0.236
0	0	0	0	0
2	0.06	0.184	0.04	0.311
4	0.134	0.404	0.08	0.73
6	0.226	0.66	0.12	1.29

(see code)   Back to top

Effect of isolation rates on the final epidemic size

Each of the isolation rates ϕ_P, and ϕ_R on the reduction of final epidemic size. The plots on the left shows the percentages of change in the final size with respect to the change in isolation rates. The Table on the right list the numerical values of the changes.

change ϕ_P only ☑
change ϕ_R only ☑
change $\phi_P = \phi_R = \phi$ ☑

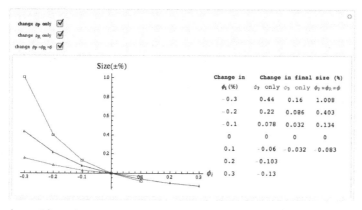

Change in ϕ_i (%)	Change in final size (%)		
	ϕ_P only	ϕ_R only	$\phi_P = \phi_R = \phi$
−0.3	0.44	0.16	1.008
−0.2	0.22	0.086	0.403
−0.1	0.078	0.032	0.134
0	0	0	0
0.1	−0.06	−0.032	−0.083
0.2	−0.103		
0.3	−0.13		

(see code)   Back to top

Appendix A

Notations and definitions

Table A.1 Notations and definitions used in some of the models.

Symbol	Definition
S	Number of susceptible individuals
E	Number of exposed (latent) individuals
I	Number of infectious individuals
R	Number of recovered individuals
β	Transmission rate of an infection
$\mathcal{D}_E$	Latent period
$\mathcal{D}_I$	Infectious period
$\mathcal{D}_{I_l}$	Isolation-adjusted infectious period
$\mathcal{R}_0$	Basic reproduction number
$\mathcal{R}_c, \mathcal{R}_v$	Control reproduction number

For continuous-time models

t	Independent variable for time
a (or θ, τ)	Independent variable for age
k (or α)	Rate of progression from latent to infectious stage
γ (or δ)	Recovery rate
μ	Natural death rate
χ	Quarantine rate
ϕ	Isolation rate
EDA	Exponential distribution assumption
GDA	Gamma distribution assumption
EDM	Exponential distribution model
GDM	Gamma distribution model

For discrete-time models

n	Independent variable for time step
α	Probability of remaining latent
δ	Probability of remaining infectious
$1 - \gamma$	Quarantine probability
$1 - \sigma$	Isolation probability
GDA	Geometric distribution assumption
PDA, BDA	Poisson, binomial distribution assumption
GDM	Geometric distribution model
PDM, BDM	Poisson, binomial distribution model

Appendix B

Examples of MATHEMATICA codes

This appendix includes the MATHEMATICA codes used to generate the computational results presented in Figures 4.1–4.4. Note that in the codes the equations are written using fractions: $s(t) = S(t)/N, i(t) = I(t)/N$, etc. These codes have been written based on version 8.0 of the software.

This code is used to generate Figures 4.1-4.3

```
Manipulate[
 Module[{t, seq0, ieq0, req0, s0, i0, r0, sol0, seqB, ieqB, reqB, solB,
    sB, iB, rB, iAB, rAB, γ, R0, ep, beta0, β, plotBi, plotBr, range, maxT},
    γ = 1 / Duration; R0 = beta0 / γ; maxT = 500; beta0 = 1.5 * γ;
    ep = 0.35; β = beta0 * (1 + ep * Cos[2 * Pi * t / 365]); range = 1;

    seq0 = (s0'[t] == -β s0[t] i0[t]);
    ieq0 = (i0'[t] == (β s0[t] - γ) i0[t]);
    req0 = (r0'[t] == γ i0[t]);
    sol0 = NDSolve[{seq0, ieq0, req0, s0[t0] == 1.0 - infect, i0[t0] == infect,
        r0[t0] == 0}, {s0, i0, r0}, {t, 0, maxT}, MaxSteps → 100 000];

    seqB = (sB'[t] == -β  sB[t] iB[t]);
    ieqB = (iB'[t] == (β  sB[t] - γ) iB[t]);
    reqB = (rB'[t] == γ iB[t]);
    solB = Quiet[NDSolve[
        {seqB, ieqB, reqB, sB[tv] == (1 - p) * s0[tv] /. sol0, iB[tv] == i0[tv] /. sol0,
        rB[tv] == r0[tv] /. sol0}, {sB, iB, rB}, {t, tv, maxT}, MaxSteps → 100 000]];
    iAB[t_] := Piecewise[{{i0[t] /. sol0, 0 ≤ t < tv}, {iB[t] /. solB, tv ≤ t ≤ maxT}}];
    rAB[t_] := Piecewise[{{r0[t] /. sol0, 0 ≤ t < tv}, {rB[t] /. solB, tv ≤ t ≤ maxT}}];

    plotBi = Plot[{i0[t] /. sol0, iAB[t]},
        {t, 0, maxT}, AxesOrigin → {0, 0}, PlotRange → {{0, maxT}, {0, 0.1}},
        PlotStyle → {{Green, Thick}, {Red, Dashed, Thick}}, AspectRatio → .8,
        ImageSize → 180, FrameLabel → {"Time (days)", "Fraction infected"},
        LabelStyle → {Bold, 10}, Frame → {{True, False}, {True, False}},
        Epilog → {{Green, Thick, Line[{{.25 maxT, .085 range}, {.35 maxT, .085 range}}]},
          {Red, Dashed, Thick, Line[{{.25 maxT, .07 range}, {.35 maxT, .07 range}}]},
          Text[Style["No vaccination", Bold, 10], {.67 maxT, .085 range}],
          Text[Style["With vaccination", Bold, 10], {.7 maxT, .07 range}]}]
    ];
```

```
plotBr = Plot[{r0[t] /. sol0, rAB[t]},
  {t, 0, maxT}, AxesOrigin → {0, 0}, PlotRange → {{0, maxT}, {0, 1}},
  PlotStyle → {{Blue, Thick}, {Purple, Dashed, Thick}}, AspectRatio → .8,
  ImageSize → 180, FrameLabel → {"Time (days)", "Cumulative cases"},
  LabelStyle → {Bold, 10}, Frame → {{True, False}, {True, False}},
  Epilog → {{Blue, Thick, Line[{{.25 maxT, .85 range}, {.35 maxT, .85 range}}]},
    {Purple, Dashed, Thick, Line[{{.25 maxT, .7 range}, {.35 maxT, .7 range}}]},
    Text[Style["No vaccination", Bold, 10], {.67 maxT, .85 range}],
    Text[Style["With vaccination", Bold, 10], {.7 maxT, .7 range}]}
  ];
Grid[{{plotBi, plotBr}}]
],

Style["Time of introduction:", Bold, Medium],
{{t0, 30, Style["t₀", 12, Bold]}, 0,
  300, 1, ImageSize → Tiny, Appearance → "Labeled"},
Style["Infectious period", 12, Bold],
{{Duration, 3, Style["1/γ", 12, Bold]},
  0.5, 20, 0.1, ImageSize → Tiny, Appearance → "Labeled"},
Style["Vaccination", 12, Bold],
{{tv, 30, Style["tᵥ", 12, Bold]}, 0,
  400, 1, ImageSize → Tiny, Appearance → "Labeled"},
{{p, 0.1, Style["p", 12, Bold]}, 0, 1, 0.05, ImageSize → Tiny,
  Appearance → "Labeled"}, Delimiter,
Style["Initial conditions", 12, Bold],
{{infect, .0000001, Style["I₀", 12, Bold]}, 0,
  1 - immune, 0.00001, ImageSize → Tiny, Appearance → "Labeled"},
{{immune, 0, Style["R₀", 12, Bold]}, 0, 0.9, 0.05, ImageSize → Tiny,
  Appearance → "Labeled"}, ControlPlacement → Left]
```

This code is used to generate Figure 4.4

```
Manipulate[Module[
  {gamma, R0, statevectornames, stotal, Infectotal, finalR, time, state, eventrate,
   totalrate, eventratio, susceptible, infec, infec2, timeatpeak, peak, meanpeaktime,
   meanpeak, meanfinal, summary, chart1, chart2, stoptime, peaklist, peaktimelist},
  Clear All;
  gamma = 1 / per; R0 = beta / gamma;
  statevectornames = {"s", "i", "r"}; stotal = {}; Infectotal = {};
  peaklist = {}; peaktimelist = {}; (*These lists are for later use*)
  finalR = Table[0, {i, 1, simuno}];
  Do[
    time = Table[0, {i, step + 1}]; state = Table[0, {i, step + 1}, {j, 3}]; time[[1]] = 0;
    state[[1]] = {totaln - I0, I0, 0}; (*initial values S(0), I(0), R(0)*)
    i = 1;
    While[time[[i]] ≥ 0 && i < step && state[[i, 1]] > 0 && state[[i, 2]] > 0,
      (*We need to add this constraint so that
        the state variables do not become negative.*)
      {
        eventrate = {beta * state[[i, 1]] * state[[i, 2]] / totaln, gamma * state[[i, 2]]};
        (*This is a list of possible events*)
        totalrate = Total[eventrate]; (*Sum of the events*)
```

```
    eventratio = eventrate / totalrate;
    (*This determines the weights for RandomChoice *)
    time[[i + 1]] = time[[i]] + If[totalrate > 0,
        Random[ExponentialDistribution[totalrate]], 0];
    state[[i + 1]] = state[[i]] + RandomChoice[
        eventratio -> {{-1, 1, 0}, {0, -1, 1}}];
    i = i + 1
    }
];
susceptible = Table[{time[[i]], state[[i, 1]] / totaln}, {i, 1, step}];
(*A list of susceptible fractions and the corresponding times *)
infec = Table[{time[[i]], state[[i, 2]] / totaln}, {i, 1, step}];
(*A list of infective fractions and the corresponding times *)
infec2 = Table[state[[i, 2]] / totaln, {i, 1, step}];
(*A list a infectives without time cordinates*)
peak = Max[infec2]; (*Peak size for a single run*)
timeatpeak = time[[First[Flatten[Position[infec2, peak]]]]];
(*Time at the position where peak is reached for a single run*)
stoptime = First[Flatten[Position[time, Max[time]]]];
(*The position where the simulation stopped*)
AppendTo[stotal, susceptible]; (*stotal includes
  susceptible from multiple runs*)
AppendTo[Infectotal, infec]; (*Infectotal includes
  infectives from multiple runs*)
AppendTo[peaklist, peak]; (*A list of peak sizes from multiple runs*)
AppendTo[peaktimelist, timeatpeak];
(*A list of peak times from multiple runs*)
finalR[[k]] = state[[stoptime, 3]] / totaln,
(*Fraction of recovered at the stoptime in k-th run*)
{k, 1, simuno}]; meanpeak = N[Mean[peaklist]];
meanpeaktime = N[Mean[peaktimelist]]; meanfinal = N[Mean[finalR]];
summary = Grid[{{Grid[{{{"Summary of", Style[simuno], "Stochastic Realizations"}}}],
    SpanFromLeft},
    {"Basic RN", Grid[{{{"R₀ =", Style[SetPrecision[R0, 2]]}}}], SpanFromLeft},
    {"Means\n", Item["Peak Size\n       " Style[SetPrecision[meanpeak, 2]],
        Background → LightYellow], "Time to Peak\n        "
        Style[SetPrecision[meanpeaktime, 2]], "Final Size\n      "
        Style[SetPrecision[meanfinal, 2]]}}, Frame → All, ItemSize → 8];
chart1 = BarChart[peaklist, PlotRange → {{1, simuno}, {0, 2 * meanpeak}},
    FrameLabel → {"Realizations", "Peak Size", "", ""}, Frame → True, FrameTicks →
    {Automatic, Automatic, None, None}, LabelStyle → {Bold, 12}, ImageSize → 250,
    Epilog → {{Red, Thick, Dashed, Line[{{0, meanpeak}, {simuno, meanpeak}}]}}];
chart2 = BarChart[peaktimelist, PlotRange → {{1, simuno}, {0, 3 * meanpeaktime}},
    FrameLabel → {"Realizations", "Peak Time", "", ""},
    Frame → True, FrameTicks → {Automatic, Automatic, None, None},
    LabelStyle → {Bold, 12}, ImageSize → 250, Epilog →
    {{Red, Thick, Dashed, Line[{{0, meanpeaktime}, {simuno, meanpeaktime}}]}}];
Grid[{{
    Grid[{{ListPlot[stotal, PlotStyle → {{Purple, PointSize[0.008]}, Red, Green},
        AxesLabel → {"time (day)", "Susceptible"}, PlotRange → {0, 1},
        ImageSize → 260, LabelStyle → {Bold, 12}, AspectRatio → .6],
        ListPlot[Infectotal, PlotStyle → {{Purple, PointSize[0.008]}, Red, Green},
        AxesLabel → {"time (day)", "Infected"}, PlotRange → {0, 2 * meanpeak},
        LabelStyle → {Bold, 12}, AspectRatio → .6, ImageSize → 260]}}},
    {Grid[{{chart1, chart2}}]}, {}, {Text @ summary}}],
```

```
Style["Transmission rate:", Bold, Medium],
{{beta, 0.45, Style[ "β", 12, Bold]},
 0.001, 1, .02, Appearance → "Labeled", ImageSize → Small},
Style["Infectious period:", Bold, Medium],
{{per, 3, Style["1/γ", Bold, Medium]},
 {1, 2, 3, 4, 5, 6, 7, 8}, ImageSize -> Small}, Delimiter,
Style["Population size:", Bold, Medium],
{{totaln, 20000, Style["N", Bold, Medium]}, 1000, 50000,
 100, Appearance → "Labeled", ImageSize -> Small}, Delimiter,
Style["Initial number of infectives:", Bold, Medium],
{{I0, 5, Style["I(0)", Bold, Medium]}, 1, 20,
 1, Appearance → "Labeled", ImageSize -> Small}, Delimiter,
Style["Number of time steps:", Bold, Medium],
{{step, 30000, Style["T", Bold, Medium]}, 100, 40000,
 100, Appearance → "Labeled", ImageSize -> Small}, Delimiter,
Style["Number of realizations:", Bold, Medium],
{{simuno, 20, Style["m", Bold, Medium]},
 5, 50, 5, ImageSize → Small, Appearance → "Labeled"},
ControlPlacement → Left, LabelStyle → Black, (*SaveDefinitions→True,*)
SynchronousUpdating → False, ContinuousAction → None]
```

Bibliography

Abu-Raddad, L. J., Magaret, A. S., Celum, C., Wald, A., Longini Jr, I. M., Self, S. G. and Corey, L. (2008). Genital herpes has played a more important role than any other sexually transmitted infection in driving hiv prevalence in africa, *PLoS One* **3**, 5, p. e2230.

ACIP (2009). H1N1 Vaccination Recommendations, `http://www.cdc.gov/h1n1flu/vaccination/acip.htm`.

Alfaro-Murillo, J. A., Towers, S. and Feng, Z. (2013). A deterministic model for influenza infection with multiple strains and antigenic drift, *Journal of Biological Dynamics* **7**, 1, pp. 199–211.

Allen, L. J. and van den Driessche, P. (2008). The basic reproduction number in some discrete-time epidemic models, *Journal of Difference Equations and Applications* **14**, 10-11, pp. 1127–1147.

Allen, S., Bennett, S., Riley, E., Rowe, P., Jakobsen, P., O'Donnell, A. and Greenwood, B. (1992). Morbidity from malaria and immune responses to defined plasmodium falciparum antigens in children with sickle cell trait in the gambia, *Transactions of the Royal Society of Tropical Medicine and Hygiene* **86**, 5, pp. 494–498.

Allison, A. C. (1954). Protection afforded by sickle-cell trait against subtertian malarial infection, *British Medical Journal* **1**, 4857, p. 290.

Alonso, W. J., Viboud, C., Simonsen, L., Hirano, E. W., Daufenbach, L. Z. and Miller, M. A. (2007). Seasonality of influenza in brazil: a traveling wave from the amazon to the subtropics, *American Journal of Epidemiology* **165**, 12, pp. 1434–1442.

Aluoch, J. (1997). Higher resistance to plasmodium falciparum infection in patients with homozygous sickle cell disease in western kenya, *Tropical Medicine & International Health* **2**, 6, pp. 568–571.

Anderson, G. W. and Amstein, M. G. (1948). Communicable disease control, *The American Journal of Nursing* **48**, 7, p. 30.

Anderson, G. W., Arnstein, M. G., Lester, M. R. *et al.* (1962). *Communicable Disease Control, A Volume for the Public Health Worker* (The Macmillan Co).

Anderson, R. and May, R. (1982a). Coevolution of hosts and parasites, *Parasitology* **85**, 02, pp. 411–426.

Anderson, R. M. (1990). *The transmission dynamics of sexually transmitted diseases: the behavioural component* (International Union for the Scientific Study of Population).

Anderson, R. M. and May, R. M. (1978). Regulation and stability of host-parasite population interactions: I. regulatory processes, *The Journal of Animal Ecology* **47**, pp. 219–247.

Anderson, R. M. and May, R. M. (1982b). Directly transmitted infectious diseases: control by vaccination, *Science* **215**, 4536, pp. 1053–1060.

Anderson, R. M. and May, R. M. (1984). Spatial, temporal, and genetic heterogeneity in host populations and the design of immunization programmes, *Mathematical Medicine and Biology* **1**, 3, pp. 233–266.

Anderson, R. M. and May, R. M. (1991). *Infectious diseases of humans: dynamics and control* (Oxford University Press).

Andreasen, V. (1995). Instability in an sir-model with age-dependent susceptibility, *Mathematical Population Dynamics* **1**, pp. 3–14.

Arino, J., Brauer, F., Van den Driessche, P., Watmough, J. and Wu, J. (2006). Simple models for containment of a pandemic, *Journal of the Royal Society Interface* **3**, 8, pp. 453–457.

Aron, J. L. and May, R. M. (1982). The population dynamics of malaria, in *The population dynamics of infectious diseases: theory and applications* (Springer), pp. 139–179.

Aron, J. L. and Schwartz, I. B. (1984a). Seasonality and period-doubling bifurcations in an epidemic model, *Journal of Theoretical Biology* **110**, 4, pp. 665–679.

Aron, J. L. and Schwartz, I. B. (1984b). Some new directions for research in epidemic models, *Mathematical Medicine and Biology* **1**, 3, pp. 267–276.

Bansal, S., Pourbohloul, B. and Meyers, L. A. (2006). A comparative analysis of influenza vaccination programs, *PLoS Medicine* **3**, 10, p. e387.

Barbour, A. D. (1978). MacDonald's model and the transmission of bilharzia, *Transactions of the Royal Society of Tropical Medicine and Hygiene* **72**, 1, pp. 6–15.

Bartlett, M. (1960). The critical community size for measles in the united states, *Journal of the Royal Statistical Society. Series A (General)* **123**, 1, pp. 37–44.

Bartlett, M. S. (1957). Measles periodicity and community size, *Journal of the Royal Statistical Society. Series A (General)* **120**, 1, pp. 48–70.

Beck, K., Keener, J. and Ricciardi, P. (1984). The effect of epidemics on genetic evolution, *Journal of Mathematical Biology* **19**, 1, pp. 79–94.

Belongia, E., Kieke, B., Coleman, L., Donahue, J., Irving, S., Meece, J., Vandermause, M., Shay, D., Gargiullo, P., Balish, A. *et al.* (2008). Interim within-season estimate of the effectiveness of trivalent inactivated influenza vaccine-marshfield, Wisconsin, 2007-08 influenza season, *Journal of the American Medical Association* **299**, 20, pp. 2381–2384.

Bloom, B. R. (1994). *Tuberculosis: pathogenesis, protection, and control*, Vol. 312 (ASM Press, Washington, DC).

Blower, S., Aschenbach, A., Gershengorn, H. and Kahn, J. (2001). Predicting the unpredictable: transmission of drug-resistant HIV, *Nature Medicine* **7**, 9, pp. 1016–1020.

Blower, S. and Dowlatabadi, H. (1994). Sensitivity and uncertainty analysis of complex models of disease transmission: an HIV model, as an example, *International Statistical Review/Revue Internationale de Statistique* **62**, 2, pp. 229–243.

Blower, S., Gershengorn, H. B. and Grant, R. (2000). A tale of two futures: HIV and antiretroviral therapy in San Francisco, *Science* **287**, 5453, pp. 650–654.

Blower, S. and Ma, L. (2004). Calculating the contribution of herpes simplex virus type 2 epidemics to increasing HIV incidence: treatment implications, *Clinical Infectious Diseases* **39**, Supplement 5, pp. S240–S247.

Blower, S., Porco, T. and Darby, G. (1998). Predicting and preventing the emergence of antiviral drug resistance in HSV-2, *Nature Medicine* **4**, 6, pp. 673–678.

Blythe, S. P. and Castillo-Chavez, C. (1989). Like-with-like preference and sexual mixing models, *Mathematical Biosciences* **96**, 2, pp. 221–238.

Bonabeau, E., Toubiana, L. and Flahault, A. (1998). The geographical spread of influenza, *Proceedings of the Royal Society of London. Series B: Biological Sciences* **265**, 1413, pp. 2421–2425.

Boots, M. and Bowers, R. G. (1999). Three mechanisms of host resistance to microparasitesavoidance, recovery and toleranceshow different evolutionary dynamics, *Journal of Theoretical Biology* **201**, 1, pp. 13–23.

Boots, M. and Haraguchi, Y. (1999). The evolution of costly resistance in host-parasite systems, *The American Naturalist* **153**, 4, pp. 359–370.

Bowers, R. G. (1999). A baseline model for the apparent competition between many host strains: the evolution of host resistance to microparasites, *Journal of Theoretical Biology* **200**, 1, pp. 65–75.

Bowers, R. G., Boots, M. and Begon, M. (1994). Life-history trade-offs and the evolution of pathogen resistance: competition between host strains, *Proceedings of the Royal Society of London. Series B: Biological Sciences* **257**, 1350, pp. 247–253.

Bowers, R. G., Hoyle, A., White, A. and Boots, M. (2005). The geometric theory of adaptive evolution: trade-off and invasion plots, *Journal of Theoretical Biology* **233**, 3, pp. 363–377.

Bowers, R. G. and Turner, J. (1997). Community structure and the interplay between interspecific infection and competition, *Journal of Theoretical Biology* **187**, 1, pp. 95–109.

Brauer, F. (2008). Age-of-infection and the final size relation, *Mathematical Biosciences and Engineering* **5**, 4, pp. 681–690.

Brauer, F. and Castillo-Chavez, C. (2012). *Mathematical Models in Population Biology and Epidemiology*, 2nd edn. (Springer-Verlag).

Brauer, F., Feng, Z. and Castillo-Chavez, C. (2010). Discrete epidemic models, *Mathematical Biosciences* **7**, p. 1.

Bremermann, H. J. and Thieme, H. (1989). A competitive exclusion principle for pathogen virulence, *Journal of Mathematical Biology* **27**, 2, pp. 179–190.

Busenberg, S. and Castillo-Chavez, C. (1991). A general solution of the problem of mixing of sub-populations and its application to risk- and age-structured epidemic models for the spread of aids, *Mathematical Medicine and Biology* **8**, 1, pp. 1–29.

Capasso, V. (1993). *Mathematical structures of epidemic systems*, Vol. 97 (Springer).

Carrat, F., Vergu, E., Ferguson, N. M., Lemaitre, M., Cauchemez, S., Leach, S. and Valleron, A.-J. (2008). Time lines of infection and disease in human influenza: a review of volunteer challenge studies, *American Journal of Epidemiology* **167**, 7, pp. 775–785.

Casella, G. and Berger, R. L. (2001). *Statistical inference* (Thomson Learning).

Castillo-Chavez, C. (1987). *Cross-immunity in the dynamics of homogenous and heterogeneous populations*, Vol. 87 (Mathematical Sciences Institute, Cornell University).

Castillo-Chavez, C., Busenberg, S. and Gerow, K. (1991). Pair formation in structured populations, in *Differential equations with applications in biology, physics and engineering* (Marcel Dekker, New York), pp. 47–65.

Castillo-Chavez, C., Castillo-Garsow, C. W. and Yakubu, A.-A. (2003). Mathematical models of isolation and quarantine, *JAMA* **290**, 21, pp. 2876–2877.

Castillo-Chavez, C. and Feng, Z. (1997). To treat or not to treat: the case of tuberculosis, *Journal of Mathematical Biology* **35**, 6, pp. 629–656.

Castillo-Chavez, C. and Feng, Z. (1998). Global stability of an age-structure model for tb and its applications to optimal vaccination strategies, *Mathematical Biosciences* **151**, 2, pp. 135–154.

Castillo-Chavez, C., Feng, Z. and Huang, W. (2002). On the computation of ∇_0 and its role on global stability, in *Mathematical approaches for emerging and reemerging infectious diseases: an introduction*, Vol. 1 (Springer), p. 229.

Castillo-Chavez, C., Hethcote, H., Andreasen, V., Levin, S. and Liu, W. M. (1989a). Epidemiological models with age structure, proportionate mixing, and cross-immunity, *Journal of Mathematical Biology* **27**, 3, pp. 233–258.

Castillo-Chavez, C., Huang, W. and Li, J. (1995). Dynamics of multiple pathogen strains in heterosexual epidemiological models, in *Differential equations and applications to biology and industry* (World Scientific Publishing Co.), pp. 289–298.

Castillo-Chavez, C., Huang, W. and Li, J. (1996). Competitive exclusion in gonorrhea models and other sexually transmitted diseases, *SIAM Journal on Applied Mathematics* **56**, 2, pp. 494–508.

Castillo-Chavez, C. and Velasco-Hernandez, J. X. (1998). On the relationship between evolution of virulence and host demography, *Journal of Theoretical Biology* **192**, 4, pp. 437–444.

Castillo-Chavez, C., Velasco-Hernandez, J. X. and Fridman, S. (1994). *Modeling contact structures in biology* (Springer).

Castillo-Chavez, C. *et al.* (1989b). *Mathematical and statistical approaches to AIDS epidemiology* (Springer).

Cauchemez, S., Carrat, F., Viboud, C., Valleron, A. and Boelle, P. (2004). A Bayesian MCMC approach to study transmission of influenza: application to household longitudinal data, *Statistics in Medicine* **23**, 22, pp. 3469–3487.

CDC (2010). Seasonal influenza vaccine supply and distribution in the united states, http://www.cdc.gov/flu/about/qa/vaxdistribution.htm.

CDC (2011). Weekly influenza surveillance report 2009-2010 influenza season, Accessed 1 June 2011, http://www.cdc.gov/flu/weekly/weeklyarchives2009-2010/weekly20.htm.

CDC (2013). The influenza (flu) viruses, http://www.cdc.gov/flu/about/viruses/.

Chin, T. D., Foley, J. F., Doto, I. L., Gravelle, C. R. and Weston, J. (1960). Morbidity and mortality characteristics of asian strain influenza, *Public Health Reports* **75**, 2, p. 149.

Chitsulo, L., Engels, D., Montresor, A. and Savioli, L. (2000). The global status of schistosomiasis and its control, *Acta Tropica* **77**, 1, pp. 41–51.

Chow, L., Fan, M. and Feng, Z. (2011). Dynamics of a multigroup epidemiological model with group-targeted vaccination strategies, *Journal of Theoretical Biology* **291**, pp. 56–64.

Chowell, G., Fenimore, P. W., Castillo-Garsow, M. A. and Castillo-Chavez, C. (2003). SARS outbreaks in Ontario, Hong Kong and Singapore: the role of diagnosis and isolation as a control mechanism, *Journal of Theoretical Biology* **224**, 1, pp. 1–8.

Cohen, M. S., Hellmann, N., Levy, J. A., DeCock, K. and Lange, J. (2008). The spread, treatment, and prevention of hiv-1: evolution of a global pandemic, *The Journal of Clinical Investigation* **118**, 4, p. 1244.

Colizza, V., Barrat, A., Barthelemy, M., Valleron, A.-J. and Vespignani, A. (2007). Modeling the worldwide spread of pandemic influenza: baseline case and containment interventions, *PLoS Medicine* **4**, 1, p. e13.

Corey, L., Wald, A., Patel, R., Sacks, S. L., Tyring, S. K., Warren, T., Douglas Jr, J. M., Paavonen, J., Morrow, R. A., Beutner, K. R. *et al.* (2004). Once-daily valacyclovir to reduce the risk of transmission of genital herpes, *New England Journal of Medicine* **350**, 1, pp. 11–20.

Couch, R. B. and Kasel, J. A. (1983). Immunity to influenza in man, *Annual Reviews in Microbiology* **37**, 1, pp. 529–549.

Cunningham, J. (1979). A deterministic model for measles. *Zeitschrift für Naturforschung. Section C: Biosciences* **34**, 7-8, p. 647.

Davies, C., Webster, J. and Woolhouse, M. (2001). Trade–offs in the evolution of virulence in an indirectly transmitted macroparasite, *Proceedings of the Royal Society of London. Series B: Biological Sciences* **268**, 1464, pp. 251–257.

Day, T., Park, A., Madras, N., Gumel, A. and Wu, J. (2006). When is quarantine a useful control strategy for emerging infectious diseases, *American Journal of Epidemiology* **163**, 5, pp. 479–485.

De Boer, R. J. and Perelson, A. S. (1998). Target cell limited and immune control models of hiv infection: a comparison, *Journal of Theoretical Biology* **190**, 3, pp. 201–214.

De Jong, M., Diekmann, O. and Heesterbeek, J. (1994). The computation of $\mathcal{R}_0$ for discrete-time epidemic models with dynamic heterogeneity, *Mathematical Biosciences* **119**, 1, pp. 97–114.

De Jong, M. C., Diekmann, O. and Heesterbeek, H. (1995). How does transmission of infection depend on population size, *Epidemic models: their structure and relation to data* **5**, 2, pp. 84–94.

Del Valle, S. Y., Hyman, J., Hethcote, H. W. and Eubank, S. G. (2007). Mixing patterns between age groups in social networks, *Social Networks* **29**, 4, pp. 539–554.

Diekmann, O. and Heesterbeek, J. (2000). *Mathematical epidemiology of infectious diseases* (Wiley Chichester).

Diekmann, O., Heesterbeek, J. and Metz, J. A. (1990). On the definition and the computation of the basic reproduction ratio $\mathcal{R}_0$ in models for infectious diseases in heterogeneous populations, *Journal of Mathematical Biology* **28**, 4, pp. 365–382.

Dietz, K. (1976). The incidence of infectious diseases under the influence of seasonal fluctuations, in *Mathematical models in medicine* (Springer), pp. 1–15.

Dietz, K., Molineaux, L. and Thomas, A. (1974). A malaria model tested in the african savannah, *Bulletin of the World Health Organization* **50**, 3-4, p. 347.

Dietz, K. and Schenzle, D. (1985a). Mathematical models for infectious disease statistics, in *A celebration of statistics* (Springer), pp. 167–204.

Dietz, K. and Schenzle, D. (1985b). Proportionate mixing models for age-dependent infection transmission, *Journal of Bathematical Biology* **22**, 1, pp. 117–120.

Dobson, A. (1988). The population biology of parasite-induced changes in host behavior, *Quarterly Review of Biology* **63**, 2, pp. 139–165.

Doedel, E. J. (1981). Auto: A program for the automatic bifurcation analysis of autonomous systems, *Congressus Numerantium* **30**, pp. 265–284.

Drakeley, C., Secka, I., Correa, S., Greenwood, B. and Targett, G. (1999). Host haematological factors influencing the transmission of plasmodium falciparum gametocytes to anopheles gambiae ss mosquitoes, *Tropical Medicine & International Health* **4**, 2, pp. 131–138.

Dushoff, J., Plotkin, J. B., Viboud, C., Simonsen, L., Miller, M., Loeb, M. and David, J. (2007). Vaccinating to protect a vulnerable subpopulation, *PLoS Medicine* **4**, 5, p. e174.

Earn, D. J., Dushoff, J. and Levin, S. A. (2002). Ecology and evolution of the flu, *Trends in Ecology & Evolution* **17**, 7, pp. 334–340.

Farrington, C. (1990). Modeling risks of infection for measles, mumps and rubella, *Stat. Med.* **9**, p. 953.

Feng, Z. (1994). *A mathematical model for the dynamics of childhood dieseases under the impact of isolation*, Ph.D. thesis, Arizona State University.

Feng, Z. (2007). Final and peak epidemic sizes for seir models with quarantine and isolation. *Mathematical biosciences and engineering: MBE* **4**, 4, p. 675.

Feng, Z., Eppert, A., Milner, F. A. and Minchella, D. (2004a). Estimation of parameters governing the transmission dynamics of schistosomes, *Applied Mathematics Letters*

17, 10, pp. 1105–1112.

Feng, Z., Huang, W. and Castillo-Chavez, C. (2001). On the role of variable latent periods in mathematical models for tuberculosis, *Journal of Dynamics and Differential Equations* **13**, 2, pp. 425–452.

Feng, Z., Iannelli, M. and Milner, F. A. (2002). A two-strain tuberculosis model with age of infection, *SIAM Journal on Applied Mathematics* **62**, 5, pp. 1634–1656.

Feng, Z., Qiu, Z., Sang, Z., Lorenzo, C. and Glasser, J. (2013). Modeling the synergy between HSV-2 and HIV and potential impact of HSV-2 therapy, *Mathematical Biosciences* **245**, pp. 171–187.

Feng, Z., Smith, D. L., Ellis McKenzie, F. and Levin, S. A. (2004b). Coupling ecology and evolution: malaria and the S-gene across time scales, *Mathematical Biosciences* **189**, 1, pp. 1–19.

Feng, Z. and Thieme, H. R. (1995). Recurrent outbreaks of childhood diseases revisited: the impact of isolation, *Mathematical Biosciences* **128**, 1, pp. 93–130.

Feng, Z. and Thieme, H. R. (2000a). Endemic models with arbitrarily distributed periods of infection i: Fundamental properties of the model, *SIAM Journal on Applied Mathematics* **61**, 3, pp. 803–833.

Feng, Z. and Thieme, H. R. (2000b). Endemic models with arbitrarily distributed periods of infection ii: Fast disease dynamics and permanent recovery, *SIAM Journal on Applied Mathematics* **61**, 3, pp. 983–1012.

Feng, Z., Towers, S. and Yang, Y. (2011). Modeling the effects of vaccination and treatment on pandemic influenza, *The AAPS Journal* **13**, 3, pp. 427–437.

Feng, Z., Velasco-Hernandez, J. and Tapia-Santos, B. (2012a). A mathematical model for coupling within-host and between-host dynamics in an environmentally-driven infectious disease, *Mathematical Biosciences* **241**, 1, pp. 49–55.

Feng, Z., Velasco-Hernandez, J., Tapia-Santos, B. and Leite, M. C. A. (2012b). A model for coupling within-host and between-host dynamics in an infectious disease, *Nonlinear Dynamics* **68**, 3, pp. 401–411.

Feng, Z. and Velasco-Hernandez, J. X. (1997). Competitive exclusion in a vector-host model for the dengue fever, *Journal of Mathematical Biology* **35**, 5, pp. 523–544.

Feng, Z., Xu, D. and Zhao, H. (2007). Epidemiological models with non-exponentially distributed disease stages and applications to disease control, *Bulletin of Mathematical Biology* **69**, 5, pp. 1511–1536.

Feng, Z., Yang, Y., Xu, D., Zhang, P., McCauley, M. M. and Glasser, J. W. (2009). Timely identification of optimal control strategies for emerging infectious diseases, *Journal of Theoretical Biology* **259**, 1, pp. 165–171.

Feng, Z., Yi, Y. and Zhu, H. (2004c). Fast and slow dynamics of malaria and the s-gene frequency, *Journal of Dynamics and Differential Equations* **16**, 4, pp. 869–896.

Ferguson, N. M., Cummings, D. A., Fraser, C., Cajka, J. C., Cooley, P. C. and Burke, D. S. (2006). Strategies for mitigating an influenza pandemic, *Nature* **442**, 7101, pp. 448–452.

Ferguson, N. M., Galvani, A. P. and Bush, R. M. (2003). Ecological and immunological determinants of influenza evolution, *Nature* **422**, 6930, pp. 428–433.

Fiore, A. E., Uyeki, T. M., Broder, K., Finelli, L., Euler, G. L., Singleton, J. A., Iskander, J. K., Wortley, P. M., Shay, D. K., Bresee, J. S. *et al.* (2009). Prevention and control of influenza with vaccines, *MMWR* http://www.cdc.gov/mmwr/preview/mmwrhtml/rr59e0729a1.htm.

Fister, K. R., Lenhart, S. and McNally, J. S. (1998). Optimizing chemotherapy in an hiv model, *Electronic Journal of Differential Equations* **1998**, 32, pp. 1–12.

Fitch, W. M., Bush, R. M., Bender, C. A. and Cox, N. J. (1997). Long term trends in the evolution of h (3) ha1 human influenza type a, *Proceedings of the National Academy of Sciences* **94**, 15, pp. 7712–7718.

Flahault, A., Letrait, S., Blin, P., Hazout, S., Menares, J. and Valleron, A. (1988). Modelling the 1985 influenza epidemic in france, *Statistics in Medicine* **7**, 11, pp. 1147–1155.

Fleming, W. H. and Rishel, R. W. (1975). *Deterministic and stochastic optimal control*, Vol. 1 (Springer New York).

Foss, A. M., Vickerman, P. T., Chalabi, Z., Mayaud, P., Alary, M. and Watts, C. H. (2009). Dynamic modeling of herpes simplex virus type-2 (HSV-2) transmission: issues in structural uncertainty, *Bulletin of Mathematical Biology* **71**, 3, pp. 720–749.

Foss, A. M., Vickerman, P. T., Mayaud, P., Weiss, H. A., Ramesh, B., Reza-Paul, S., Washington, R., Blanchard, J., Moses, S., Lowndes, C. M. *et al.* (2011). Modelling the interactions between herpes simplex virus type 2 and HIV: implications for the HIV epidemic in southern India, *Sexually Transmitted Infections* **87**, 1, pp. 22–27.

Fox, J. P., Cooney, M. K., Hall, C. E. and Foy, H. M. (1982). Influenzavirus infections in Seattle families, 1975–1979 II. Pattern of infection in invaded households and relation of age and prior antibody to occurrence of infection and related illness, *American Journal of Epidemiology* **116**, 2, pp. 228–242.

Frank, S. A. (1992). A kin selection model for the evolution of virulence, *Proceedings of the Royal Society of London. Series B: Biological Sciences* **250**, 1329, pp. 195–197.

Fraser, C., Riley, S., Anderson, R. M. and Ferguson, N. M. (2004). Factors that make an infectious disease outbreak controllable, *Proceedings of the National Academy of Sciences of the United States of America* **101**, 16, pp. 6146–6151.

Gao, L. Q., Mena-Lorca, J. and Hethcote, H. W. (1995). Four sei endemic models with periodicity and separatrices, *Mathematical Biosciences* **128**, 1, pp. 157–184.

Gerberding, J. L. (2003). Faster... but fast enough? responding to the epidemic of severe acute respiratory syndrome, *New England Journal of Medicine* **348**, 20, pp. 2030–2031.

Germann, T. C., Kadau, K., Longini, I. M. and Macken, C. A. (2006). Mitigation strategies for pandemic influenza in the united states, *Proceedings of the National Academy of Sciences* **103**, 15, pp. 5935–5940.

Gilchrist, M. A. and Sasaki, A. (2002). Modeling host–parasite coevolution: a nested approach based on mechanistic models, *Journal of Theoretical Biology* **218**, 3, pp. 289–308.

Glass, R. J., Glass, L. M., Beyeler, W. E., Min, H. J. *et al.* (2006). Targeted social distancing design for pandemic influenza, *Emerging Infectious Diseases* **12**, 11, pp. 1671–1681.

Glasser, J., Feng, Z., Moylan, A., Del Valle, S. and Castillo-Chavez, C. (2012). Mixing in age-structured population models of infectious diseases, *Mathematical Biosciences* **235**, 1, pp. 1–7.

Glasser, J., Taneri, D., Feng, Z., Chuang, J.-H., Tüll, P., Thompson, W., McCauley, M. M. and Alexander, J. (2010). Evaluation of targeted influenza vaccination strategies via population modeling, *PLoS One* **5**, 9, p. e12777.

Glezen, W. P. (1996). Emerging infections: pandemic influenza, *Epidemiologic Reviews* **18**, 1, pp. 64–76.

Glezen, W. P., Couch, R. B. and Six, H. R. (1982). The influenza herald wave, *American Journal of Epidemiology* **116**, 4, pp. 589–598.

Govaert, T. M., Thijs, C., Masurel, N., Sprenger, M., Dinant, G. and Knottnerus, J. (1994). The efficacy of influenza vaccination in elderly individuals, *JAMA* **272**, 21, pp. 1661–1665.

Gower, C. M. and Webster, J. P. (2004). Fitness of indirectly transmitted pathogens: restraint and constraint, *Evolution* **58**, 6, pp. 1178–1184.

Gower, C. M. and Webster, J. P. (2005). Intraspecific competition and the evolution of virulence in a parasitic trematode, *Evolution* **59**, 3, pp. 544–553.

Grossman, Z. (1980). Oscillatory phenomena in a model of infectious diseases, *Theoretical Population Biology* **18**, 2, pp. 204–243.

Guckenheimer, J. and Holmes, P. (1983). *Nonlinear oscillations, dynamical systems, and bifurcations of vector fields* (Springer).

Gupta, S., Swinton, J. and Anderson, R. M. (1994). Theoretical studies of the effects of heterogeneity in the parasite population on the transmission dynamics of malaria, *Proceedings of the Royal Society of London. Series B: Biological Sciences* **256**, 1347, pp. 231–238.

Haber, M. J., Shay, D. K., Davis, X. M., Patel, R., Jin, X., Weintraub, E., Orenstein, E. and Thompson, W. W. (2007). Effectiveness of interventions to reduce contact rates during a simulated influenza pandemic, *Emerging Infectious Diseases* **13**, 4, p. 581.

Hadeler, K. (1982). An integral equation for helminthic infections: Stability of the non-infected population, *Trends in theoretical and practical nonlinear differential equations, Lecture Notes in Pure and Applied Mathematics* **90**.

Hadeler, K. and Dietz, K. (1983). Nonlinear hyperbolic partial differential equations for the dynamics of parasite populations, *Computers & Mathematics with Applications* **9**, 3, pp. 415–430.

Hadeler, K. and Müller, J. (1996). Vaccination in age structured populations ii: optimal strategies, in *Models for infectious human diseases: Their structure and relation to data* (Cambridge University Press), p. 102.

Hamer, W. H. (1906). *The Milroy lectures on epidemic disease in England: The evidence of variability and of persistency of type* (Bedford Press).

Hannoun, C., Megas, F. and Piercy, J. (2004). Immunogenicity and protective efficacy of influenza vaccination, *Virus Research* **103**, 1, pp. 133–138.

Hasting, I. (2000). Models of human genetic disease: how biased are the standard formulae? *Genetical Research* **75**, 1, pp. 107–114.

Heesterbeek, J. (2000). *Mathematical Epidemiology of Infectious Diseases: Model Building, Analysis and Interpretation*, Vol. 5 (John Wiley & Sons).

Helton, J. C., Johnson, J. D., Sallaberry, C. J. and Storlie, C. B. (2006). Survey of sampling-based methods for uncertainty and sensitivity analysis, *Reliability Engineering & System Safety* **91**, 10, pp. 1175–1209.

Hernandez-Ceron, N., Feng, Z. and Castillo-Chavez, C. (2013a). Discrete epidemic models with arbitrary stage distributions and applications to disease control, *Bulletin of Mathematical Biology* **75**, 10, pp. 1716–1746.

Hernandez-Ceron, N., Feng, Z. and van den Driessche, P. (2013b). Reproduction numbers for discrete-time epidemic models with arbitrary stage distributions, *Journal of Difference Equations and Applications* **19**, 10, pp. 1671–1693.

Hethcote, H. (1996). *Modeling heterogeneous mixing in infectious disease dynamics* (Cambridge University Press).

Hethcote, H., Zhien, M. and Shengbing, L. (2002). Effects of quarantine in six endemic models for infectious diseases, *Mathematical Biosciences* **180**, 1, pp. 141–160.

Hethcote, H. W. (1976). Qualitative analyses of communicable disease models, *Mathematical Biosciences* **28**, 3, pp. 335–356.

Hethcote, H. W. (2000). The mathematics of infectious diseases, *SIAM Review* **42**, 4, pp. 599–653.

Hethcote, H. W. and Levin, S. A. (1989). Periodicity in epidemiological models, *Applied*

Mathematical Ecology **18**, pp. 193–211.

Hethcote, H. W., Stech, H. W. and Van Den Driessche, P. (1981). Nonlinear oscillations in epidemic models, *SIAM Journal on Applied Mathematics* **40**, 1, pp. 1–9.

Hethcote, H. W. and Tudor, D. W. (1980). Integral equation models for endemic infectious diseases, *Journal of Mathematical Biology* **9**, 1, pp. 37–47.

Hethcote, H. W. and Yorke, J. A. (1984). *Gonorrhea transmission dynamics and control*, Vol. 56 (Springer Berlin).

Hopewell, P. C. (1994). Overview of clinical tuberculosis, in *Tuberculosis: Pathogenesis, protection, and control* (ASM, Washington, DC), pp. 25–46.

Hoppensteadt, F. (1974). An age structured epidemic model, *J. Franklin Inst.* **197**, pp. 325–333.

Hoppensteadt, F. (1975). *Mathematical theories of populations: demographics, genetics and epidemics*, Vol. 20 (SIAM).

Hoyle, A. and Bowers, R. G. (2008). Can possible evolutionary outcomes be determined directly from the population dynamics? *Theoretical Population Biology* **74**, 4, pp. 311–323.

Hsieh, Y.-H., King, C.-C., Chen, C. W., Ho, M.-S., Lee, J.-Y., Liu, F.-C., Wu, Y.-C. and JulianWu, J.-S. (2005). Quarantine for SARS, Taiwan, *Emerging Infectious Diseases* **11**, 2, p. 278.

Hurford, A., Cownden, D. and Day, T. (2010). Next-generation tools for evolutionary invasion analyses, *Journal of the Royal Society Interface* **7**, 45, pp. 561–571.

Jacquez, J. A., Simon, C. P., Koopman, J., Sattenspiel, L. and Perry, T. (1988). Modeling and analyzing hiv transmission: the effect of contact patterns, *Mathematical Biosciences* **92**, 2, pp. 119–199.

Jung, E., Lenhart, S. and Feng, Z. (2002). Optimal control of treatments in a two-strain tuberculosis model, *Discrete and Continuous Dynamical Systems Series B* **2**, 4, pp. 473–482.

Keeling, M. J. and Eames, K. T. (2005). Networks and epidemic models, *Journal of the Royal Society Interface* **2**, 4, pp. 295–307.

Kent, J. (1993). The epidemiology of multidrug-resistant tuberculosis in the united states. *The Medical Clinics of North America* **77**, 6, pp. 1391–1409.

King, J. C., Cummings, G. E., Stoddard, J., Readmond, B. X., Magder, L. S., Stong, M., Hoffmaster, M., Rubin, J., Tsai, T., Ruff, E. *et al.* (2005). A pilot study of the effectiveness of a school-based influenza vaccination program, *Pediatrics* **116**, 6, pp. e868–e873.

King, J. C., Haugh, C. J., Dupont, W. D., Thompson, J. M., Wright, P. F. and Edwards, K. M. (1988). Laboratory and epidemiologic assessment of a recent influenza b outbreak, *Journal of Medical Virology* **25**, 3, pp. 361–368.

King Jr, J. C., Stoddard, J. J., Gaglani, M. J., Moore, K. A., Magder, L., McClure, E., Rubin, J. D., Englund, J. A. and Neuzil, K. (2006). Effectiveness of school-based influenza vaccination, *New England Journal of Medicine* **355**, 24, pp. 2523–2532.

Kirschner, D. (1999). Dynamics of Co-infection with *M. tuberculosis* and HIV-1, *Theoretical population biology* **55**, 1, pp. 94–109.

Kirschner, D., Lenhart, S. and Serbin, S. (1997). Optimal control of the chemotherapy of HIV, *Journal of Mathematical Biology* **35**, 7, pp. 775–792.

Kochi, A. (2001). The global tuberculosis situation and the new control strategy of the world health organization, *Bulletin of the World Health Organization* **79**, 1, pp. 71–75.

Kolata, G. (1995). First documented case of TB passed on airliner is reported by the US, Accessed 10 January 2014, http://www.nytimes.com/1995/03/03/us/

`first-documented-case-of-tb-passed-on-airliner-is-reported-by-the-us.`
`html`.

Le Hesran, J. Y., Personne, I., Personne, P., Fievet, N., Dubois, B., Beyemé, M., Boudin, C., Cot, M. and Deloron, P. (1999). Longitudinal study of plasmodium falciparum infection and immune responses in infants with or without the sickle cell trait. *International Journal of Epidemiology* **28**, 4, pp. 793–798.

Lee, N., Hui, D., Wu, A., Chan, P., Cameron, P., Joynt, G. M., Ahuja, A., Yung, M. Y., Leung, C., To, K. *et al.* (2003). A major outbreak of severe acute respiratory syndrome in hong kong, *New England Journal of Medicine* **348**, 20, pp. 1986–1994.

Lee, V. J., Yap, J., Ong, J. B., Chan, K.-P., Lin, R. T., Chan, S. P., Goh, K. T., Leo, Y.-S., Mark, I. and Chen, C. (2009). Influenza excess mortality from 1950–2000 in tropical singapore, *PLoS One* **4**, 12, p. e8096.

Lell, B., May, J., Schmidt-Ott, R. J., Lehman, L. G., Luckner, D., Greve, B., Matousek, P., Schmid, D., Herbich, K., Mockenhaupt, F. P. *et al.* (1999). The role of red blood cell polymorphisms in resistance and susceptibility to malaria, *Clinical Infectious Diseases* **28**, 4, pp. 794–799.

Leo, Y., Chen, M., Heng, B., Lee, C., Paton, N., Ang, B., Choo, P., Lim, S., Ling, A., Ling, M. *et al.* (2003). Severe acute respiratory syndrome–Singapore, 2003, *MMWR* **52**, pp. 405–411.

Lessler, J., Reich, N. G., Brookmeyer, R., Perl, T. M., Nelson, K. E. and Cummings, D. A. (2009). Incubation periods of acute respiratory viral infections: A systematic review, *The Lancet Infectious Diseases* **9**, 5, pp. 291–300.

Levin, S. and Pimentel, D. (1981). Selection of intermediate rates of increase in parasite-host systems, *American Naturalist* **117**, 3, pp. 308–315.

Lewis, M. A., Rencławowicz, J., van Den Driessche, P. and Wonham, M. (2006). A comparison of continuous and discrete-time west nile virus models, *Bulletin of Mathematical Biology* **68**, 3, pp. 491–509.

Lipsitch, M., Cohen, T., Cooper, B., Robins, J. M., Ma, S., James, L., Gopalakrishna, G., Chew, S. K., Tan, C. C., Samore, M. H. *et al.* (2003). Transmission dynamics and control of severe acute respiratory syndrome, *Science* **300**, 5627, pp. 1966–1970.

Lipsitch, M., Cohen, T., Murray, M. and Levin, B. R. (2007). Antiviral resistance and the control of pandemic influenza, *PLoS Medicine* **4**, 1, p. e15.

Lipsitch, M. and Nowak, M. A. (1995). The evolution of virulence in sexually transmitted HIV/AIDS, *Journal of Theoretical Biology* **174**, 4, pp. 427–440.

Liu, W.-m., Hethcote, H. W. and Levin, S. A. (1987). Dynamical behavior of epidemiological models with nonlinear incidence rates, *Journal of Mathematical Biology* **25**, 4, pp. 359–380.

Liu, W.-m., Levin, S. A. and Iwasa, Y. (1986). Influence of nonlinear incidence rates upon the behavior of SIRS epidemiological models, *Journal of Mathematical Biology* **23**, 2, pp. 187–204.

Lloyd, A. L. (2001a). Destabilization of epidemic models with the inclusion of realistic distributions of infectious periods, *Proceedings of the Royal Society of London. Series B: Biological Sciences* **268**, 1470, pp. 985–993.

Lloyd, A. L. (2001b). Realistic distributions of infectious periods in epidemic models: changing patterns of persistence and dynamics, *Theoretical Population Biology* **60**, 1, pp. 59–71.

Lo, J. Y., Tsang, T. H., Leung, Y.-H., Yeung, E. Y., Wu, T., Lim, W. W. *et al.* (2005). Respiratory infections during sars outbreak, hong kong, 2003, *Emerging Infectious Diseases* **11**, 11, p. 1738.

Loeb, M., Russell, M. L., Moss, L., Fonseca, K., Fox, J., Earn, D. J., Aoki, F., Horsman,

G., Van Caeseele, P., Chokani, K. *et al.* (2010). Effect of influenza vaccination of children on infection rates in Hutterite communities, *JAMA* **303**, 10, pp. 943–950.

London, W. P. and Yorke, J. A. (1973). Recurrent outbreaks of measles, chickenpox and mumps i. seasonal variation in contact rates, *American Journal of Epidemiology* **98**, 6, pp. 453–468.

Longini, I. M., Halloran, M. E., Nizam, A. and Yang, Y. (2004). Containing pandemic influenza with antiviral agents, *American Journal of Epidemiology* **159**, 7, pp. 623–633.

Longini, I. M., Nizam, A., Xu, S., Ungchusak, K., Hanshaoworakul, W., Cummings, D. A. and Halloran, M. E. (2005). Containing pandemic influenza at the source, *Science* **309**, 5737, pp. 1083–1087.

Luk, J., Gross, P. and Thompson, W. W. (2001). Observations on mortality during the 1918 influenza pandemic, *Clinical Infectious Diseases* **33**, 8, pp. 1375–1378.

Ma, J. and Earn, D. J. (2006). Generality of the final size formula for an epidemic of a newly invading infectious disease, *Bulletin of Mathematical Biology* **68**, 3, pp. 679–702.

Macdonald, G. (1957). *The epidemiology and control of malaria* (London, Oxford Univ. Press).

Manuel, O., Pascual, M., Hoschler, K., Giulieri, S., Alves, D., Ellefsen, K., Bart, P.-A., Venetz, J.-P., Calandra, T. and Cavassini, M. (2011). Humoral response to the influenza a h1n1/09 monovalent as03-adjuvanted vaccine in immunocompromised patients, *Clinical Infectious Diseases* **52**, 2, pp. 248–256.

Marino, S., Hogue, I., Ray, C. and Kirschner, D. (2008). A methodology for performing global uncertainty and sensitivity analysis in systems biology, *Journal of theoretical biology* **254**, 1, pp. 178–196.

Marsh, K. and Snow, R. (1999). Malaria transmission and morbidity. *Parassitologia* **41**, 1-3, p. 241.

Massara, C. L., Peixoto, S. V., Barros, H. d. S., Enk, M. J., Carvalho, O. d. S. and Schall, V. (2004). Factors associated with schistosomiasis mansoni in a population from the municipality of Jaboticatubas, state of Minas Gerais, Brazil, *Memórias do Instituto Oswaldo Cruz* **99**, pp. 127–134.

May, R. M. (2004). Uses and abuses of mathematics in biology, *Science* **303**, 5659, pp. 790–793.

May, R. M. and Anderson, R. (1983). Epidemiology and genetics in the coevolution of parasites and hosts, *Proceedings of the Royal society of London. Series B. Biological sciences* **219**, 1216, pp. 281–313.

May, R. M. and Nowak, M. A. (1994). Superinfection, metapopulation dynamics, and the evolution of diversity, *Journal of Theoretical Biology* **170**, 1, pp. 95–114.

May, R. M. and Nowak, M. A. (1995). Coinfection and the evolution of parasite virulence, *Proceedings of the Royal Society of London. Series B: Biological Sciences* **261**, 1361, pp. 209–215.

McKenzie, F. E., Ferreira, M. U., Baird, J. K., Snounou, G. and Bossert, W. H. (2001a). Meiotic recombination, cross-reactivity, and persistence in plasmodium falciparum, *Evolution* **55**, 7, pp. 1299–1307.

McKenzie, F. E., Killeen, G. F., Beier, J. C. and Bossert, W. H. (2001b). Seasonality, parasite diversity, and local extinctions in plasmodium falciparum malaria, *Ecology* **82**, 10, pp. 2673–2681.

McNeill, W. H. (1976). Plagues and peoples. anchor, *Garden City, NY* **51**.

Medlock, J. and Galvani, A. P. (2009). Optimizing influenza vaccine distribution, *Science* **325**, 5948, pp. 1705–1708.

Mena-Lorca, J., Velasco-Hernandez, J. X. and Castillo-Chavez, C. (1999). Density-dependent dynamics and superinfection in an epidemic model, *Mathematical Medicine and Biology* **16**, 4, pp. 307–317.

Metz, J., Nisbet, R. and Geritz, S. (1992). How should we define fitness for general ecological scenarios? *Trends in Ecology & Evolution* **7**, 6, pp. 198–202.

Miller, B. (1993). Preventive therapy for tuberculosis. *The Medical clinics of North America* **77**, 6, p. 1263.

Miller, M., White, A. and Boots, M. (2005). The evolution of host resistance: tolerance and control as distinct strategies, *Journal of Theoretical Biology* **236**, 2, pp. 198–207.

Miller, R. K. (1971). *Nonlinear Volterra integral equations*, 48 (WA Benjamin Menlo Park, California).

Minchella, D. (1985). Host life-history variation in response to parasitism, *Parasitology* **90**, 01, pp. 205–216.

Minchella, D. J. and Loverde, P. T. (1981). A cost of increased early reproductive effort in the snail biomphalaria glabrata, *The American Naturalist* **118**, 6, pp. 876–881.

MMWR, M. (1999). Mortality weekly report, *CDC Surveillance Summaries* **48**.

Molinari, N.-A. M., Ortega-Sanchez, I. R., Messonnier, M. L., Thompson, W. W., Wortley, P. M., Weintraub, E. and Bridges, C. B. (2007). The annual impact of seasonal influenza in the us: measuring disease burden and costs, *Vaccine* **25**, 27, pp. 5086–5096.

Monath, T. P. *et al.* (1989). *The arboviruses: epidemiology and ecology. Volume V.* (CRC Press, Inc.).

Monto, A. S., Davenport, F. M., Napier, J. A. and Francis, T. (1970). Modification of an outbreak of influenza in tecumseh, michigan by vaccination of schoolchildren, *Journal of Infectious Diseases* **122**, 1-2, pp. 16–25.

Moscona, A. (2005). Neuraminidase inhibitors for influenza, *New England Journal of Medicine* **353**, 13, pp. 1363–1373.

Mosquera, J. and Adler, F. R. (1998). Evolution of virulence: a unified framework for coinfection and superinfection, *Journal of Theoretical Biology* **195**, 3, pp. 293–313.

Mossong, J., Hens, N., Jit, M., Beutels, P., Auranen, K., Mikolajczyk, R., Massari, M., Salmaso, S., Tomba, G. S., Wallinga, J. *et al.* (2008). Social contacts and mixing patterns relevant to the spread of infectious diseases, *PLoS Medicine* **5**, 3, p. e74.

Mukandavire, Z. and Garira, W. (2007a). Age and sex structured model for assessing the demographic impact of mother-to-child transmission of HIV/AIDS, *Bulletin of Mathematical Biology* **69**, 6, pp. 2061–2092.

Mukandavire, Z. and Garira, W. (2007b). Sex-structured HIV/AIDS model to analyse the effects of condom use with application to Zimbabwe, *Journal of Mathematical Biology* **54**, 5, pp. 669–699.

Nakajima, S., Nishikawa, F., Nakamura, K. and Nakajima, K. (1992). Comparison of the ha genes of type b influenza viruses in herald waves and later epidemic seasons, *Epidemiology and Infection* **109**, pp. 559–559.

Nardell, E. A. (1995). Interrupting transmission from patients with unsuspected tuberculosis: a unique role for upper-room ultraviolet air disinfection, *American Journal of Infection Control* **23**, 2, pp. 156–164.

Naresh, R. and Tripathi, A. (2005). Modelling and analysis of HIV-TB co-infection in a variable size population, *Mathematical Modelling and Analysis* **10**, 3, pp. 275–286.

Newman, M. E. (2003). The structure and function of complex networks, *SIAM Review* **45**, 2, pp. 167–256.

Newton, E. A. and Kuder, J. M. (2000). A model of the transmission and control of genital herpes, *Sexually Transmitted Diseases* **27**, 7, pp. 363–370.

Nichol, K. L., Nordin, J. D., Nelson, D. B., Mullooly, J. P. and Hak, E. (2007). Effectiveness of influenza vaccine in the community-dwelling elderly, *New England Journal of Medicine* **357**, 14, pp. 1373–1381.

Nold, A. (1980). Heterogeneity in disease-transmission modeling, *Mathematical Biosciences* **52**, 3, pp. 227–240.

Nowak, M. and May, R. M. (2000). *Virus dynamics: mathematical principles of immunology and virology* (Oxford University Press).

Nowak, M. A. and May, R. M. (1994). Superinfection and the evolution of parasite virulence, *Proceedings of the Royal Society of London. Series B: Biological Sciences* **255**, 1342, pp. 81–89.

Ntoumi, F., Mercereau-Puijalon, O., Ossari, S., Luty, A., Reltien, J., Georges, A. and Millet, P. (1997a). *Plasmodium falciparum*: Sickle-cell trait is associated with higher prevalence of multiple infections in gabonese children with asymptomatic infections, *Experimental Parasitology* **87**, 1, pp. 39–46.

Ntoumi, F., Rogier, C., Dieye, A., Trape, J.-F., Millet, P. and Mercereau-Puijalon, O. (1997b). Imbalanced distribution of Plasmodium falciparum MSP-1 genotypes related to sickle-cell trait, *Molecular Medicine* **3**, 9, p. 581.

Nuño, M., Feng, Z., Martcheva, M. and Castillo-Chavez, C. (2005). Dynamics of two-strain influenza with isolation and partial cross-immunity, *SIAM Journal on Applied Mathematics* **65**, 3, pp. 964–982.

Peiris, J., Chu, C., Cheng, V., Chan, K., Hung, I., Poon, L., Law, K., Tang, B., Hon, T., Chan, C. *et al.* (2003). Clinical progression and viral load in a community outbreak of coronavirus-associated SARS pneumonia: A prospective study, *The Lancet* **361**, 9371, pp. 1767–1772.

Plant, R. E. and Wilson, L. (1986). Models for age structured populations with distributed maturation rates, *Journal of Mathematical Biology* **23**, 2, pp. 247–262.

Pontrëïiagin, L. S. (1962). *The mathematical theory of optimal processes*, Vol. 4 (CRC Press).

Porco, T. C., Small, P. M., Blower, S. M. *et al.* (2001). Amplification dynamics: predicting the effect of HIV on tuberculosis outbreaks, *JAIDS* **28**, 5, pp. 437–444.

Qiu, Z. and Feng, Z. (2010). Transmission dynamics of an influenza model with vaccination and antiviral treatment, *Bulletin of Mathematical Biology* **72**, 1, pp. 1–33.

Raimundo, S. M., Engel, A. B., Yang, H. M. and Bassanezi, R. C. (2003). An approach to estimating the transmission coefficients for AIDS and for tuberculosis using mathematical models, *Systems Analysis Modelling Simulation* **43**, 4, pp. 423–442.

Rambaut, A., Pybus, O. G., Nelson, M. I., Viboud, C., Taubenberger, J. K. and Holmes, E. C. (2008). The genomic and epidemiological dynamics of human influenza a virus, *Nature* **453**, 7195, pp. 615–619.

Reichert, T. A., Sugaya, N., Fedson, D. S., Glezen, W. P., Simonsen, L. and Tashiro, M. (2001). The japanese experience with vaccinating schoolchildren against influenza, *New England Journal of Medicine* **344**, 12, pp. 889–896.

Renshaw, E. (1993). *Modelling biological populations in space and time*, Vol. 11 (Cambridge University Press).

Robert, V., Read, A., Essong, J., Tchuinkam, T., Mulder, B., Verhave, J.-P. and Carnevale, P. (1996). Effect of gametocyte sex ratio on infectivity of *plasmodium falciparum* to *anopheles gambiae*, *Transactions of the Royal Society of Tropical Medicine and Hygiene* **90**, 6, pp. 621–624.

Roeger, L.-I. W., Feng, Z., Castillo-Chavez, C. *et al.* (2009). Modeling TB and HIV co-infections, *Mathematical Biosciences and Engineering* **6**, 4, pp. 815–837.

Rong, L., Feng, Z. and Perelson, A. S. (2007a). Emergence of HIV-1 drug resistance during antiretroviral treatment, *Bulletin of Mathematical Biology* **69**, 6, pp. 2027–2060.

Rong, L., Feng, Z. and Perelson, A. S. (2007b). Mathematical analysis of age-structured hiv-1 dynamics with combination antiretroviral therapy, *SIAM Journal on Applied Mathematics* **67**, 3, pp. 731–756.

Rong, L., Gilchrist, M. A., Feng, Z. and Perelson, A. S. (2007c). Modeling within-host HIV-1 dynamics and the evolution of drug resistance: trade-offs between viral enzyme function and drug susceptibility, *Journal of Theoretical Biology* **247**, 4, pp. 804–818.

Rosner, F. (1995). *Medicine in the Bible and the Talmud*, Vol. 5 (Ktav Pub Incorporated).

Russell, C. A., Jones, T. C., Barr, I. G., Cox, N. J., Garten, R. J., Gregory, V., Gust, I. D., Hampson, A. W., Hay, A. J., Hurt, A. C. et al. (2008). The global circulation of seasonal influenza a (h3n2) viruses, *Science* **320**, 5874, pp. 340–346.

Salyers, A., Abigail and Whitt, D. D. (1994). *Bacterial Pathogenesis: A molecular approach* (ASM Press, Washington, DC).

Sandland, G. J. and Minchella, D. J. (2003a). Costs of immune defense: an enigma wrapped in an environmental cloak, *Trends in Parasitology* **19**, 12, pp. 571–574.

Sandland, G. J. and Minchella, D. J. (2003b). Effects of diet and echinostoma revolutum infection on energy allocation patterns in juvenile lymnaea elodes snails, *Oecologia* **134**, 4, pp. 479–486.

Sandland, G. J. and Minchella, D. J. (2004). Life-history plasticity in hosts (Lymnaea elodes) exposed to differing resources and parasitism, *Canadian Journal of Zoology* **82**, 10, pp. 1672–1677.

Schaffer, W. M. and Kot, M. (1985). Nearly one dimensional dynamics in an epidemic, *Journal of Theoretical Biology* **112**, 2, pp. 403–427.

Schenzle, D. (1984). An age-structured model of pre-and post-vaccination measles transmission, *Mathematical Medicine and Biology* **1**, 2, pp. 169–191.

Schiller, J. S. and Euler, G. (2009). Vaccination coverage estimates from the national health interview survey: United states, 2008, *National Center for Health Statistics and National Center for Immunization and Respiratory Diseases* **July**, http://www.cdc.gov/nchs/data/hestat/vaccine_coverage/vaccine_coverage.pdf.

Schinazi, R. B. (1999). Strategies to control the genital herpes epidemic, *Mathematical Biosciences* **159**, 2, pp. 113–121.

Schulzer, M., Radhamani, M., Grzybowski, S., Mak, E. and Fitzgerald, J. M. (1994). A mathematical model for the prediction of the impact of HIV infection on tuberculosis, *International Journal of Epidemiology* **23**, 2, pp. 400–407.

Schwartz, I. B. and Smith, H. (1983). Infinite subharmonic bifurcation in an seir epidemic model, *Journal of Mathematical Biology* **18**, 3, pp. 233–253.

Shim, E., Feng, Z., Martcheva, M. and Castillo-Chavez, C. (2006). An age-structured epidemic model of rotavirus with vaccination, *Journal of Mathematical Biology* **53**, 4, pp. 719–746.

Simmerman, J. M., Thawatsupha, P., Kingnate, D., Fukuda, K., Chaising, A. and Dowell, S. F. (2004). Influenza in thailand: a case study for middle income countries, *Vaccine* **23**, 2, pp. 182–187.

Smith, H. (1983a). Multiple stable subharmonics for a periodic epidemic model, *Journal of Mathematical Biology* **17**, 2, pp. 179–190.

Smith, H. (1983b). Subharmonic bifurcation in an sir epidemic model, *Journal of Mathematical Biology* **17**, 2, pp. 163–177.

Smith, P. and Moss, A. (1994). *Epidemiology of tuberculosis*, Vol. 47 (ASM Press, Washington, DC).

Smith, T. A., Leuenberger, R. and Lengeler, C. (2001). Child mortality and malaria transmission intensity in africa, *Trends in Parasitology* **17**, 3, pp. 145–149.

Soper, H. (1929). The interpretation of periodicity in disease prevalence, *Journal of the Royal Statistical Society* **92**, 1, pp. 34–73.

Stirnadel, H. A., Felger, I., Smith, T., Tanner, M., Beck, H.-P. *et al.* (1999). Malaria infection and morbidity in infants in relation to genetic polymorphisms in tanzania, *Tropical Medicine & International Health* **4**, 3, pp. 187–193.

Stoer, J., Witzgall, C., Stoer, J. and Stoer, J. (1970). *Convexity and optimization in finite dimensions*, Vol. 1 (Springer-Verlag Berlin).

Sturrock, R. (2001). Schistosomiasis epidemiology and control: how did we get here and where should we go? *Memórias do Instituto Oswaldo Cruz* **96**, pp. 17–27.

Styblo, K. (1991). Selected papers. vol. 24, epidemiology of tuberculosis, *Hague, The Netherlands: Royal Netherlands Tuberculosis Association* .

Sullivan, A., Agusto, F., Bewick, S., Su, C., Lenhart, S. and Zhao, X. (2012). A mathematical model for within-host toxoplasma gondii invasion dynamics, *Mathematical Biosciences and Engineering* **9**, 3, pp. 647–662.

Taber, L., Paredes, A., Glezen, W., Couch, R. *et al.* (1981). Infection with influenza A/Victoria virus in Houston families, 1976, *Journal of Hygiene* **86**, 3, pp. 303–313.

Tan, C. C. (2008). Public health response: A view from Singapore, in *Severe acute respiratory syndrome: A clinical guide* (Wiley), p. 139.

Taylor, H. and Karlin, T. (1998). *An introduction to stochastic modeling* (Academic Press).

Thacker, S. B. (1986). The persistence of influenza a in human populations, *Epidemiologic Reviews* **8**, 1, pp. 129–142.

Thieme, H. R. (1991). Stability change of the endemic equilibrium in age-structured models for the spread of SIR type infectious diseases, in *Differential equations models in biology, epidemiology and ecology* (Springer), pp. 139–158.

Thieme, H. R. (2003). *Mathematics in population biology* (Princeton University Press).

Thieme, H. R. and Castillo-Chavez, C. (1993). How may infection-age-dependent infectivity affect the dynamics of HIV/AIDS, *SIAM Journal on Applied Mathematics* **53**, 5, pp. 1447–1479.

Towers, S. and Feng, Z. (2009). Pandemic H1N1 influenza: Predicting the course of vaccination programme in the United States, *Eurosurveillance* **14**, http://www.eurosurveillance.org/ViewArticle.aspx?ArticleId=19358.

Trager, W. and Gill, G. S. (1992). Enhanced gametocyte formation in young erythrocytes by plasmodium falciparum in vitro, *Journal of Eukaryotic Microbiology* **39**, 3, pp. 429–432.

Trager, W., Gill, G. S., Lawrence, C. and Nagel, R. L. (1999). *Plasmodium falciparum*: Enhanced gametocyte formation *in vitro* in reticulocyte-rich blood, *Experimental Parasitology* **91**, 2, pp. 115–118.

Tsang, K. W., Ho, P. L., Ooi, G. C., Yee, W. K., Wang, T., Chan-Yeung, M., Lam, W. K., Seto, W. H., Yam, L. Y., Cheung, T. M. *et al.* (2003). A cluster of cases of severe acute respiratory syndrome in hong kong, *New England Journal of Medicine* **348**, 20, pp. 1977–1985.

Velasco-Hernandez, J. X. (1994). A model for Chagas disease involving transmission by vectors and blood transfusion, *Theoretical Population Biology* **46**, 1, pp. 1–31.

Viboud, C., Alonso, W. J. and Simonsen, L. (2006a). Influenza in tropical regions, *PLoS Medicine* **3**, 4, p. e89.

Viboud, C., Bjørnstad, O. N., Smith, D. L., Simonsen, L., Miller, M. A. and Grenfell, B. T. (2006b). Synchrony, waves, and spatial hierarchies in the spread of influenza, *Science* **312**, 5772, pp. 447–451.

Vynnycky, E. and Edmunds, W. (2008). Analyses of the 1957 (asian) influenza pandemic in the united kingdom and the impact of school closures, *Epidemiology and Infection*

136, 2, pp. 166–179.

Wald, A., Langenberg, A. G., Link, K., Izu, A. E., Ashley, R., Warren, T., Tyring, S., Douglas Jr, J. M. and Corey, L. (2001). Effect of condoms on reducing the transmission of herpes simplex virus type 2 from men to women, *JAMA* **285**, 24, pp. 3100–3106.

Wallinga, J. and Teunis, P. (2004). Different epidemic curves for severe acute respiratory syndrome reveal similar impacts of control measures, *American Journal of Epidemiology* **160**, 6, pp. 509–516.

Wallinga, J., Teunis, P. and Kretzschmar, M. (2006). Using data on social contacts to estimate age-specific transmission parameters for respiratory-spread infectious agents, *American Journal of Epidemiology* **164**, 10, pp. 936–944.

Wallinga, J., van Boven, M. and Lipsitch, M. (2010). Optimizing infectious disease interventions during an emerging epidemic, *Proceedings of the National Academy of Sciences* **107**, 2, pp. 923–928.

Wearing, H., Rohani, P. and Keeling, M. (2005). Appropriate models for the management of infectious diseases, *PLoS Medicine* **2**, 7, p. e174.

Webster, J. P., Gower, C. M. and Norton, A. J. (2008). Evolutionary concepts in predicting and evaluating the impact of mass chemotherapy schistosomiasis control programmes on parasites and their hosts, *Evolutionary Applications* **1**, 1, pp. 66–83.

Wesley, C. L., Allen, L. J., Jonsson, C. B., Chu, Y.-K. and Owen, R. D. (2009). A discrete-time rodent-hantavirus model structured by infection and developmental stages, *Advanced Studies in Pure Mathematics* **53**, pp. 387–398.

White, R., Freeman, E., Orroth, K., Bakker, R., Weiss, H., OFarrell, N., Buvé, A., Hayes, R. and Glynn, J. (2008). Population-level effect of HSV-2 therapy on the incidence of HIV in sub-Saharan Africa, *Sexually transmitted infections* **84**, Suppl 2, pp. ii12–ii18.

WHO (2003). Consensus document on the epidemiology of severe acute respiratory syndrome (SARS), `http://www.who.int/iris/handle/10665/70863#sthash.yUFbbsKV.dpuf`.

WHO (2009). Influenza fact sheet no. 211, `http://www.who.int/mediacentre/factsheets/fs211/en/`.

Wiggins, S. (2003). *Introduction to applied nonlinear dynamical systems and chaos* (Springer).

Wodarz, D. (2007). *Killer cell dynamics: mathematical and computational approaches to immunology*, Vol. 32 (Springer).

Wu, L.-I. and Feng, Z. (2000). Homoclinic bifurcation in an SIQR model for childhood diseases, *Journal of Differential Equations* **168**, 1, pp. 150–167.

Xu, D., Sandland, G. J., Minchella, D. J. and Feng, Z. (2012). Interactions among virulence, coinfection and drug resistance in a complex life-cycle parasite, *Journal of Theoretical Biology* **304**, pp. 197–210.

Yan, P. and Feng, Z. (2010). Variability order of the latent and the infectious periods in a deterministic seir epidemic model and evaluation of control effectiveness, *Mathematical Biosciences* **224**, 1, pp. 43–52.

Yang, C. K. and Brauer, F. (2008). Calculation of $\mathcal{R}_0$ for age-of-infection models, *Math. Biosci. Eng.* **5**, 3, pp. 585–599.

Yang, Y., Feng, Z., Xu, D., Sandland, G. J. and Minchella, D. J. (2012). Evolution of host resistance to parasite infection in the snail–schistosome–human system, *Journal of Mathematical Biology* **65**, 2, pp. 201–236.

Zagheni, E., Billari, F. C., Manfredi, P., Melegaro, A., Mossong, J. and Edmunds, W. J. (2008). Using time-use data to parameterize models for the spread of close-contact infectious diseases, *American Journal of Epidemiology* **168**, 9, pp. 1082–1090.

Zhang, P., Feng, Z. and Milner, F. (2007a). A schistosomiasis model with an age-structure in human hosts and its application to treatment strategies, *Mathematical Biosciences* **205**, 1, pp. 83–107.

Zhang, P., Sandland, G. J., Feng, Z., Xu, D. and Minchella, D. J. (2007b). Evolutionary implications for interactions between multiple strains of host and parasite, *Journal of Theoretical Biology* **248**, 2, pp. 225–240.

Zhou, J. and Hethcote, H. W. (1994). Population size dependent incidence in models for diseases without immunity, *Journal of Mathematical Biology* **32**, 8, pp. 809–834.

Printed in the United States
By Bookmasters